Transforming the Indonesian Uplands

Studies in Environmental Anthropology
edited by Roy Ellen, University of Kent at Canterbury, UK

This series is a vehicle for publishing up-to-date monograph studies on particular issues in particular places which are sensitive to both socio-cultural and ecological factors (i.e. sea level rise and rain forest depletion). Emphasis will be placed on the perception of the environment, indigenous knowledge and the ethnography of environmental issues. While basically anthropological, the series will consider works from authors working in adjacent fields.

Volume 1
A Place Against Time
Land and Environment in Papua New Guinea
Paul Sillitoe

Volume 2
People, Land and Water in the Arab Middle East
Environments and Landscapes in the Bilâd ash-Shâm
William Lancaster and Fidelity Lancaster

Volume 3
Protecting the Arctic
Indigenous Peoples and Cultural Survival
Mark Nuttall

Volume 4
Transforming the Indonesian Uplands
Marginality, Power and Production
edited by Tania Murray Li

This book is part of a series. The publisher will accept continuation orders which may be cancelled at any time and which provide for automatic billing and shipping of each title in the series upon publication. Please write for details.

Transforming the Indonesian Uplands
Marginality, Power and Production

Edited by

Tania Murray Li
*Dalhousie University
Halifax, Nova Scotia
Canada*

harwood academic publishers
Australia • Canada • China • France • Germany
India • Japan • Luxembourg • Malaysia
The Netherlands • Russia • Singapore • Switzerland

Copyright © 1999 OPA (Overseas Publishers Association) N.V. Published by license under the Harwood Academic Publishers imprint, part of The Gordon and Breach Publishing Group.

All rights reserved.

No part of this book may be reproduced or utilized in any form or by any means, electronic or mechanical, including photocopying and recording, or by any information storage or retrieval system, without permission in writing from the publisher. Printed in Singapore.

Amsteldijk 166
1st Floor
1079 LH Amsterdam
The Netherlands

British Library Cataloguing in Publication Data

Transforming the Indonesian Uplands: marginality, power
  and production. – (Studies in environmental anthropology;
  v. 4)
  1. Rural development – Indonesian 2. Indonesia – Economic
  conditions – 1945– 3. Indonesia – Politics and government –
  1966–
  I. Li, Tania, 1959–
  333.7'3'09598

ISBN: 90-5702-401-2 (softcover)
ISSN: 1025-5869

FRONT COVER: KARO BATAK GROUP IN KAMPONG LAU TEPU, C. 1885 (G.R. LAMBERT, COLLECTION ROYAL TROPICAL INSTITUTE AMSTERDAM).

CONTENTS

List of Maps, Figures and Tables — vii

Notes on Contributors — ix

Acknowledgements — xi

Introduction — xiii
Tania Murray Li

Chapter 1 Marginality, Power and Production: Analysing 1
          Upland Transformations
          Tania Murray Li

Section I Constituting the Uplands: Economies and Traditions

Chapter 2 Maize and Tobacco in Upland Indonesia, 45
          1600–1940
          Peter Boomgaard

Chapter 3 Culturalising the Indonesian Uplands 79
          Joel S. Kahn

Chapter 4 "It's not Economical": The Market Roots of a 105
          Moral Economy in Highland Sulawesi
          Albert Schrauwers

Section II Representing the Uplands: Traditional Knowledge and Environments Reconsidered

Chapter 5 Forest Knowledge, Forest Transformation: 131
          Political Contingency, Historical Ecology and the
          Renegotiation of Nature in Central Seram
          Roy Ellen

Chapter 6 Becoming a Tribal Elder, and Other Green 159
          Development Fantasies
          Anna Lowenhaupt Tsing

Chapter 7 Representations of the "Other" by Others: The 203
          Ethnographic Challenge Posed by Planters' Views
          of Peasants in Indonesia
          Michael Dove

Section III	**Changing Agrarian Relations: Commodity Production and State Agendas**	
Chapter 8	Nucleus and Plasma: Contract Farming and the Exercise of Power in Upland West Java *Ben White*	231
Chapter 9	From Home Gardens to Fruit Gardens: Resource Stabilisation and Rural Differentiation in Upland Java *Krisnawati Suryanata*	257
Chapter 10	Agrarian Transformations in the Uplands of Langkat: Survival of Independent Karo Batak Rubber Smallholders *Tine G. Ruiter*	279
Index		311

LIST OF MAPS, FIGURES AND TABLES

MAPS

Frontispiece	Indonesia	xii

Chapter 5
Map 1	The Eastern Part of the Amahai Sub-district, Seram	133

Chapter 6
Map 1	Mangkiling Village Territory	192
Map 2	Mangkiling Village	193
Map 3	Mangkiling Village	197

FIGURES

Chapter 4
Figure 1	Household Boundaries	115
Figure 2	Household Development Cycles	121

Chapter 5
Figure 1	Videotape of Komisi Soumori	147

Chapter 9
Figure 1	Distribution of Upland Fields in Tumpakpuri	269

Chapter 10
Figure 1	Karo Batak Group in Kampong Lau Tepu	289

TABLES

Chapter 6
Table 1	Excerpt from "Inventory Lost of Flora", Yayasan Borneo	187

Chapter 9
Table 1	Annual Growth Rates of Fruit Production in East Java	259
Table 2	Share-tenacy Terms After Tree Planting in Tumpakpuri	272

NOTES ON CONTRIBUTORS

Peter Boomgaard is Director of the Royal Institute of Linguistics and Anthropology (KITLV), Leiden, The Netherlands and Professor of Economics and Environmental History of Southeast Asia at the University of Amsterdam.

Michael Dove is Professor of Social Ecology at the Yale University School of Forestry and Environmental Science, USA.

Roy Ellen is a Professor in the Department of Anthropology, University of Kent at Canterbury, UK.

Joel S. Kahn, formerly at La Trobe University, Melbourne, Australia has recently taken up the Chair in Social Anthropology at Sussex University, Brighton, UK.

Tania Murray Li is an Associate Professor in the Department of Sociology and Social Anthropology of Dalhousie University, Halifax, Canada.

Tine G. Ruiter is a Ph.D. candidate in social anthropology at the University of Amsterdam, The Netherlands.

Albert Schrauwers is a Temporary Lecturer in the Department of Anthropology at the London School of Economics, UK.

Krisnawati Suryanata is an Assistant Professor in the Department of Geography and Department of Urban and Regional Planning at the University of Hawaii at Manoa, Honolulu.

Anna Lowenhaupt Tsing is a Professor in the Department of Anthropology at the University of California, Santa Cruz, USA.

Ben White is Professor of Rural Sociology at the Institute of Social Studies, The Hague and Professor of Social Science at the University of Amsterdam, The Netherlands.

ACKNOWLEDGEMENTS

The idea for this book was inspired by *Agrarian Transformations: Local Processes and the State in Southeast Asia* (Hart et al., 1989). Since that study focused exclusively on the wet rice lowlands, it seemed to me that a counterpart volume, which explores transformations in Southeast Asia's upland areas, still needed to be written. Like its model, this book draws upon local studies to expose underlying processes, and much of its analysis is relevant across the region. At the same time, by focusing on a single country, it has been possible to give a fuller account of the political and cultural dimensions of agrarian change.

I would like to thank Bob Hefner for his warm support and his direct contributions to this project. He co-authored the initial proposal and helped to organize the 1995 conference at which the papers were first discussed. The conference involved ten scholars in two days of unusually intense and productive discussions, a source of many of the ideas explored in Chapter One. I would like to thank the participants for their willingness to engage in the task of searching, collectively, for a conceptual framework capable of generating fresh insight and making sense of the whole. Henri Bastaman and Peter Brosius contributed to the conference; Anna Tsing and Tine Ruiter were not present at the conference but prepared chapters subsequently. I am especially grateful to the contributors for undertaking two sets of revisions, refining their arguments in relation to the themes that crystallized during the three year process of moving from conference to book. Albert Schrauwers generously offered assistance with the final stages of the editorial process. Dalhousie staff and students who have helped in various ways include Donna Edwards, Louise Uhryniuk, Michael Dickinson, Lucas Sorbara and Dawn Aeron-Wason. As always, I thank Victor Li for his support and advice throughout.

Funding for the conference was provided by the Canadian International Development Agency (CIDA) under the Environmental Management Development in Indonesia Project (a joint project of Dalhousie University and The Ministry of State for Environment, Government of Indonesia). Much of the work on Chapter One and the editing tasks were carried out while I held a fellowship at the Institute for Southeast Asian Studies, Singapore, with financial support from CIDA through the Canada-ASEAN Fund. Additional support was provided by the Canadian Social Science and Humanities Research Council. None of these parties is in any way responsible for the results presented here.

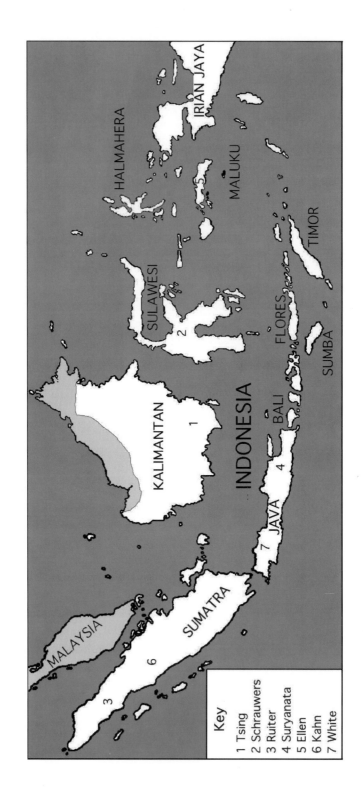

INTRODUCTION*

This book aims to examine and re-assess the transformations occurring in the Indonesian uplands. The goal in doing so is, first and foremost, to draw attention to an area of Indonesia undergoing far-reaching political, economic, and social change, but about which there has been, thus far, very little synthetic and comparative discussion. Drawing upon current theoretical debates in social anthropology, development studies, and political ecology, the book addresses the changing histories and identities of uplanders, particularly as they relate to new modes of livelihood, and shifting relationships to the natural resource base, to markets, and to the state.

During the past twenty years, numerous studies have explored political-economic changes in Indonesia's lowlands, focusing especially upon the impact of the "green revolution" in wet rice agriculture (Hart *et al.*, 1989; Heyzer 1987; Stoler 1977). During the late 1960s and early 1970s, the widespread introduction of new strains of high-yielding rice and industrially-manufactured chemical inputs made possible extraordinary increases in foodcrop production. These programmes depended on more than the diffusion of socially-neutral technical inputs (White 1989). Access to green revolution technologies required new and administratively expensive credit programs, which often worked to the advantage of middle- and large-farmers, thus enhancing rather than diminishing rural inequality. Similarly, efforts to maximise farm profits encouraged farmers in many lowland regions to introduce more restrictive forms of labour organisation, which disadvantaged landless workers, especially women (Stoler 1977). The efficient distribution of inputs and the quick marketing of harvests also required massive investments in the construction and improvement of roads. This in turn facilitated the diffusion of new consumer goods, the movement of investors into rural agriculture, and the dissemination of new lifestyles. Finally, and perhaps most fundamentally, the success of "green revolution" initiatives was accompanied by, and indeed dependent upon, a significant expansion in state capacity. New programmes required government intervention into rural communities on a scale and with a duration not previously seen in the independence era.

---

* Parts of this introduction draw from the conference statement "Agrarian Transformation in Upland Indonesia", Robert Hefner and Tania Li, 1995. References cited in this chapter are found after Chapter 1.

Throughout the 1970s and the 1980s, as these changes (and others) were occurring in lowland Indonesia, upland areas in the hilly or mountainous interiors of most of Indonesia's provinces were undergoing economic, political, and social changes of no less significance. Vast populations gain their livelihoods in upland regions through a mix of swidden farming, tree-crop cultivation, forest extraction, fixed-field permanent agriculture, and wage labour. As in the lowlands, recent years have witnessed road construction, crop intensification, capital investment, deforestation, and a movement of people and ideas on a scale unparalleled in contemporary times. With these developments have come fundamental changes in upland economy, polity, and morality, as rural people respond to new pressures and take advantage of new opportunities.

For a variety of practical and political reasons, the number of people living in Southeast Asia's upland areas has routinely been underestimated (Poffenberger 1990: xxi). There is no official category under which upland populations are enumerated, but the number of people living in and around forests or directly dependent upon them may serve as a proxy: an official source (Bappenas 1993) acknowledges twelve million people while another source (Lynch and Talbott 1995: 22, 55) estimates that the number is upwards of 60 million people. Many have lived in upland regions for generations, while others have migrated more recently in search of land and livelihoods.

Uplanders have been viewed variously as innocent tribals, maintaining distinctive traditional ways of life; as peasant farmers, albeit perhaps rather inefficient ones; as destroyers of the environment and illegal squatters; and, more recently, as expert environmentalists, holding the secrets to sustainable and equitable community based resource management systems. As these widely divergent perceptions indicate, there are several potentially and actually conflicting interests at work in modern Indonesia's upland regions. For the Indonesian state, the primary concern has been to bring order, control and "development" to upland regions, while deploying upland resources to serve national goals. Para-statal and private interests view upland forests as a harvestable resource, and are also attracted by the possibilities for the expansion of commercial agriculture, often on a large scale. For many anthropologists, ecologists and social activists, the concern has been to preserve unique and diverse ways of life, while helping upland people to gain more secure control over their resources and balance livelihood and sustainability concerns. Environmentalists lobby for conservation to protect biodiversity potential. Ethnic or regional politics are virtually impossible in Indonesia, and there are no visible groups representing the interests of upland peoples, but it can be assumed that their interests and concerns are diverse, and not identical with those of the other stakeholders.

A re-assessment of upland transformation must move beyond these pre-established agendas, and ask some searching questions about how contradictory pressures are experienced locally and how they are worked out in the context of every day lives and practices. If we are able to rid ourselves of the image of uplanders as innocents, victims or villains, and treat them seriously as agents contributing to the making of history, the questions become: What future is desired? How exactly is it being made? Under what constellation of opportunities and constraints? What, in short, are the processes through which upland change is occurring? These are the concerns at the core of this volume.

Reacting against the simplified portrayals of social transformation developed by modernisation theorists in the 1950s and 1960s, many historians and anthropologists suspect that comparative research is inherently and unacceptably reductionist. But the dearth of comparative projects has created difficulties of its own: "[S]ynthetic and comparative work is lagging well behind the production of detailed empirical studies. We face the danger that social researchers, disillusioned with the old theoretical certainties … will become very good at producing detailed case studies but rather bad at communicating the general implications of their work to a wider academic audience, not to speak of a wider public of development practitioners." (Booth 1993: 59). For the discipline of anthropology, Nicholas Thomas (1991: 316) argues the need to "refuse the bounds of conveniently sized localities through venturing to speak about regional relations and histories". He proposes a reinvigoration of comparative analysis on a regional scale as an appropriate framework in which to discuss processes of social change and differentiation, and the use of ethnography to generate wider argument.

In order to meet the challenges posed by comparative, policy-relevant analysis, this book adopts a rather distinctive approach. A feature of this approach is the attempt to bring uplands and lowlands, community and state, tradition and transformation, within a single field of vision. Therefore, while the social characterisation of the uplands could proceed (and most commonly has proceeded) by distinguishing Java from the Outer Islands, discussing Eastern Indonesia as a separate sphere, and emphasizing the difference between "traditional" or "tribal" people and the diverse mass of migrants who live alongside them, our emphasis is upon the processes which have constituted both differences and similarities in upland

and lowland ways of life. Further, while much of the existing literature about the uplands is centrally concerned with the ecological implications of agricultural production on sloping land, we are equally interested in the association of uplands with various forms of political and social marginality. In this volume, therefore, we link the conventional concerns of ecology with those of political economy and cultural studies, exploring the uplands as components of national and global systems of meaning, power and production.

As a physical category, the term uplands is rather loosely defined. Nevertheless, it is an indigenous category in much of the Archipelago, and is in common use in both the practitioner and scholarly literature on Southeast Asia written in English. According to Allen (1993: 226, citing Spencer 1949: 28), whose usage we follow: "uplands could be defined as containing a core of 'hilly to mountainous landscapes of steeply inclined surfaces and the table lands and plateaus lying at higher elevations'. It might be added that the discussion concerns land which is not flood-irrigated, not the immediate coastal fringe, estuarine or alluvial plains and swampland, nor is it seasonally flooded." The chapters in this volume relate to physical landscapes which fit within the broad category uplands so defined, while drawing attention to the ways in which the social and political dimensions of being upland, up-river and "of the interior" often overlap. There is no single English or Indonesian term which captures both the social and physical dimensions of the upland landscape. The social aspects resonate with the Indonesian term *pelosok*, roughly translatable as "the boonies", while the term *pedalaman*, "the interior" is commonly used in Indonesia to refer to areas far from centres of government. As the unnamed reviewer of this volume noted, many of the "processes, histories, relationships and ... discourses" explored in this book "are not unique to either the "uplands" or Indonesia". Difficulties of definition notwithstanding, we hope that this observation makes the work presented here more, rather than less, compelling.

The premise of this book is that a broadly based examination of the uplands, one that cross-cuts geographic, social and conceptual spaces, and which pools insights from a range of disciplines and academic styles, has the potential to break new ground. There is no expectation that a totalising, definitive, truth about the uplands will emerge from the present endeavour, but only that selected issues will become more visible, and more complex. This is, to adopt Michael Dove's term, an "engaged study", which addresses real-world issues. It does so not by proposing solutions to upland problems, but rather by filling some of the analytical gaps that surround upland development agendas and practices, and placing practical concerns in a broader political and economic context. In order to do this, and to accomplish the task of comparison and synthesis, it is necessary to attend to patterns, processes and mechanisms of change (White 1989) at various points of intersection between the local and the global. It is also necessary to focus the inquiry upon selected themes, and search for concepts capable of generating fresh insights and explanations.

Three central analytical themes run through the volume. The first concerns the nature and effects of marginality and the related processes of "traditionalisation" and transformation. The Indonesian uplands, we suggest, have been constituted as a marginal domain through a long and continuing history of political, economic and social engagement with the lowlands. Marginality must therefore be understood in terms of relationships, rather than simple facts of geography or ecology. Dichotomous conceptions and related evolutionary models which assume tradition, stasis, subsistence orientation, and general backwardness to be natural features of upland populations are both theoretically moribund and empirically unsupported. Transformations (past or present) cannot be viewed in terms of the familiar impact myth, which proposes that all was quiet before change arrived (market, state, new crops, technologies, migrants and projects). Furthermore "tradition" and the emergence and maintenance of distinctive ways of life have to be viewed historically, as the outcomes of processes of marginalization, "traditionalization" and, in some cases, ethnogenesis. Modern processes of state formation (both colonial and post-colonial) and capitalist expansion were stimulus to, and indeed part of, the emergence of many of the institutions and practices usually regarded as "traditional".

Second, there is a concern with the question of power, and the characteristic ways in which it operates in upland settings. The territorialisation processes through which modern states have attempted to order and control upland resources and populations are of central importance. So too are the informal, personalized lines of patronage which tie upland and lowland elites to each other, and tie upland farmers and labourers to those with access to state protection, authority or sources of capital. Both these forms of power hinge upon the real or supposed ignorance and isolation of uplanders. The definition of uplanders as backward people has legitimated harsh measures such as land expropriation and forced resettlement as well as more or less benevolent forms of paternalism and control. Assumptions about isolation have been central to the construction of visions of "development", both conventional and green varieties. To the extent that they ignore uplanders' historical experiences and current aspirations, "development" policies and programmes produce results which are often problematic, if not actually perverse. Most upland people wish to have access to the promised benefits of "development" programmes, but contest the disadvantageous terms upon which their participation in "development" is currently arranged.

Finally, there is a concern with the forms of production that take shape under the varied political-economic and ecological conditions prevailing in the uplands. These include shifting cultivation on "traditional" land or newly cleared or logged forest land; production of cash crops, such as vegetables and tobacco; tree crop holdings, large or small in scale; contract farming arrangements; plantation agriculture; and wage labour in extractive industries. These systems of production and the livelihoods they provide are more often found in combination than as separate pure forms. They do not coincide with, and indeed invalidate, the idea of clear cut social divisions ("traditional" shifting cultivators are often part time wage workers; estate labourers have small holdings; farmers working along logging roads include those indigenous to the area as well as newcomers). Each of these forms of production has been affected by increased accessibility to markets (an outcome of major road building and logging activities in the 1970s and 80s), and the introduction of chemical inputs, hybrids and new technologies, sometimes state sponsored but more often, in the uplands, adopted under local initiative. Yet much of the diversity, dynamism and productivity of upland environments, and the insight and creativity of upland populations, is overlooked in official "development" programmes and policies which assume a starting point at or near zero, and assert that the most pressing task is to bring change to areas where it has not (yet) occurred. Some "green" approaches also assume a subsistence orientation and degree of detachment from market-oriented production which was rarely present in the past, and is even less likely to be encountered under current conditions.

The attempt to identify a conceptual repertoire capable of exposing the processes at work in Indonesia's uplands is pursued in Chapter 1. In that chapter, I elaborate upon the three themes summarized above, locating them in the theoretical literature and in the specifics of upland history, ethnography and "development" agendas. The chapter is not a blueprint or prescription for the chapters that follow, but rather an attempt to synthesise and make sense of the rich and diverse material presented in this book and other sources. Each of the subsequent chapters addresses a component of the overall problematic. They draw upon original ethnographic, historical and case study material from several regions of the country, but they are not designed to "cover" a representative sample of upland scenes. Rather, each attempts to identify and illuminate one or more underlying processes, as they are encountered in the activities and predicaments of upland life. The authors do not adhere to a single theoretical perspective. Drawing upon different disciplinary traditions and styles of inquiry, they offer a range of perspectives. Terms treated as relatively unproblematic in one chapter (state, uplands, tradition, culture, accumulation) are picked up and scrutinised in another so that, piece by piece, a deeper understanding of upland transformations can emerge. The chapters relate to each other in the form of a constellation - each occupying a separate space but in dialogue with one or more of the other contributions and with the volume's central themes. Such diversity seems to be inevitable in interdisciplinary work, and, in terms of the rather ambitious goals of the volume, both productive and necessary.

Contrary to the myth of unpeopled and unproductive upland terrains, and also to the administrative and scholarly obsession with lowland rice and colonial cash crops, Peter Boomgaard describes the early and spontaneous transformation of upland agriculture initiated around 1600 with the adoption of a New World staple, maize, and small holder tobacco. Through a meticulous examination of the archival record, he establishes a correlation in time and space of these two crops, together with livestock rearing. He uses this data to outline a complex, productive, and relatively sustainable agrarian system that was found in many areas of the Archipelago and persisted for several hundred years. While the full social and political implications of this system cannot be discerned from the sources available, he highlights some key points. Maize cultivation increased the carrying capacity of upland terrain, permitting more people to live at high altitudes. It was politically significant, therefore, in enabling some people to escape the oppression and insecurity of lowland polities and reconstitute themselves as "highlanders" or "tribes" on the peripheries of state control. At the same time, the co-existence of maize production with tobacco indicates that uplanders, for all their physical remove, did not seek cashless isolation: they were tied into lowland markets, as well as credit and taxation arrangements. The linkages between uplands and lowlands in the period discussed by Boomgaard were clearly complex, and involved varying combinations of resistance and collaboration in the service of both political and economic agendas.

Joel Kahn draws attention to the culturalising idiom favoured by officials and elites in characterisations of Indonesia's diversity. Through this idiom, "relations between uplands and lowlands, core and periphery, inner and outer Indonesia, rich and poor, powerful and marginal," which might be understood in terms of unequal access to resources and/or power, become cast in cultural terms as relations between distinct cultural groups. He examines the historical conditions under which this idiom emerged in highland Sumatra at the turn of the century. These conditions included colonial programs to intensify rule, the efforts of colonial officials and scholars of the "ethical" persuasion to preserve native "traditions", and the activities of Minangkabau elites negotiating issues of modernity and identity while repositioning themselves in relation to the colonial regime. Kahn makes power an important focus of his investigation. Challenging the view that power in Indonesia can be understood in terms of apparently unchanging cultural preoccupations, such as hierarchy defined in terms of distance from fixed centers, Kahn exposes the embeddedness of "traditional" forms of power in thoroughly modern processes of state formation and foreign-investment. In upland settings, local power brokers draw their authority and privilege not from monopoly of land (characteristic of lowland societies) but first and foremost from their connections with the colonial and post-colonial state, and/or large corporations, such as estates and multinationals. So positioned, they dispense favours, jobs, permits, contracts, support and, where necessary, protection from petty forms of legal and extra-legal rent-seeking and harassment. These kinds of relations (among others) are commonly read, or misread, through the optic of cultural difference.

Albert Schrauwers' study illustrates, in rich detail, some of the culturalising processes outlined by Kahn in more general terms. Schrauwers challenges the assumption that the apparent "traditionalism" of upland peasantries results from a failure to adapt and change. Reporting on his study in the highlands of Central Sulawesi, he demonstrates that To Pamona "traditions" of complex, multi-family households, elaborate feasts and labour exchanges, and a moral economy of mutual assistance can best be understood in terms of the rational calculus of the market. Rather than signaling a retreat from or opposition to a market economy, these institutions co-emerged with it, under conditions of colonial tutelage. Forced down from their hilltop hamlets, settled in narrow valleys and required to develop sawah, the To Pamona were subject to a process of "peasantisation". Inadequate land and capital, unevenly distributed, led to strategies of maximising inputs of unpaid family labour, accessed through an elaboration of "traditional" institutions, relationships and claims. Yet these practices are read by the state through a cultural lens as primordial attributes, thereby obscuring both their recent nature and the role of state policies in framing the context in which they emerged. Official rhetoric celebrates the communal character of Pamona society, while simultaneously denigrating it as backward and using it as a prime explanation for Pamona poverty and immiseration. Schrauwer's study highlights the active engagement of upland people and state agencies in defining (and contesting) the terms under which agrarian transformations occur and the idioms through which they are managed and understood. It illustrates one of the many diverse forms that capitalism can take in upland settings and confirms that there are a range of modern pathways to "culture" and "tradition".

The Nuaulu discussed by Roy Ellen, like the To Pamona, are officially designated *masyarakat terasing* (estranged peoples). They too were resettled under pressure away from their ancestral lands, moving in this case to the Moluccan coast. Yet the Nuaulu have retained, in their own minds and those of observers, the characteristics of an upland, forest oriented people. Conventional wisdom might label this a case of persistent tradition. Yet there is much in Ellen's account to disrupt the construction of Nuaulu as unchanging, traditional "others". He shows that they became Nuaulu, consolidating a distinctive identity and tribal structure not in the "natural" domain of the forest but in interaction with coastal polities, colonial power and the spice trade. Moreover, their forest is not in fact "natural", but a product of generations of human intervention and modification, blurring the stark distinction between forests and farms characteristic of bureaucratic schemes. Their views of nature and the environment are not fixed but have changed historically as Nuaulu have encountered different political and economic conditions. Having initially welcomed logging and transmigration as developments which increased their access to ancestral lands in the hilly interior and facilitated their hunting and cash cropping endeavours, they have since become disenchanted. To articulate their protest, they have adopted a new rhetoric, one which is recognizably environmentalist, and anticipates both in its tone and the technology of its delivery an urban, and possibly global audience. Nuaulu pleas for assistance could be read in terms of the impact myth: "isolated tribal uplanders invaded and threatened by outside forces" but such a reading would not capture the transformations in which they have participated over many centuries. Nor would it focus attention appropriately on the cause of Nuaulu frustration: not development itself, but the price they have had to pay for a development process which has enriched others but benefited them very little.

Anna Tsing's account focuses upon the contemporary production of a tribal identity by innovative leaders engaging with "development" (both state and "green" varieties) in the Meratus Mountains of South Kalimantan. She discusses the strategies through which community leaders have drawn attention and resources to their locale. They have played, simultaneously, to a globally-constituted "green" fantasy formulated as tribal wisdom and to a state rhetoric that promises roads, facilities and other benefits to those who are primitive and backward, yet open to change and compliant enough to deserve favour. The context is one in which incursions by loggers, migrants, plantations, and transmigration schemes, plus the threat of involuntary resettlement, make the issue of Meratus control over their lives and resources a pressing matter. Rather than a story of victimization, however, she demonstrates the strength, limitations, and above all the ingenuity of Meratus participation in the reconstruction of the upland social formation. News reports, lists of tree species and medicinal herbs, and the production of maps are vehicles for the construction of a "tribal situation" which promises opportunities for collaboration and sets up arenas of ambiguity or "room for maneuver" within which Meratus advance their claims.

Outside "green" circles, tribal identity and the cultural difference attributed to uplanders continue to be viewed negatively, as signs of a development deficit. Michael Dove analyses the world view of plantation managers, and explores the logic through which problems over labour and land come to be represented as problems of primitive culture and irrationality. Para-statal and private sector plantations occupy an increasingly large space in the upland landscape and in "development" plans. Their expansion provokes a contest between state and local representations of upland realities, a struggle over meanings corresponding to the competition over resources and benefit streams. Dove identifies a remarkably consistent, pan-Indonesian tendency for managers to label uplanders tribal, primitive, ignorant, or simply backward and strange. The generality of this discourse indicates a language of power, which operates by apprehending economic conflicts as matters of culture and world view. By contrast, locals attribute planter behavior to familiar, potentially shared attributes — namely, greed and self interest. For plantations and their state sponsors, the primitive label provides a rationale for expropriation and justifies harsh discipline, as well as continuous efforts to direct, persuade and, if necessary, enforce conformity to the social and spatial requirements of "development".

As an alternative to plantation agriculture, contract farming schemes are an increasingly significant way in which Indonesia's upland cultivators are being incorporated into wider economic circuits and state-defined "development" agendas. Besides signaling a shift from mixed farming to monoculture, they reconfigure the physical, social and political landscape. Ben White's study examines the experience of contract farmers in the hilly southern region of West Java, focusing upon a hybrid coconut scheme run by a large nationalized plantation corporation. He shows that contract farming communities deviate markedly from the neo-populist vision of homogeneous, modernising family farms which characterize official rhetoric. They reflect instead the uneven distribution of power and resources that preceded the scheme, and the new forms of patronage, resistance and accommodation that emerge in the context of its implementation. In the scheme he examines, profits overall fell short of economic projections, but some groups made significant economic gains: notably the government officials who were assigned the most fertile and accessible plots. Rich farmers whose valuable tree crops were bulldozed to make way for the scheme lost out. So too did poor farmers who, lacking connections, were allocated inadequate land. Most seriously marginalised were those excluded from membership of the scheme for expressly political reasons. Wage labour for absentee landlords was common, and poorer farmers also tended to be absent working for wages elsewhere. Women and children did most of the work demanded of contract farmers, even while official membership, resources and decision making authority were vested exclusively in men. White's analysis offers some insight into the patterns and processes that can be expected to emerge as contract farming expands over the vast upland areas outside Java.

The expansion of plantation production and contract farming notwithstanding, smallholdings are still the predominant form of upland agriculture. Krisna Suryanata describes the introduction of intensive fruit production on Java's upland fields, as farmers respond to improved access to increasingly affluent urban markets. Although discussed by officials in the environment- and community-friendly rhetoric of "home gardens", "agroforestry" and the restoration of "degraded uplands" commercial fruit production is still, as Suryanata argues, a "strategy of private accumulation". Groves of temperate fruit require major investments of capital in terracing, seed stock and maintenance prior to the first yield. Interplanted annuals cushion the transition period, but soon yield to the more profitable crop. She compares two communities, one in which land owners have gradually lost control of their trees through leasing arrangement with capital-rich "apple lords", and another in which lower capital requirements and a labour shortage favour share cropping arrangements. Despite the concentration of tree tenure, farmers in the apple area have still experienced a general prosperity resulting from high labour demand. In the orange district, in contrast, ownership is more equitable but production is unstable and income gains have not been sufficient to draw outmigrants back to the community. Both examples indicate that proponents of sustainable, subsistence-oriented livelihood systems for upland smallholders need to anticipate the ways in which farmers' land use and investment decisions reflect opportunities, priorities and expectations framed within the broader national context.

Addressing issues of politics and identity as well as production, and covering both the colonial period and the present day, Tine Ruiter's contribution connects many of the themes explored in other chapters. She investigates the formation of a Karo Batak peasantry on the borders of foreign-owned plantations in upland Sumatra. She shows how patterns of production, as well as patterns of leadership, hierarchy and "tradition", were shaped through a process of interaction with the plantations and with the colonial and post colonial states. Colonial state "protection" for Karo land rights and traditions was half hearted. The authorities were forced to weigh the advantages of meeting Karo aspirations (and thereby securing their loyalty vis a vis the rebellious Achenese) versus the profits to be gained from the plantation sector. The Karo, for their part, have steadily pursued their economic goals, turning first to rubber and then to cloves, coffee and investments in education, all without state assistance and, to a considerable extent, contrary to plans made for them by the state. Ruiter focuses particularly on processes of class differentiation within Karo villages, and the factors which contribute to or mitigate against it. Although they are not the centre of her analysis, she reveals also the extreme marginalisation of the upland's poorest residents: landless Javanese, former plantation workers cast adrift in territory not their own.

Chapter 1

MARGINALITY, POWER AND PRODUCTION:
ANALYSING UPLAND TRANSFORMATIONS

Tania Murray Li

The Indonesian uplands have been defined, constituted, imagined, managed, controlled, exploited and "developed" through a range of discourses and practices. These include academic work, government policies, national and international activism, and various popular understandings. Common to all of them is a perception of the uplands as a marginal domain, socially, economically and physically removed from the mainstream, "traditional", undeveloped, left behind. Rather than accept the marginality of the uplands as a "natural" fact, I seek here to locate the constitution of marginality historically and in specific processes of knowledge, power and production.

I argue that the gap between the assumptions driving upland "development" (whether commercial, sustainable or conservationist in orientation) and the actual conditions that pertain in these areas requires some explanation. Following Dove (1985b), I suggest that the reasons for this gap are primarily economic and political. They have to do with instituting and sustaining, or critiquing and opposing, systems of accumulation and control. Representations of the uplands in terms of marginality serve particular agendas, and have real effects in the shaping of upland development. They cannot simply be dismissed as incorrect, and replaced with more nuanced historical and ethnographic accounts. They need to be examined in relation to the contexts in which they are generated and the purposes that they serve.[1]

MARGINALITY, TRADITION AND TRANSFORMATION: CONSTITUTING THE UPLANDS

The Social Construction of Marginality

The concept of marginality provides a point of departure from which some key dimensions of upland transformation can be exposed.[2] Analysing the uplands in terms of marginality has three implications. First, uplands and lowlands are brought within a single analytical frame, and treated as one integrated system (cf Burling 1965). Margins are an essential part of the whole, not separate, complete objects of study in themselves. Second, marginality is clearly a relational concept, involving a social construction, not merely a natural one (Shields 1991). Finally there is an obvious asymmetry between margin and centre: the two do not stand in a relationship of two equal parts of an encompassing whole. As Anna Tsing has observed, marginality is always "an ongoing relationship with power" (1993:90). The constitution of margins and centres is best understood therefore as a hegemonic project, subject to contestation and reformulation. Seldom if ever is it a completed, hegemonic accomplishment (Roseberry 1994). The cultural, economic and political projects of people living and working in the uplands are constituted in relation to various hegemonic agendas, but never are they simple reflections of them.

The social construction of marginality (as a hegemonic project) involves a process through which particular spaces become subject to descriptions which are simplified, stereotyped and contrastive, and then rated according to criteria defined at the centre (Shields 1991). The rating may be positive or negative: often it is disputed, or the subject of ambivalence. Margins are characteristically the site of nostalgia and fascination as well as derision. Regardless of the positive or negative valuation, "imagined spaces" or "place myths" become enacted; they form the prejudices of people designing policies, making decisions and interpreting outcomes, underpinning both the rhetoric and the substance of interventions made by the centre in the periphery (Shields 1991: 47).

In Indonesia, the uplands are negatively associated with backwardness, poverty, ignorance, disorderliness and a stubborn refusal to live as "normal" citizens. A counterposing set of images, ones relating to freedom and the integrity of people, communities and environment, also exist. But the positive side of the dichotomy is rather weakened by the ambiguous character of the uplands environment. Most upland areas of Indonesia do not fit neatly into the contrastive schemes commonly encountered for the social organisation of space (cf. Short 1991). They are neither wild nor tame. The potentially positive attributes of wilderness (unspoiled hills and forests) are sullied by the presence of people and agricultural activities. Yet the agricultural activities that take place in the uplands seem too precarious to satisfy the nostalgic advocate of peasant life, iconified in Indonesia by the padi farmer. For those who support the idea of "traditional", ecologically benign agriculture in forested areas, and are therefore prepared to merge the concepts of wilderness and farm, the commercial nature of many upland enterprises poses another problem. As an "imagined country" (Short 1991), the uplands are complex. There being little consensus on how to characterise the uplands and their deficiencies, proposed models of change (conservation, forest extraction, smallholder support, plantation agriculture, transmigration) are, not surprisingly, contradictory.

Recognising similar complexities in India's "agrarian environments", Agrawal and Sivaramakrishnan (forthcoming) point to hybrid landscapes and explore the consequences of partitioning landscape into the environment (ideally untouched by humans), and the agrarian (clearly the product of human agency). One consequence has been a division of scholarship, studies of agrarian change focusing upon the agricultural heartlands, especially those affected by the green revolution, while environmental scholarship has been preoccupied with mountains, forests, tribal populations and semi-arid regions. This division obscures the connections between environmental change and agrarian structures, as well as the source of many "environmental" struggles in typically agrarian concerns. It obscures the role of humans in producing nature, even in apparently remote locales (see Ellen, Tsing this volume). It also lends itself to the construction of social typologies, the environment evoking an array of naturalised or exoticized terms such as "woman", "indigenous", "community" and "local", reifications at odds with the diversity and flux of actual social formations. Their analysis offers insights into the ways in which upland terrains in India and also in Indonesia, as I will show, have been simultaneously "environmentalised" and "culturalised"[3] in ways which mark them as both different and deficient.

Constituting Margins and Centres in Indonesian Geography and History

Although there are exceptions, to go upland, inland, towards forests, away from coasts and from sawah cultivation is, in many parts of contemporary Indonesia to move from domains of greater to lesser power and prestige, from centres towards margins. The association of the uplands with cultural difference, and the negative rating of that difference, has a very long history in Indonesia, and reflects changes taking place in the lowlands, especially along the coasts. Early converts to Islam, such as the thirteen century Samudra-Pasai dynasty on the North Sumatran coast, still felt the need for political and spiritual legitimation by the "existing powers of the interior" (Hall 1978:223). Ancient residents of the Tengger Highlands in Java were both feared and admired for their alleged spiritual powers (Hefner 1990). Over time, however, the locus of power throughout the archipelago shifted towards Islam and the coast (with the important exceptions of Mataram and the Minangkabau kingdom, both inland centres). There was a split between those who chose to live as Muslims on the coast or alongside the rivers, and those who preferred the uplands, jungles and interiors. Upland populations and their habitats became quite consistently despised. In Sulawesi for example, the terms Toraja (Bugis, "people of the interior") and Halefuru or Alifuru (Ternatan, wilderness, forest), both disparaging, were in common use in coastal areas in the sixteenth century. They were subsequently taken up and used by Europeans (Henley 1989:n54). From a coastal perspective, crucial political relationships were those connecting one coastal settlement to another through hierarchies of dependence and obligation. Rather than emphasise commonalties of culture across a given land mass, there was a marked social distinction between "people of the littoral kingdoms and the barbarian population of the tribal interior" (Henley 1989:8).

Across the archipelago coastal centres continue to relate both to their interior domains and also to other coastal centres. There are different models for apprehending the resulting relationships. For administrative purposes, the whole of Indonesia is linked by a singular hierarchy centred on Jakarta. In terms of culture, the official model of "unity in diversity" proposes a non-hierarchical reading of difference, and highlights culture (rather than, say, class) in the definition of the nation (Kahn, this volume). The model is belied by the numerous ways in which cultural standards defined in Jakarta frame official interventions in all fields (health, education, agriculture, housing, administration). Needless to say these standards are reconstructed in multiple ways according to local conditions, priorities, and room for manoeuver (Schrauwers, Tsing this volume).

Besides the standards imposed by adherence to world religions, many of the cultural norms or standards by which the uplands have been judged deficient are a relatively recent invention. According to John Pemberton (1994) "Javanese Culture" was fashioned through a nineteenth century dialogue between the royal courts of Surakarta and the colonial presence. Against this culture the practices of minor aristocrats, urban commoners, lowland Javanese peasants, upland Javanese, all those off-Java, and uplanders off Java most of all, came to be judged both distinct and deficient. Thus, he argues, the cultural project of "majoritising" lowland Java has been, simultaneously, a project which "minoritises" and marginalises others on the social and geographic periphery of that self-defined centre.

Countering Evolutionary Imaginings

A feature of marginality as a social construction is the elision of differences in culture with differences in time, conceptualised in evolutionary terms.

By virtue of living in "out-of-the-way-places" (Tsing 1993) uplanders have also (wrongly) been characterised "peoples without history" (Wolf 1982).[4] From the perspective of those at the self-defined centre, the uplands are deemed marginal because they have failed to change, "develop" or modernise.[5] By implication, the difference observed at the margins confirms the distance that the centre has progressed. But the uplands (and other places and peoples normally characterised as "traditional") have endured the same period of historical time as those places and peoples deemed "modern". By rejecting the evolutionary assumption of a unilinear trajectory from traditional to modern, it becomes possible to look at difference as a historical product. Instead of assigning unfamiliar cultural forms to the category "tradition", and reading them as relics from the past, attention can be turned to the processes through which diverse cultural forms are generated and maintained.

In the case of the uplands in Indonesia, renewed attention to regional histories reveals that centuries of interaction with the lowlands, with state programmes, and with national and international markets have been central to the formation and reformation of their cultures and practices, and to their very identities as "communities". Distinctive traditions have been an outcome of change, not its antithesis. Moreover, change is not a new predicament faced by "traditional" people confronted by dichotomous choices between community and state, subsistence and market, past and future. Nor is it imposed unilaterally by outsiders, as the "impact myth" proposes.[6] It is, rather, the result of creative engagement and cultural production.

Upland histories in Indonesia have been remarkably non-linear. They have involved a variety of crops falling in and out of favour according to the broader political and economic context; periods of strong engagement with markets, followed by disengagement; and periods of focused state attention succeeded by periods of relative neglect. The state of being a "traditional" tribesperson or upland peasant turns out to be not a starting point for, but a product of these complex histories.

According to Anthony Reid (cited in Colombijn *et al.*, 1996), major population concentrations in pre-colonial Indonesia were located not on the coastal plains, but in the interior, and particularly the upland valleys and plateaux. The reasons for this were both economic and political. Sixteenth century reports (Reid 1988:19) describe livelihood systems in the interior of some of the eastern islands that were diverse and complex. Rice, for example, was planted on hillside swiddens, broadcast on flood plains, and transplanted into ploughed, bunded fields. Where terrain permitted, swidden was especially favoured because of its high productivity in relation to labour inputs. Surplus hill rice was exported through well established trade relations with lowland districts, while fish, salt and other items were imported (Reid 1988:28). Counter to the received evolutionary wisdom, hill rice, once thought to be a wild, "original" cultigen, turns out to be a strain developed from swamp rice, selected for its productivity and the upland lifestyle it made possible (Helliwell 1992 citing Chang 1984, 1989).

If high productivity made upland livelihoods attractive, so too did the possibility of escape from oppressive systems of lowland rule, debt bondage and slavery. Some sanctuary was afforded by the difficult, forested terrain, although uplanders and residents of the small islands of Eastern Indonesia were still captured, victims of slave raids or of internal warring and disunity (Reid 1988:122). With the spread of Islam, which forbade the enslavement of fellow Muslims, pressure was increased on the animist frontiers. In Java, non-Muslim Tengger highlanders were enslaved by Mataram forces during the seventeenth century. The survivors retreated from the accessible, midslope region to the high mountain zones, where they built tightly clustered settlements on steep ridges well suited for defence (Hefner 1990:37–38). Likewise in Northern Maluku, early settlements were built in the hilly interiors for security reasons, and movement to the coast began only with the temptations of the spice trade (Andaya L. 1993; Ellen 1979).

Some of the production systems developed by those who have elected to live in upland terrains are especially adapted to conditions of political uncertainty and insecurity. Specialists in forest collection, such as the Penan of Borneo have long traded their products to settled agriculturists in return for staples and other valuables (Hoffman 1988). They have continued to balance limited agricultural production with forest-products trade over a period of centuries — not because of a failure to advance up the evolutionary ladder, but because of the positive advantages that such combinations provide. These include, most especially, the ability to retreat into the forests and subsist on sago when their relationships with farmers, trade partners and patrons threaten unacceptable levels of subordination (Sellato 1994).

Boomgaard (this volume) argues that it was the early and rapid adoption of maize, beginning in Eastern Indonesia, which permitted larger populations to subsist in upland zones. Maize facilitated the upward flight of the Tengger Javanese after the Muslim conquest of Majapahit (Hefner 1990:57). Political insecurity may also account for the adoption of maize and other new crops in the Lesser Sundas prior to European arrival.[7] The development of lowland sawah can be explained similarly by a political, rather than an evolutionary logic. It was promoted or imposed by lowland or coastal lords not because of its productivity as such, but because it was suitable for cultivation by subjugated populations who were forced to concentrate at important trading centres and in other locales where they could be monitored, subjected to corveé obligations, and taxed.[8]

Capturing Upland Wealth

Both the possibility of escape from subordination and the productivity of upland farming systems drew people towards or encouraged them to remain in the uplands. Besides agriculture, the uplands offer other sources of livelihood and profit, including forest products traded on international markets and mineral resources. In Borneo, according to Padoch and Peluso (1996:4), such activities have supported interior populations and attracted in-migrants for at least two thousand years, linking the most remote areas to regional and international systems of trade. A central concern of coastal powers has been to devise mechanisms for capturing some (or most) of the wealth generated in upland or hinterland settings.

Different political dynamics prevailed at different periods and in different regions. Some upland societies, especially in Eastern Indonesia (Timor) were themselves complex, stratified states. In others, for example Toraja in the nineteenth century, upland chiefs colluded with coastal powers in the oppression and enslavement of upland populations. Alliances between upland and lowland élites were shaped by struggles for control over the lucrative coffee trade. Coffee was also a source of rivalry between upland communities, setting off the skirmishes from which victims or slaves were generated (Bigalke 1983). Few pre-colonial states were powerful enough to exert systematic control over populations living in the uplands, and they did not even attempt to control territory (Bentley 1986). Where coastal powers were strong, for example, Jambi in the seventeenth and eighteenth centuries (Andaya, B. 1993), extraction of rice, pepper, and other interior produce was accomplished through coercive mechanisms such as levies and tolls, forced indebtedness and enslavement, as well as through enticements such as reduced corvée obligations. But downriver control was continuously contested. Resistance included boycotts and the transfer of business to more hospitable ports, withholding produce, refusing to accept credit, and eventually, abandoning the production of goods which tied people to unfavourable terms of trade (Andaya, B. 1993). Where highlanders had several choices of trade route, by trail or river, their autonomy was enhanced; it declined to the extent they were dependent upon a single riverine system (Bronson 1977).[9] Coastal states in Kalimantan faced another dilemma: hinterland populations too well controlled tended to become Muslim and lose interest in forest collection (Healey 1985:18). Those determined to retain their autonomy could simply reject trade relations (Rousseau 1989:49) or bypass middlemen and debt bondage by developing an alternative mechanism to access imported prestige goods — the labour migration of young men (Healey 1985:5, 22). They also resorted to large-scale migrations involving whole communities moving into the interior or closer to alternative, more hospitable trading centres (Healey 1985:16, 28).

Thus in economic and political matters the relationship between upland and lowland systems has long been marked by tension. The uplands were, in many parts of the archipelago, central to the prosperity of lowland and coastal states, yet their populations were treated with unmitigated derision and contempt. The difficulty of controlling the interior was construed (ironically but unsurprisingly) as confirmation of the cultural inferiority of its inhabitants. The uplands and interiors of Indonesia's islands were complex and contradictory spaces: they could be domains of freedom and autonomy from oppressive lowland systems, or of powerlessness so extreme that people became entrapped and subordinated; they could be domains of high productivity, envied and sought-after by lowland profiteers, or of harsh and sparse livelihoods gained on the run. Regardless of which dynamic prevailed, people who lived in the uplands did so not by default, by-passed by history, but for positive reasons of economy, security, and cultural style formed in dialogue with lowland agendas.

Tradition and Cultural Diversity as Regimes of Knowledge and Power

Recent historical ethnographies have emphasised that the emergence of cultural distinctiveness among upland populations has been a two-way process — not the unilateral product of centralised lowland power, nor the unilateral product of tradition-bound people without history. According to Tsing, for example, the Meratus in the mountains of Southeast Kalimantan have "lived on the border of state rule and Banjar regionalism for centuries and have elaborated a marginality that has developed in dialogue with state policies and regional politics. Indeed ... it is this elaboration of marginality that regional officials mistake for an isolated, primitive tradition" (1993:29). Kahn, Ruiter and Schrauwers (this volume) describe the elaboration of "traditional" forms of peasant production in the modern era. Hefner (1990:23) argues similarly that centuries of interaction and local creativity went into the formation of a "different way of being Javanese" in the Tengger Highlands. The early history of Tengger included a prominent role in state-sponsored religious cults of Majapahit, and a period as a refuge for Hindus fleeing the imposition of Islam, but it was in the context of compulsory colonial coffee production in the mid nineteenth century that highland culture was definitively shaped. In this period the physical landscape was transformed with the complete clearing of midslope forests. The population increased greatly with the influx of landless lowland migrants fleeing the even harsher conditions of compulsory sugar cultivation (Hefner 1990:43, citing Onghokham 1975:215). Most importantly, the entire population was settled on the land, organised for production on a smallholder basis, and subjected to land taxes payable in coffee. A product of colonial policies, independent smallholding became the hallmark of Tengger cultural identity, a positively espoused and strongly defended way of life which highlanders contrasted with the subservience and indebtedness, as well as the airs and graces, of the rural Javanese lowlands. Tengger "tradition" thus emerged in the context of the regional and transnational political-economic histories of which Tengger has always been a part.

The disjunction between accounts which highlight the historical constitution of difference and marginality, and conventional accounts which characterise the uplands as places with "different" cultures but without history can be explained, I argue, not as a failure of knowledge, but rather in terms of regimes of truth and power. As noted earlier, a discourse about difference, cast as savagery, facilitated the enslavement and exploitation of upland peoples in the pre-colonial period. In the late colonial period the label "traditional" was deployed to explain (and justify) the underdeveloped state of Indonesia's upland regions. It was a term that served to obscure the extent of change brought about by colonial rule. For example in Sumatra, where a lively indigenous commerce in gold, iron, textiles, cattle and tobacco had been displaced by enforced coffee cultivation, the slump resulting from coffee's decline was interpreted by Dutch observers as evidence of a "natural", traditional economy (Kahn 1993:173–179; Bowen 1991; Sherman 1990). So too in upland Java where the retreat of smallholders into a "grim subsistence agriculture" (Hefner 1990:11; Palte 1989) could be read as evidence of a failure to "develop", as if poverty, tiny, marginal plots of land, and "traditional" practices of labour exchange and feasting had endured from time immemorial. Similar narratives are deployed, under parallel conditions, in contemporary upland Sulawesi, as Schrauwers (this volume) so clearly demonstrates.

The notion of "tradition" was also used quite deliberately to legitimate colonial policies of indirect rule. Early this century, various traditions and customs (*adat*) were codified by scholars and officials, and used as the basis for rule by Dutch-appointed "traditional" leaders.[10] These frozen codes helped to obscure the extent and nature of changes brought about by the colonial regime (Barber 1989:95). The concept of the "adat community" assumed, as it simultaneously sought to engineer, a rural population separated into named ethnic groups with "traditions" stable enough from person to person and context to context to serve as definitions of group identity, and centralised political structures with recognised leaders capable of articulating a single "tradition" on behalf of the whole. These were assumptions that fitted some areas better than others. In Minangkabau, for example, they conformed better to the identities that emerged during colonial rule than with those that preceded it (Kahn, this volume).[11] As a result of differences in "fit" and, more importantly, different degrees of colonial interest in asserting control over the people and resources of particular locales, the process of "traditionalisation" or "adatisation" occurred unevenly across the archipelago. Those groups that underwent the adat codification process found their customs abstracted from daily practice, catalogued and held up as laws to be enforced by adat chiefs. In other regions, where this process did not occur or was incomplete, identities, practices and authority in matters of custom remained more flexible and diffuse.[12]

The process of inventing or creating traditions was not only a colonial imposition. It has been a central and continuing feature of Indonesian modernity as those who are mobile (socially or geographically) seek to secure a sense of identity (Kahn 1993:122; Gibson 1994:160; Volkman 1985), while others turn to "tradition" in the attempt exert some control over their lives in changing times (George 1991; Schrauwers; Ellen; Tsing; this volume). Images of a traditional "other" are generated also by urban idealists and romantics (both Indonesian and foreign) seeking "traditional people" in whom to place their faith (Tsing, this volume). They are further enhanced as an urban imaginary becomes translated into a tourist agenda.

Finally, processes of traditionalisation or, to use Joel Kahn's term "culturalisation", are generated by the New Order regime, which needs a defined and controlled degree of "diversity" in order to promote its form of "unity" (Kahn, this volume; Pemberton 1994). As the Dutch had previously acknowledged, a minority becomes more manageable when it has been appropriately defined and pinned down spatially and socially.[13] In Indonesia today, people without an acceptable "culture" or "tradition" cannot be readily incorporated within the state's framework for "unity in diversity". Whether they are advantaged or disadvantaged by this exclusion depends upon an assessment of the ways in which "tradition" and its associated images are taken up by various players, including uplanders themselves, in the formation and contestation of policies, practices and identities.

Reiterating a point made earlier: to acknowledge cultural difference is not necessarily to concede to a singular hierarchy that arrays difference in terms of centres and margins. The question of marginality is always an empirical one; it has to do with hegemonic claims and their local resonances, refractions and outcomes. Webb Keane (1997:38–39) makes this point: "To the extent that people understand themselves to be "marginal," or simply "local," they may be accepting at least some of the authority that makes somewhere else — the capital city, the nation, the state, the global economy — a proper, even foundational, frame of reference". He shows how the authority of the nation insinuates itself into people's self-understanding, channelled as it is through the mundane disciplines of schools, village meeting halls and bureaucratic encounters. His analysis of local language and culture in Sumba indicates a hegemonic project which, for this locale at least, is quite far advanced.

Policy makers, administrators, activists and critics are primarily concerned with shaping the present and the future of the uplands; yet it is the past, represented in terms of "culture" and "tradition", that dominates their discourse and shapes current practices and alternatives. For state agencies, environmentalists and social activists alike, assumptions about the marginality of upland ecologies, livelihoods and people form the basis of "development narratives" or "cultural scripts for action" (Hoben 1995). That is, they define problems, filter out contradictory data, and structure options in ways which enable and legitimise specific forms of intervention.[14] It is, as I will show, essentially the *same* image of the "traditional" and/or marginal uplander that informs both the construction of state policies and also, paradoxically, their critique.

POWER, TERRITORIALISATION AND "DEVELOPMENT" AGENDAS

To define particular regions, peoples and practices as marginal, disorderly, traditional, and/or in need of "development" is not simply to describe the social world: it is to deploy a discourse of power. In this section I focus upon the creation and reconfiguration of margins and minorities as part of the dynamic of modern state formation. Central to understanding state formation (following Philip Corrigan 1994) is an account of *how rule is accomplished*. This is a matter more fundamental than the question of who rules, or even whose interests are served, and it takes us into the heart of state-systems, institutions and processes.[15]

Of particular relevance in the contemporary uplands is the process of territorialisation defined by Vandergeest and Peluso (1995:387) as the process through which "all modern states divide their territories into complex and overlapping political and economic zones, rearrange people and resources within these units, and create regulations delineating how and by whom these areas can be used". Territorialising initiatives have been undertaken by both colonial and post-colonial regimes prompted by a search for profit by favoured élites, for tax revenues to support administrative systems, or by the need to assert state authority in areas that, although they may lie clearly within national boundaries, are not fully enmeshed in state-defined institutions and processes. Ongoing and incomplete, territorialising initiatives are commonly contested by the populace. Moreover they involve many government departments, each with different and possibly conflicting approaches. Strategies for increased control may include privatising natural resources (within state-defined frameworks) or direct management by state-agencies; encouraging settlement in unpopulated areas or forbidding settlement; centralising administrative authority or devolving authority to lower levels. The making of maps, the conduct of censuses, the drawing up of village boundaries and lists, the classification and staking of forests can all be seen as mechanisms to define, regulate and assert control over the relationship between population and resources.

Territorialising initiatives in the uplands of Indonesia (as well as Thailand, the focus of Vandergeest and Peluso's analysis) have historically been less intense than those in the lowlands, but this situation is changing. State projects to intensify territorial control have been of central importance in transforming the uplands in recent decades.[16] Their justification or, to use Hoben's (1995) phrase, their "cultural script for action" draws upon a discourse about marginality which emphasises the unruliness and deficiencies of the populations and landscapes to which state-defined order and "development" are the common-sense, apparently a-political, response.

Territorialisation Projects of the Colonial State

As noted earlier, territorial control or the direct attempt to regulate the relationship between population and resources was not a feature of rule in pre-colonial state systems. Even in the densely populated pre-colonial *negara* described by Clifford Geertz (1980), critical relations of power and control were organised along interpersonal rather than territorial lines. There were territorially organised systems for village governance and the regulation of irrigation and production but these were designed to accomplish practical goals and were not marshalled in ways that helped to accomplish rule. Indeed there was little concern with rule as such, hence Geertz's conclusion (1980:13) that the Balinese negara was a theatre state directed "not towards tyranny, whose systematic concentration of power it was incompetent to effect, and not even very methodically toward government, which it pursued indifferently and hesitantly, but rather towards spectacle, toward ceremony, toward the public dramatisation of the ruling obsessions of Balinese culture: social inequality and status pride."

Modern state formation, and with it the process of territorialisation, began under colonial rule. An early goal was to increase state control over labour through the imposition of a territorialised system for village administration. The pre-colonial rural system in Java, characterised by personalised allegiances and channels of extraction, had been "unable to bind people to existing settlements for any length of time" (Breman 1988:26). Fleeing debts and excessive corvée demands, people frequently moved off in search of less oppressive conditions with another master, or autonomy (often temporary) on a forest frontier. It was in the early nineteenth century that the colonial regime pinned people down into households and villages, surveyed land, fixed and enforced *desa* boundaries, and represented the result in maps, lists and censuses (Breman 1980, 1988). Contrary to the assumption that orderly, homogenous villages are a natural feature of the Javanese landscape, Breman argues that it was colonial policy which *created* the peasant village (1980:9–14).[17] By the end of the nineteenth century, all the cultivable land in lowland Java had been brought into production, and "everybody was set in their place", freezing the agrarian social structure in the "harnessed construction" of the desa community (1980:41).

In the uplands, territorialisation was initiated as an attempt to increase colonial control over natural resources and facilitate the release of land for large scale commercial agriculture. To this end the 1870 Agrarian Law restricted customary rights to land in favour of the state, and made swidden farming on extensive terrain, the management of lucrative tree crop groves, and long standing patterns of forest use illegal (Peluso 1990; Kahn 1993). To the extent that uplanders lost control over resources, or were forced to access them illegally, this law also served as a mechanism to gain control over labour. In the Javanese uplands colonial laws constituted villagers on the fringe of the island's teak forests as poachers and thieves on state land. While not effectively excluded from the forests, their illegal status nevertheless made them vulnerable to state sanctions, facilitating their co-optation as forest labour on the most minimal terms and subjecting them to multiple supplementary forms of extraction and harassment (Peluso 1990:33–35).

In the Sumatran uplands at the turn of the century, state control over territory was incomplete. Villagers were able to reform themselves on the fringes of colonial plantations and retain a foothold on the land — particularly on the steeper, less accessible slopes. So long as upland populations had access to the means of production, however marginal the terrain, labour for the European plantations and resin forests had to be imported (Kahn 1993: 246; Bowen 1991:79; Ruiter, this volume). Plantation labour was drawn from the landless, indebted, disciplined lowland Javanese population while native uplanders were, predictably, branded as lazy and deficient (Alatas 1972; Dove, this volume). In the indirectly governed territories the Dutch assumed that native rulers were in control over the so-called wastelands, and western entrepreneurs had to negotiate with them for access to land. Their arrangements led, however, to similar sets of restrictions on peasant access, and in some cases to protracted conflict (Ruiter, this volume).

Territorialisation and Development in the Post-Colonial Era

In the post-colonial period, state interventions in the uplands have come increasingly to be framed through a discourse of marginality and the need for "development". Disorder in the relationship between people and the resource base is seen to stem, in part at least, from the deficiencies of upland culture and personality (Dove, this volume). To bring order where it is lacking, current mechanisms include the official designation of land as "forest"; the development of plantations and transmigration sites which bring well-ordered, usually lowland populations into unruly areas; the regularisation of spontaneous land settlement; and the resettlement of the so-called *masyarakat terasing* (isolated or backward tribes) into properly administered villages. These four approaches are somewhat contradictory, indicating the unfinished nature of territorialisation processes (Vandergeest and Peluso 1995:391), as well as the internal complexity of the governing regime. While they each have a territorialising dimension, they are framed in more particular terms, as initiatives designed to bring about one or other aspect of "development".

In contrast to the lowlands, cadastral surveys and land titles are almost unknown in the uplands and territorial control is exercised most directly through forest law and policy (Barber 1989:5; Peluso 1992). Under the Basic Forestry Law of 1967, nearly three quarters of Indonesia's total land area, including much of the uplands, is defined as "forest", regardless of its current vegetation. This legal framework implies, from the outset, that most agrarian activities in and around the "forest" are illegitimate. But rather than confront the economic issues directly, transgressions are interpreted within a "development" rhetoric stressing personal characteristics of backwardness, ignorance and indiscipline. Forest-dependent villagers are said to have "low levels of awareness", lack "advanced patterns of thought", and have "inadequate consciousness" of the forest's functions (Barber 1989:137, 172, 282) thereby legitimising intervention to provide appropriate "guidance". Control is also exercised directly through military and police interventions to protect the "forest" from illegal incursions (Barber 1989:131–137; Tjitradjaja 1993). Java's forests continue to be associated in state discourse with criminals and outlaws (Peluso 1990).

The significance of "forests" to the process of extending state control over the uplands goes considerably beyond any concern with trees. On crowded Java, according to Barber (1989:3–20), 50% of the population is landless yet 23% of the total land area is classified as national forest, off limits to agriculture or other uses. Within this area only traces of natural forest remain and state teak and pine plantations are largely unproductive. The landscape is characterised by secondary forest, degraded land, and farms with more or less permanent crops. On the borders of this "forest" are six thousand villages, with 30 million inhabitants, mainly drawing their livelihoods from the forest as labourers and through illegal extraction and farming. With a burgeoning landless population, forest villages are placing increased pressure on upland watersheds, as farming extends and is intensified in the effort to meet subsistence needs. Since Java's forests now contribute only a "minuscule portion of the nation's forestry income" (Barber 1989:124), control over the population occupying this upland terrain, rather than the search for revenue or profit, is the major logic at work in Java's forest management.

Political, economic and administrative control over a mass of impoverished and dispossessed "forest" villagers on Java, and a dispersed, often inaccessible population off Java, has not been easy to impose. Administratively, much power over "forest" areas, and therefore over most of the uplands, is concentrated in the Department of Forestry. For "forest" villagers on Java this one department dominates their interactions with the state, and has major influence over political, economic and security matters (Barber 1989:148). In the vast "forests" off Java, concessionaires with forest leases are delegated the same extensive powers. For the resident population, the only "room to manoeuver" arises in the spaces between the mandates of different government departments as these are interpreted by various parties at the local level (see Tsing, this volume).

In competition with the Department of Forestry, the Department of Agriculture has an interest in the massive logged-over (or burned-over) lands of Kalimantan for conversion to oil palm or other plantations, offering an alternative way to bring "development" to the huge land areas and populations under Forestry control. Social forestry programs can be seen, in part at least, as an attempt to forestall this alternative. Under the banner of "development", "environment" and "participation", social forestry programs promise to address the needs of the people by permitting limited agricultural activities to take place in "forests" under the control and guidance of the Forest Department (Barber 1989:229, 410–11; Djamaludin Suryohadikusumo 1995). Long time opponents, forest villagers are now to become allies of the Forest Department in its project to retain control over its domain. At the same time, coercive removal of those practising unregulated smallholder agricultural activities within "forest" zones continues to be government policy. Reinvigorated enforcement measures were announced in 1993 through a joint Decree of the Ministers of Forestry, Home Affairs and Transmigration dealing with so-called shifting cultivators and forest destroyers (Barber et al. 1995:15). People who farm within "forests" are to identified and inventoried (located, listed, classified); they are then liable to be relocated into official transmigration settlements (Departmen Kehutanan 1994; Barber et al. 1995:14).

The Transmigration Department is also a potential competitor when it seeks to have "forest" released for conversion to farms and new settlements. In the past, transmigration sites were mostly in the lowlands and their goal was irrigated rice production. Since suitable lowland sites have become scarce, however, the focus of the program has shifted to upland areas which would be suitable for either timber plantations (under Forestry control) or for commercial tree crop estates (under the Department of Agriculture)(Brookfield et al. 1995:89, 105; Hidayati 1991:43). The current mandate of Transmigration is to furnish (suitably disciplined) labour to exploit and develop underutilised resources, and thereby promote economic growth while also bringing political and administrative order to peripheral areas. Schemes for large scale agricultural development such as plantations and nucleus estates, sometimes pursued in conjunction with transmigration, have a similar effect: they enmesh territory and people in frameworks of state control, simultaneously reshaping the landscape, developing "model" settlements, and providing a clear rationale for the extension of roads, services and the administrative machinery of government. They include some of those displaced to make room for "development" and further marginalise others (White, Dove, this volume).

In economic terms, many of the large scale state schemes to realign people and resources are expensive, inefficient, and unprofitable (SKEPHI and Kiddel-Monroe 1993:245–259; White this volume). Where they disrupt people's livelihoods, they are also unpopular, and must be implemented through coercive mechanisms (White, Dove, this volume). Yet they have important effects. While these projects may or may not bring in "development" (improved livelihoods, greater productivity, roads and services) there is no doubt that "development" brings with it the administrative and coercive machinery of the state. Moreover, the logic is self-confirming: to the extent that livelihoods are not improved, projects fail to produce, services are poor, and people recalcitrant, this only proves that a stronger state presence and more "development" are needed (cf. Ferguson 1994).[18]

In less grandiose fashion than transmigration and estate schemes, and often more peacefully, the intensification of state control over population and resources can also be accomplished by regularising the spontaneous incursion of migrants into frontier zones, especially where these have been made accessible by logging roads.[19] Once newcomers have been organised into administrative units (desas), their daily activities can be monitored and regulated through the various village committees and institutions specified in law. As newcomers trying to make their way outside the formal structure of a project, they may be especially eager to transform themselves into model communities and thereby legitimate their presence and consolidate their hold over resources. They want and need to be enmeshed in state systems in order to claim their place as citizens and as clients of state-officials and institutions. Order along the frontier is not guaranteed, however, as the processes of settlement and expanding state territorial control themselves routinely reproduce conditions for conflict. By ignoring the local populations and indigenous resource management institutions already in place, they provoke the events, cases, or situations which draw media attention and evince further rounds of state intervention. As critics of the Village Administrative Law No.5/1979 have often pointed out (Moniaga 1993), state administrative policies in frontier areas have been designed to dismantle local forms of order and regulation in order to substitute government ones, but this transition is not easily accomplished.

Some of the responsibility for refashioning indigenous institutions falls upon the Department of Social Welfare, through its programs for the so-called isolated and estranged people (masyarakat terasing). Tsing suggests that such people play an ideological role disproportionate to their rather modest numbers (about one million, Departmen Sosial 1994a:1) as they:

> have quietly become icons of the archaic disorder that represents the limit and test of state order and development. From the perspective of the elite, "primitives", unlike communists, are not regarded as seriously dangerous but rather as wildly untutored — somewhat like ordinary village farmers, but much more so. Disorderly yet vulnerable, primitives are relatively scarce, and their taming becomes an exemplary lesson in marginality through which the more advanced rural poor can be expected to position themselves nearer the centre (1993:28).

The continued presence of (a few) primitives on the periphery could be viewed as evidence that state projects to extend order and control remain incomplete. Yet, as Tsing suggests, their presence also serves as an opportunity to "re-state" and justify "development" agendas. Officials contrast the strangeness of masyarakat terasing to the positively valued, comforting homogeneity of the "ordinary, average, Indonesian" (*bangsa Indonesia secara rata-rata dalam keseluruhan*", (Haryati Soebardio, Minister for Social Affairs 1993:vii).[20] Guidelines for fieldworkers list the backward characteristics of masyarakat terasing and specify the physical, social and administrative elements needed to render them "normal" in national terms[21] (Departmen Sosial 1994b). Increasingly, they are permitted to remain in their current locations (reorganised into newly built, standardised, more orderly settlements) but the model of "development" to be pursued is pre-defined.[22] Thus the presence of the "other" helps to affirm the characteristics of the "normal developed Indonesian person", overriding, for a moment, the vast regional, class and other differences that characterise the Archipelago.

Local Hierarchies and Channels for State Power

For most Indonesians living in the cities or lowlands of Java, upland shifting cultivators or forest product collectors are very remote. TV images of wild people and places might confirm to viewers the superiority of their own modernity, but the groups most affected by their presence are those who encounter them directly: managers, officials, workers, transmigrants and farmers living in the uplands. These groups compete with them for resources and struggle to assert control over their land and labour. Among these diverse groups of outsiders (some of whom have been present for decades) the ease with which the local or "indigenous" people can be labelled primitive permits their claims to be ignored or dismissed (Dove, Ellen, this volume). Heterogeneous groups of newcomers can define for themselves an identity which centres on being, at the very least, more "developed" than the people already in place.[23]

Ethnic stereotypes apart, both newcomers and long time residents of the uplands are divided along class lines, and their relationships with each other can be quite complex. Imported plantation labourers often seek to re-establish themselves as peasant farmers on the borders of estate land (Stoler 1985; Ruiter, this volume) and transmigrants unable to survive and prosper on the farmland allotted to them seek additional land (Hidayati 1991, 1994). Those with powerful patrons may simply take over, but many acknowledge that the locals have prior and legitimate land claims and seek to establish alliances or relations of clientship with them (Ruiter, this volume). In these interactions they reverse the state-sponsored ethnic hierarchy which asserts the superiority of the Javanese (and lowlanders) over uplanders (and natives). The position of "indigenous" leaders may also be complex. Some become brokers of land to outsiders and may betray the interests of their kin and co-villagers in favour of building alliances with newcomers and with patrons promising access to jobs, resources, or state power (Tjitradjaja 1993). In the uplands there is no simple correlation of ethnic identity with class status; nor is there a single trajectory of change prefigured by state plans and programs.

In frontier conditions such as those along the logging roads criss-crossing the uplands and interior of Kalimantan and other islands, the machinery of the state is often weak, and its representatives few. It is local patrons who act as brokers of state-derived power and largesse (land rights, credit, subsidies, licenses, cf Hart 1989). Whether patrons are members of well-established local élites, wealthy newcomers, officials drawn from or posted to the regions, or some of these in combination, it is their links to specific state institutions, to senior officials, and to the idea of "the state" as a locus of legitimate authority that provides them with power in and over the periphery (Kahn, Ruiter, this volume). They help to operationalise, while at the same time they personify, various forms of institutional power and control. Patronage is quite clearly reinforced by "development", as some profit from state hand-outs while others need more assistance to negotiate the increased array of rules and bureaucratic procedures that are part and parcel of territorialising processes in general and "development" in particular (White, this volume; Ferguson 1994). As Kahn argues (this volume), patronage is one of the characteristic ways that power works in upland settings. The fact that particular laws may be enforced or may instead be waived by "lenient" officials; the ways in which upland people are able to play one agency or official off against another; and the various forms of accommodation as well as conflict that arise are not accidents, but rather *features* of the relationship between "the state" and its citizens.[24] Patronage is one way in which the categories centre and margin are articulated.

Strengths, limits and contradictions in state power, "development", and green agendas

Not surprisingly, in view of the range of state agencies, corporations and local interests competing for pieces of the upland terrain, mapping efforts have intensified but they have not produced either clarity or consensus. In Kalimantan a major mapping effort with foreign co-operation was undertaken with a view to defining areas available for "development", especially transmigration and oil and rubber estates (Repprot, cited in Peluso 1995:389–391). The report associated with these maps makes a claim for interpreting the "unclassified" portion in a manner favourable to the Department of Agriculture but disadvantageous both to the Department of Forestry and to customary land users.[25] It is an interpretation which can (and probably will) be disputed piecemeal, as tussles occur over initiatives to realise the Repprot vision in particular places, conjuring up alliances and oppositions of a varied nature indicative of "room for manoeuver", rather than the unilateral imposition of order as defined by a singular locus of state power (cf Tsing, this volume).[26]

Significant as state programs are, change in the uplands is not engineered from a single source. Some of the effects of state programs are unintended, indirect, or unforeseen. State programs for the uplands can also be actively disrupted and resisted through evasion, breaches and non-compliance. This is the case in Thailand, where Vandergeest and Peluso (1995:416) observe,

> Government agencies are continually reclassifying and remapping territory to account for how people have crossed earlier paper boundaries. State land management agencies are forced to recognize local rights deriving from local classification, modes of communication, and enforcement mechanisms. Programs such as those awarding limited land rights to cultivators in reserve forest

areas are simultaneously a state attempt to contain people's activities and a state response to what people had done to undermine previous such policies.

In Indonesia, the emergence of social forestry programs, the inclusion of forest-villagers in transmigration programs, renewed attempts to enforce forest boundaries, moves to bring administrative order to spontaneous land settlement, failures of resettlement schemes and increased willingness to service so-called *masyarakat terasing* in their ancestral areas all indicate that territorialisation projects of various state agencies have met with considerable degrees of resistance, and are being revised and reformed accordingly.

Just as the power of state institutions is not absolute, it must be stressed that it is not necessarily malevolent: as noted earlier, territorialisation is a normal activity of governments, not one peculiar to oppressive regimes. Environmentalists and supporters of peasant struggles who assume that "traditional" communities are inclined to oppose "the state" in order to preserve "their own" institutions and practices may overlook the extent to which uplanders seek the benefits of a fuller citizenship.[27] Their demands commonly include access to roads, education, and health facilities. The oppositional characterisation of "virtuous" peasants and "vicious" states (Bernstein 1990:71) fails to do justice to the complexities of state-local relations and associated class structuring processes (Nugent 1994; Hart 1989). It neglects also the claims upon state institutions and programs for *access* to modernity which characterise many peasant and indigenous people's movements (Schuurman 1993:27), just as others reject and resist state imperatives.

In the case discussed by Tsing (this volume), Meratus people did not oppose the territorial strategies reflected in mapping and road building; rather, they wanted to ensure that their community was *on* official maps and roads, a regularised and recognised component of the national framework. Similarly, administrative changes which subdivided previously unwieldy units into smaller ones reflecting actual patterns of settlement and interaction were experienced by Meratus not *only* as intensifications of rule, but also as opportunities to form communities capable of interacting more effectively with state programs and obtaining access to development benefits. In the case described by White (this volume) villagers joining a contract farming scheme saw as its principal benefit the provision of official land titles to replace previous tenure arrangements that had proven insecure (even though, in this case, the insecurity had been caused by a state-sponsored program which had reallocated "their" land for new political and economic purposes). Complaints commonly encountered in the Indonesian uplands relate not to the concept of "development" as articulated by "the state", but to particular, localised experiences with a development which removes sources of livelihood without providing viable alternatives, fails to bring promised benefits or distributes benefits unevenly (Peluso 1992; Ellen, this volume).

The evidence from these local studies seems to indicate that critiques of Indonesian state policies which propose that "the state" should retreat from the uplands, and leave indigenous people to manage their own affairs in their "traditional" ways, misrepresent the nature of upland life just as seriously as the "development" agenda they oppose. Promoted by national and international élites, and well intentioned (just as many government efforts are well intentioned), "green" agendas also depend upon simplifications and stereotypes. Ironically, the discourses of development and anti-development seem to rely upon the very same simplified images to advance diametrically opposed agendas. As Tsing (1993:32) points out: "Ecological activists argue for the conservation of Bornean rainforests based on images of nature-loving primitive tribes. Such images of primitive conservatism are also used by developers to prove the necessity for progress in the form of forced resettlement and export-oriented resource appropriation".

Implicating the Indonesian state in forest destruction, proponents of "green" development models have declared indigenous or forest-based communities to be the appropriate keepers of this resource. Such communities are imagined to possess a set of characteristics counterposed to those of state agencies and other forest destroyers: they are tribal, or at least backed by centuries-old environmental wisdom; they have long been located in one place, to which they have spiritual as well as pragmatic attachments; they are relatively homogenous, without class divisions; they are not driven by motives of exploitation and greed; they have limited consumption requirements; and their collective desires focus upon the long-term sustainable management of forest resources for the benefit of future generations.[28] For example, the environmental advocacy group SKEPHI emphasises the harmony between people and with nature that characterises "traditional" communities: "Under the traditional *adat* land rights system, land is regarded as the common property of the community ... it is inalienable ... people do not own the land on which they live and work; they merely control it ... land belongs to God as the creator ... In this way, respect for the land and its resources is maintained for the benefit of the community as a whole" (SKEPHI and Kiddell-Monroe, 1993:231–

2). The implication of their account is that "tradition" (unlocated in time, geography or context) remains present in Indonesia's "indigenous" communities, ready to be revitalised by enlightened state and NGO backing for "community forestry under the control of existing traditional institutions" (1993:262).[29] The same account notes problems created by state policies in the uplands, such as displacement by forest concessions, transmigration schemes and nucleus estates. Yet the changes these much-critiqued programs have brought to the uplands over the past five decades — new people, new landscapes, new communities, new classes, new infrastructure, new needs and desires, new engagements with market and state — are not factored into the solution proposed, one that looks principally to "tradition" for salvation.[30]

Similarly, those concerned with the "right to survival of endangered cultures and tribal peoples" note the vulnerability of such people to the penetration of exploitative capitalism supported by state interventions to "modernise" and "develop" them. Yet they still imagine a group of "tribals, aborigines, natives, minorities, highlanders, forest people" sufficiently exempt from these processes to be "autochtonous minorities", authentic candidates for protection (Lim and Gomes 1990:2). Scholars and activists promoting common property regimes in the region also base their arguments upon a "communal presumption" rooted in the image of intact tribes.[31] But tenure patterns have shifted historically with changing forms and relations of production in ways which cannot be captured by visions of an apocalyptic collapse of autonomous communities facing market and state "intrusions".[32] Development models emphasising sustainable development, community based resource management, and social forestry also assume that upland people have a natural environmental ethic, limited current market engagements, strong community bonds and minimal class inequalities (e.g. Kepas 1985). While some acknowledge the political and economic complexity of the uplands (for example, the introductory chapter of Poffenberger 1990), they proceed quickly to the practical matters of "tools and techniques of participatory management" and "empowering communities through social forestry" (Poffenberger 1990), assuming a consensus on the future of the uplands which is superficial at best.

The simplifications noted in the practitioner literature occur in part through the elision of categories. Uplands become equated with forests while farms are ignored, especially commercial farms and plantations. Upland people become equated with indigenous people, tribal people, traditional people, forest-dependent people or shifting cultivators, belying the enormous diversity of ethnic groups and social classes that occupy upland terrain, and the alliances and identities formed in situ.[33] An overwhelming image is that of victim, struggling against "outside forces" (capitalism, plantations, concessions, state development schemes) to maintain something old, but hardly engaged in creating something new. These elisions are not merely oversights, or the result of inadequate data. They are simplifications necessary to the critique of state policies, which rely, as noted above, on equally simplified representations of upland lifestyles and ecology. Instead of a dialogue between the state and its critics, a mirror effect simply inverts the categories (wise swiddener/ destructive swiddener, valuable traditions/backward traditions) leaving the categories themselves essentialised and fundamentally unchanged. In between these opposing camps, uplanders must invent especially creative strategies in order to defend their livelihoods and advance their own agendas, attempting to turn both state and "green" discourse to their own ends (see Tsing, Ellen, Dove, this volume).

Like tradition and culture, environment and development turn out to be deeply embedded in questions of representation and power with both local and global dimensions (Bryant 1992, Moore 1993, Blaikie and Brookfield 1987). In the context of struggles over resources, "people may invest in meanings as well as in the means of production — and struggles over meaning are as much a part of the process of resource allocation as are struggles over surplus or the labour process" (Berry 1988:66). Such struggles take place within households, between communities, between communities and the state and now, increasingly, between communities and a global environmental lobby.[34] They frequently invoke divergent images of community as a remembered past, a contested present, and imagined future.

PRODUCTION, ACCUMULATION, AND IMAGINED FUTURES

The familiar but indispensable investigation of the social relations of production and accumulation provides the third analytical frame through which upland transformations can be exposed. Also important is an examination of the culturally formed wants (Hefner 1990) pursued by uplanders and reflected in their patterns of production, investment and consumption. As Hefner observes, "Economic change is never just a matter of technological diffusion, market rationalization, or 'capitalist penetration'. Deep down, it is also a matter of community, morality and power" (1990:2).

Production

In keeping with the general assumption of marginality, there is a widespread perception in Indonesia that livelihoods in the uplands are inadequate and poverty widespread. Yet, as indicated earlier, large populations have been attracted to the uplands for their agricultural potential both for subsistence and for commerce. As is the case in the lowlands (Alexander *et al.* 1991), some of the more lucrative livelihoods have not been agrarian at all, but focused rather on extractive industries, trade and wage work. There has been much discussion of the ecological limits of upland agriculture and the fact that agriculture sometimes takes place in locations where, according to some technical or administrative criteria, it should not, has obscured consideration of the types of production that actually take place in upland settings. The "environmentalisation" of the uplands, together with the "culturalisation" already explored, strongly colour the ways in which upland landscapes and production systems are apprehended. Both the forms of upland production and the distortions commonly encountered in their characterisation are the subject of this section.

One reason for the assumed inadequacy of upland livelihoods is the consistent underestimation of both the extent and the productivity of upland agriculture. The conceptual hierarchy prioritising rice is such that upland food crops (maize, cassava) scarcely figure in archival records and are still under-recorded in official statistics (Boomgaard, this volume). Off Java, colonial administrators abhorred the apparent disorder of upland farms and sought to impose Javanese style sawah, a system of cultivation requiring intensive management, water control, and a diligence reminiscent of the Dutch polder (Colombijn 1995). They ignored or underestimated the smallholder cash crop production which has been a pervasive and long-standing characteristic of swidden systems (Dove 1985b).[35] Throughout the colonial period, according to Henley and Colombijn (1995:3),

> Where an innovation was indeed viable ... the local population often recognised its value long before the Dutch did. Coffee, pioneered by indigenous farmers in both Minahasa and West Sumatra, is a case in point. The main effects of the subsequent Dutch interference in coffee cultivation ... were simply to lower profits, to increase the element of compulsion correspondingly, and to shift production away from mixed swidden fields and home gardens into monocultural plantations which both interfered with subsistence farming and promoted soil erosion.

The significance of even large scale production within the uplands has sometimes been underestimated. The labour mobilised for upland coffee at the height of the Cultivation System was two to three times that mobilised for lowland sugar (White 1983:28, cited in Hefner 1990:41). Colonial profits, ecological impacts and social dislocations associated with coffee were immense (Hefner 1990:42), yet they have received less scholarly attention.

The proportion of Indonesia's agriculture that takes place in the uplands is considerable. On Java, an island strongly associated in official discourse and the popular imagination with lowland sawah, rainfed lands and house gardens, located primarily in the uplands, account for more than half of the cultivated land (Hefner 1990:16). The extensification of agriculture in Java's uplands is continuing as a result of pressure from growing populations and also as a result of the government goal of rice self-sufficiency, which has pushed non-rice annuals (*palawija*) onto upland fields (Hardjono 1994:183). Another pressure is the attractive profit to be gained from temperate fruit and vegetable crops, increasingly in demand in urban markets (Hardjono 1994:184; Suryanata, this volume). Outside Java, agricultural extensification has occurred to meet national export targets, with three million hectares converted to tree crops alone between 1971–86 (Booth 1991:54 cited in Hardjono 1994:201). Plantation lands in Indonesia already cover 35 million hectares, and it is estimated that a further 2.8–5.6 million hectares will be needed by the year 2000 (Bappenas 1993:12).

The area devoted to dryland agriculture, including shifting cultivation, has also increased in the past few decades. In the hilly interior of East Kalimantan, for example, plans to expand sawah during the period 1963–1983 fell short, but the dryland farm area increased from 96,000 hectares to 155,000 hectares (Hidayati 1991:38). Much of the expansion was the work of in-migrants, including transmigrants who fled failed schemes or expanded beyond the edges of their designated two-hectare plot as they adopted the extensive swidden techniques of their indigenous neighbours (Hidayati 1991:43, 1994). Some found themselves stranded by downturns in the logging or oil sectors which first drew them to the province while others were attracted to the province specifically for the opportunity to grow lucrative cash crops such as pepper, while meeting food needs through swiddening (Hidayati 1991:38, 45). In Sumatra too, smallholder cultivation has expanded into forested areas as a result of improved communications, current and expected land pressure due to transmigration, and the recognition that land claims are strengthened under both adat and national law by the presence of perennials (Angelsen 1995).

Some of the agrarian changes that occurred in East Kalimantan and Sumatra were not the developments which were planned;[36] nor are in-migrants and indigenous farmers who are interested in claiming land and making profits from trees the kinds of people normally highlighted in discussions of shifting cultivation. Agricultural sector analyses tend to focus more on changes that are policy generated than those that are the unplanned, "illegal" or, from a state planner's point of view, perverse outcome of millions of separate, smallholder decisions (Barbier 1989). Yet smallholder tree crops, mostly located in the uplands, contributed 12.3% to Indonesia's gross domestic product from agriculture in 1992, while tree crops grown on estates (concentrated in Sumatra and West Nusa Tenggara) contributed only 4.9% (Barlow 1996:8). Para rubber was for some time the country's third largest foreign exchange earner after oil and timber, 76% of it "produced in tiny gardens of a hectare or so, with century-old technology, by so-called "smallholders"" (Dove 1996:43). Because their operations are small in scale, and because many of them are carried out within "forest" boundaries and without state control or assistance, it is the low technology, errant location and indiscipline of smallholders that are observed, rather than their significant contribution to the national economy. As Dove (1996:47) suggests, estate development has been given priority because it "suits the general, overarching governmental imperative of centralised control and extraction of resources, whereas smallholder development only frustrates this imperative". Smallholders reject estates or contracting schemes for these very reasons, as illustrated by Ruiter, Dove and White (this volume).

Characteristically, several types of activity and several sources of livelihood are combined in upland settings. Yet the models found in government reports are misleadingly simple. For example, the official document outlining the identification criteria for shifting cultivators and forest squatters envisages two quite distinct types of farmer. 1) The shifting cultivator, strongly traditional, resistant to change, engaged in production oriented primarily towards subsistence, who uses tree crops only as markers of land ownership and control. 2) The forest destroyer, whose main purpose is the illegal control of land for the purpose of growing high value crops such as pepper, coffee and cocoa (Departemen Kehutanan 1994).[37] A similar dichotomy is found in much academic and advocacy literature, which distinguishes integral swiddeners (the traditional, ecologically benevolent ones) from shifting cultivators, assumed to be newcomers, interested in only in cash and unconcerned about environmental damage (Hardjono 1994:202; Kartawinata and Vayda 1984; Barber *et al.* 1995:10).

The dichotomous classification of farm types and farmers tends to reproduce the more general traditional/modern dualism critiqued earlier. It shares the problem of stereotyping, and underestimates the diversity, complexity and productivity of upland farming systems. In terms of farms, it emphasises extremes (sawah versus swidden, or plantation versus swidden), but glosses over much of the significant middle ground: tegal, dryland, tree grove/swidden combinations, commercial crops on smallholdings or in fallows, farming strategies which include both upland and sawah components.[38] It also misses seasonal combinations that include fishing or wage labour, the activities of wage workers who farm around the edges of plantations or on logged over land, structural shifts in which farmers abandon their fields for a few years when better incomes are available in logging or construction work, and many other livelihood options that upland people have identified and pursued (Brookfield *et al.* 1995; Lian 1993). Finally, it tends to treat the upland farmer as a singular male subject, ignoring the diverse livelihood activities pursued by women, and the ways that these are restructured in the context of broad shifts in the regional economy, technical innovations, male wage migration, and official schemes and projects such as the contract farming scheme described by White, or the commercial fruit ventures described by Suryanata (this volume).[39]

In social terms, the dichotomous model has only a limited space for recognition of the indigenous farmer who is innovative, dynamic, aware of market prices, very much interested in cash, and making investment decisions accordingly. The presence of such a character is somewhat acknowledged in Java. Hardjono (1994:188) notes that upland farmers are pragmatic and opportunistic, seeking maximum returns. But her remark is not intended to highlight how familiar and normal are the motivations of upland farmers; rather she observes that upland farmers are pursuing profits at the expense of the environment.[40] Off Java, the pragmatic profit-seeking upland farmer is even less visible. At best, it is noted that the categories defined by dichotomous models can become "blurred". But the strongly marked image is that of the uplander as victim of market and state, as when Barber *et al.* (1995) discuss swiddeners "forced to intensify" through planting tree crops, but hardly entertain the possibility that these farmers may in fact choose to plant these profitable tree crops for all the normal economic reasons.

Ecology

In the past two or three decades, research and policy have both tended to focus upon the ecological limitations of upland production. Observers note the poor quality of some of Java's upland soils (Hardjono 1994:188) and the serious consequences of intensification upon a fixed land base: "Upland *tegal* lacks the ecological resilience of irrigated sawah. Worked intensively, it loses its fertility. Exposed to winds and rain, it erodes. Cultivated without fallow, it becomes an ideal medium for fungi and insect pests" (Hefner 1990:16). Green revolution chemicals have reversed some of the effects of fertility decline, but have not reversed erosion, and these chemicals have had social and ecological impacts of their own (Hefner 1990:81–112). When pushed past a certain point, the regenerative potential of upland ecologies is limited, or is a very long term prospect.

It is also possible to exaggerate the risks of agricultural intensification in upland settings. Estimates of average fallows in Minahasan subsistence farming indicate that a "surprisingly intensive swidden cycle was maintained over a period of several decades without apparent loss of soil fertility" (Henley and Colombijn 1995:4). Land in Sulawesi reported by colonial officials to be "worn out" in the 19th century is still in productive use today. In Kalimantan, according to Brookfield *et al.* (1995:29–30, 135–136), large areas of what now appears to be primary forest were occupied in the past, and population in many interior areas was more dense 1–300 years ago than it is today. Brookfield *et al.* (1995:229) conclude that "trajectories can be reversed when driving forces and conditions undergo major change". Potter (1996:25) suggests that the number and impact of spontaneous settlers, the "land-hungry migrants" observed practising shifting cultivation alongside logging roads may also have been exaggerated. Rather than consuming ever-greater amounts of forest and spreading out into the interior, they tend to stay close to the road itself as their point of access to markets. Similarly, when people in inaccessible villages are attracted to live near new roads, they increase resource pressure in the road corridor but presumably relieve pressure in their areas of origin (Potter 1996:26).

Thus the history of upland agriculture cannot be reduced to a story of ecological ruin. Nor is ecological change necessarily a story of impoverishment. Smallholdings carved out of the forest may become stable and profitable. The various state programs transforming the uplands (estate crop plantations, timber concessions, transmigration etc.) have different rates of economic return overall as well as for the various social groups affected by them. In terms of their capacity to support people over the long term, however, the verdict of Brookfield *et al.* on both the forest industry and large scale land conversion for agriculture in Kalimantan is guarded but positive (1995:83–111).

Accumulation

Directly or indirectly, simplified images of upland production and upland producers shape policies and programs, and also shape the characteristic patterns of accumulation that arise in upland settings. The image of upland farming as marginal or "near zero" facilitates its displacement by official schemes which assign the land to other purposes (Dove 1987; Hardjono 1994:203; Lynch and Talbott 1995:98). Once displaced, farmers are forced to reorganise their productive activities (yet again) "on the margins" of official schemes and agendas (Schrauwers, Kahn, Ruiter, White this volume). In many cases they are pushed onto higher or more sloping terrain, confirming the impression that they disregard the environment. Robbed of their assets, they meet the expectations of poverty and disorganisation that require further "development" intervention.

Assumptions about the cultural otherness of upland people may also be translated into development initiatives which disadvantage them. Dove critiques conservationist agendas which propose that "minor forest products" be promoted to meet the (apparently limited) income needs of forest-dwelling people. It is lack of power which excludes such people from enjoyment of the most profitable forest product (timber), and they are regularly punished for converting land to profitable tree crops. Should their land or some "minor forest product" become especially profitable, these too are usually taken away from them: their poverty renders them ineligible as beneficiaries. Therefore, according to Dove, the "search for "new" sources of income for "poor forest dwellers" is often, in reality, a search for opportunities that have no other claimants — a search for unsuccessful development alternatives" (1993:18). Poverty, powerlessness and exclusion from valuable resources are integrally related. Such economic and political linkages are obscured when forest communities are viewed through a lens that stresses cultural difference and prioritises conservation, while implying that marginality is an elected way of life.

Agroforestry programs have been especially favoured in recent years, because they seem to offer a singular solution to the problematic of upland marginality and a script for action acceptable to many parties: they satisfy environmentalists by conserving soil and planting trees to intensify production on already-cleared land; they meet state-defined development objectives by stabilising farm locations, increasing market production and relieving poverty; and they satisfy social activists concerned to see the land rights of upland populations regularised and secured, and their local knowledge and capacities respected. Missing from this picture, however, is an appreciation of uplanders' interest in, and vulnerability to, processes of accumulation. On this point several of the enabling assumptions (about priorities for conservation over production, egalitarian patterns of resource access and labour investment, and limited consumer desires) founder. Agroforestry technologies routinely fail to deliver the increased incomes promised, especially in relation to the labour invested, and some programs have resorted to coercion (see Lee 1995).[41] When economically successful, programs may not conserve forests but rather encourage an expansion of agriculture into forested area as local farmers seize new opportunities rather than sitting back when their (supposedly limited) needs are met (see Tomich and Noordwijk 1995; Angleson 1995). Another scenario has emerged in upland Java (Suryanata, this volume) where a technically successful transition to fruit-based "agroforestry" set up patterns of accumulation in favour of established village élites. These are not exceptional situations, and it is not clear that they can be rectified by better technologies and program incentives. They are the predictable outcomes of changing patterns of production and the dynamics of culture and class in contemporary upland settings. They tend to be ignored, overlooked or explained away in order to protect the "strategic simplifications" embedded in agroforestry initiatives.

The uneven reach of government programs also impacts on patterns of accumulation in the uplands, as it did in the lowland green revolution (Hart *et al.* 1989). Relatively small amounts of government credit are allocated to upland farmers (Palte 1989:208, cited in Hardjono 1994:184). Lack of subsidised credit and tenure insecurity in turn increase the problems of predation or exploitation by patrons and brokers noted earlier. In particular, high capital requirements and long waiting periods for the establishment of tree crops provide avenues for those with capital to acquire a stake in upland production, and sharecropping arrangements and indebtedness leading to loss of control over resources are typical (Poffenberger 1983; Li 1996a, 1997; Suryanata this volume).

Low prices for upland products are both a cause and a symptom of farmer vulnerability. Unlike lowland rice, which is destined for national, price-controlled markets, much upland produce is destined for international markets where prices are characteristically unstable. They are subject to boom-bust cycles which cannot be controlled at the national level (Hardjono 1994:185). Even when prices are high, however, the marketing chain reflects power imbalances which ensure that upland farmers receive only a small percentage of the gains. A recent reduction in the farmgate price for cloves, for example, reflected the increased profit margin officially assigned by the government to the clove marketing board (Sondakh 1996:161).

In Java, official sanctions and harassment of those living in and around "forest" areas reduce the bargaining power of both farmers and labourers (Peluso 1992). Despite the promises of social forestry, processes of class structuring proceed apace. According to Barber (1989) the predominant local labour relations (including unofficial absentee ownership and sharecropping) are replicated on "forest" lands, and familiar patterns of state patronage are well entrenched. On the most fertile "forest" lands, forester's dispensations and connections to local government and the military permit favoured élites to engage in intensive commercial agriculture, thus positioning them to capture any subsidies or benefits. Only the most degraded land, yielding the lowest possible returns to labour and capital, is generally made available to the poorest people under social forestry schemes. Thus the patterns of accumulation taking place in the "forest" zone in Java are masked by the "forestry" rhetoric in general, and social forestry rhetoric in particular, with its suggestion of reduced inequalities and "sustainable" production of trees and other crops.

In the "forests" off Java, it is common for powerful individuals to claim large areas along logging roads and then "sell" the land to latecomers, or have them work it as sharecroppers (Barber *et al.* 1995:3, 49). These forms of land-grabbing and predation exemplify the extreme inequalities of power that shape production relations in some upland settings. On the other hand, as noted earlier, where upland people are in control of their land, their labour is comparatively expensive or unavailable to would-be employers, placing a limit on some accumulative schemes. Contrary to the assumption that "traditional" swidden cultivators are the poorest people inhabiting the uplands, Brookfield *et al.* find good indications from contemporary Kalimantan that many of these people can and do adapt, intensifying their agriculture in quite productive and sustainable ways (1995:112–142). They suggest that the most vulnerable people currently in the uplands are relative newcomers: transmigrants who find themselves in impossible sites, or migrant workers left stranded by the timber industry's decline. In many cases, they cannot return whence they came, and must find new work on terms which reflect their poor bargaining position. If they cross borders and become illegal migrant workers on Malaysian territory, they encounter further jeopardy (1995:235). Ruiter (this volume) describes impoverished, landless Javanese former plantation workers drifting around the Sumatran hills, looking for work, land and security with varying degrees of success.

Through these various mechanisms, some centrally designed and others unplanned, land ownership, land and tree tenure, work opportunities, wages and profits from upland production accumulate unevenly. It is necessary to restate, however, that uplanders may consider that they have much to gain, as well as the potential to lose, from their changing engagements with the market. There are both push and pull factors at work. The extent to which the remarkable transformation of upland agriculture which has occurred over the past twenty years has been smallholder initiated (rather than the result of direct or indirect compulsion) strongly suggests that the image of victim is too simple. As with "development" and state schemes, the complaint is not against market oriented production, but against the terms under which such production takes place, terms which reflect the uneven distribution of power.

Imagined Futures

Corresponding to the political and economic changes in which upland peoples have participated in the past few decades are changed notions of community and identity, new desires and aspirations, and revised images of what the future may hold. These are not easy to apprehend directly, since public statements about both community and identity change more slowly than the underlying sets of practices and relationships (Hefner 1990). Changing consumption styles and investment priorities provide some indication. Ruiter and Suryanata (this volume) find profits from upland agriculture being invested in improved housing, education and, in Muslim areas, the Haj pilgrimage. Villagers in the upland Citanduy area studied by Henri Bastaman (1995) have been acquiring the equipment for more elaborate, lowland-style food preparation, making trade in these items an important source of off-farm income. Well connected and wealthy villagers (who, incidentally, benefited disproportionately from a project designed to make upland farming more sustainable) were the first to seek investment opportunities in trade, transport and other sectors with higher status and more potential for profit than their upland fields. In upland Sulawesi, women and men have expressed to me their desire to be able to eat rice and fish and drink sugared coffee on a regular basis (like coastal people), and to sit back in between weekly harvests of cocoa rather than working and weeding endlessly in the swiddens. Yet they also remark that they cannot understand how the majority of coastal people, who are landless labourers, are able to survive on a daily basis without food gardens (Li 1996a, 1997).

Hefner (1990) illustrates some of the more subtle political-economic dimensions of intensified upland agriculture in the Tengger highlands. New crops and their associated technologies have tied people to markets to an extent unprecedented in the long history of upland market production, and cash crops have replaced food crops in many areas. They have required costly inputs, and the environmental damage resulting from some of them rules out the possibility of reversion to former crops and modes of subsistence should market opportunities recede. Yet, contrary to the "green" assumption that traditional/upland/peasant farmers would "naturally" give priority to subsistence and sustainability concerns, Tengger farmers and those studied by Suryanata and Bastaman have indicated a rather different vision. Many do not desire or anticipate a future for their children on the land, making it logical to invest profits, including those which result from environmentally unsound practices, in education and urban lifestyles. They thereby seek to equip their children with both the economic and cultural capital to leave the uplands and integrate into the urban/lowland mainstream. Others, reduced to the status of wage labourers, may also exit the uplands but for different reasons and on different terms. Associated with greater inequalities, therefore, are new upland identities and aspirations, and new visions of self and community.

There is no single model for the transformation of upland livelihoods. Some upland areas have become more populated, drawing in wage workers and land hungry migrants. In others population has declined, a consequence of ecological collapse, limited opportunities or the pull of higher incomes obtainable elsewhere. Upland populations have different degrees of attachment to their current locales and different degrees of commitment to an agrarian future.

CONCLUSION

In this chapter I have argued that various aspects of real and supposed marginality are conflated in the uplands. They are regarded as distant places, fragile ecosystems, poverty stricken villages inhabited by "different" kinds of people and, from a political and military perspective, trouble zones. The conflation of overlapping dimensions of marginality produces an apparently "natural" fact, masking social and economic processes and the operation of power. In attempting to account for the marginality of the Indonesian uplands I have linked cultural practices and the production of images and representations to political-economic systems of accumulation and control. I have proposed that such an approach raises questions capable of illuminating the processes at work in Indonesia's uplands, and generating explanations of a comparative nature.

Drawn from this perspective, a list of questions might include the following: How does the image of upland terrain as unproductive (physically and economically marginal) sustain a definition of the uplander as backward (socially marginal)? How does an assumed economic and social marginality promote and legitimise specific systems of accumulation (logging, conversion to estates, usury, price fixing)? When plantations prove unprofitable, "development" efforts "fail", or displaced farmers become unruly, how does this serve to confirm initial assumptions about the overwhelming marginality of the uplands? How is it that the intensification of rule reproduces or intensifies marginality as one of its effects? How do margins look from the perspective of people who live or work in "out of the way places" (transmigrants, plantation workers, expatriate mine managers, officials), yet who consider themselves to belong to "centres" located elsewhere? How is marginality constructed culturally by the people whom outsiders take to be marginal by nature, and what are its political and economic consequences? How do margins and centres, communities and identities, become redefined in the context of changing aspirations and commitments? What dimensions of change become visible once we move away from the restricted vocabulary counterposing an unexamined "tradition" to a narrowly envisaged "modernity"? Some of these questions are addressed in the contributions to this volume, while others remain to be explored.

ACKNOWLEDGEMENTS

I would like to thank Tim Babcock, Peter Boomgaard, Roy Ellen, Ruth McVey and an anonymous reviewer for detailed comments on an earlier draft of this chapter; also Diana Wong, Phillip Eldridge, Philip Morrison, Neil Byron and Donald Moore for their helpful suggestions.

NOTES

1. See Ferguson (1994), Davies (1994).
2. See King and Parnwell (1990) for a parallel examination of marginality in Malaysia.
3. The term coined by Kahn, this volume; see also Padoch and Peluso (1996) on the predominance of pristine nature and exoticized culture in historical and contemporary images of Borneo; a similar point is made by Ellen (this volume).
4. See the converging debates on this topic in anthropology (Comaroff and Comaroff 1992; O'Brien and Roseberry 1991; Gupta and Ferguson 1992) and development studies (Pieterse 1992; Pred and Watts 1992; Watts 1993).
5. H. Geertz (1963: 12) offers a classic statement on the unchanging uplands.
6. See Wilk's (1997) critique of the "impact myth", and references cited earlier for problems with its dependency theory variant.
7. Fox (1992) describes the agronomic changes, but not their social and political dimensions.

8   See Henley (1994: 44); Dove (1985b); de Koninck and McTaggart (1987). Healey (1985: 8) mentions irrigated wet rice production in Kalimantan in highland areas beyond the reach of state authority but does not explain its rationale.
9   The riverine model of extractive relations works for much of the western archipelago, but not Timor or Maluku. See Andaya, L. (1993), Ellen (1979).
10 See Kahn (1993: 78–110), Hooker (1983) and Ellen (1976) on the intellectual, economic and political rationales for colonial era adat law.
11 Li (1996a) describes Sulawesi highlanders who do not stress an ethnic identity nor envisage themselves as a bounded tribe; Li (nd) discusses historical and contemporary processes of "indigenous" identity formation.
12 See, for example, the discussion of ad hoc adat-making processes in Tsing (1990).
13 Under the New Order cultural difference has been bureaucratically "rationalised" at the provincial level, while local distinctions are ignored.
14 As Pigg (1992) shows in relation to Nepal, such scripts are not created nor imposed unilaterally but generated through cultural processes at a range of levels and sites.
15 See Joseph and Nugent (1994). Abrams (1988) usefully distinguishes between the idea of "the state" as a unified source of intention and power, which is an ideological construct or mask, and the state-system, the institutions of political and executive control and their key personnel. The state-system, through its everyday operations, produces (and disguises) the relations of power on which the reified idea of "the state" is based. See also Mitchell (1991).
16 I focus here on disciplines which are territorially based but see Mitchell (1991) on other state initiatives which intensify rule while rendering power internal to everday practices. See also Ferguson (1994).
17 The desa reforms through which the colonial state claimed to be restoring villages to their "traditional" condition before the advent of despotic princely rule instead merely recreated despotism through the land monopolies and corvée rights granted to village officers; the difference was that the villagers were now "subjects" rather than personal dependents (Breman 1980: 26–27). Boomgaard (1991) argues, contra Breman, that the pre-colonial Javanese village was a significant moral community, if not a unit of rule.
18 De Koninck and McTaggart (1987: 350–1) argue that state directed land settlement schemes have a circular logic, routinely recreating the inequalities and impoverishment they are designed to redress since they produce not stable peasant environments but dynamic commercial ones in which labour and land are commoditized.
19 De Koninck (1996) makes this argument for Vietnam.
20 The cover design and many of the photos in Koentjaraningrat (ed.)1993 show near-naked Irianese engaged in exotic dances and unfamiliar tasks, emphasizing the alien and primitive nature of masyarakat terasing. Questions of history and political economy are hardly mentioned; each chapter presents an apparently isolated "tribe" recently encountering change; current land struggles are not discussed.
21 Normality in Indonesia includes adherance to a world religion. As Gibson (1994) points out, it is religious conviction as much as administrative fiat which motivates officials and others to redress the spiritual poverty and pollution, as well as material poverty and disease, of those they envisage as primitives.
22 State Ministry for Population and the Environment, Act Number 10, 1992 acknowledges the right of indigenous people to retain cultural diversity as well as traditional land, but guidelines for implementation have yet to be developed.
23 See, for example, Robinson's (1986) description of the class and ethnic dynamics of an upland mining town.
24 Relations between villagers and NGOs or corporations (especially major employers) may share in some of these features; see Kahn and Tsing, this volume.
25 In keeping with internationally acceptable rhetoric, and also some interpretations of Indonesian land and forest law, the report observes that customary land claims should be studied to ensure that land slated for development is unencumbered by prior claims. In

effect, however, the report ignores such rights by deeming shifting cultivation a non-permanent land use, thereby finding a much larger land area to be "convertible" than previous planning efforts.

26 See Peluso (1996) on "countermapping" by communities and their NGO supporters in response to the territorialising strategies of various state agencies. Of central importance is the attempt to locate cultural groups on a spatial grid, in order to stake claims and contest the influx of newcomers. This type of information is "important but scarce" in Indonesia, as census data does not reveal ethnic/cultural affiliations (Peluso 1995: 399).

27 See for example Banuri and Marglin (1993); Ghai (1994).

28 See, for example, Poffenberger (1990); Skephi and Kiddell-Monroe (1993); Colchester (1994); Moniaga (1993). Lynch and Talbott (1995: 128) recognize women as especially sound resource managers.

29 I discuss the national debate on indigenous people and land rights in Li (nd).

30 Lian (1993: 333) disputes the viability and attractiveness of "tradition" as a solution to the contemporary problems of Orang Ulu in Sarawak.

31 See Lynch and Talbott (1995); Colchester (1994); Moniaga (1993).

32 See Dove (1985a); Kahn (1993); Li (1996a, 1997).

33 Peluso, Vandergeest and Potter (1995) observe that literature on forest-dependent people in Thailand focuses overwhelmingly on the "hill tribes", who comprise only 2% of the population, and are a small minority of Thailand's forest-dependent and upland peoples.

34 Rangan's (1993) dissident account of the Chipko movement describes people driven to defend their livelihoods *against* environmentalists.

35 Similarly, C. Geertz (1963: 116) characterizes the uplands and interiors of Indonesia in the colonial period as a "monotonous expanse of enduring stability" and "essentially unchanged swidden-making", in which were "scattered" some dynamic, productive enclaves.

36 One major change in the structure of agriculture in East Kalimantan was a more direct result of government planning: the promotion of large scale plantations increased the allotted land area from 3716 hectares in 1963 to 256,162 in 1987 (Hidayati 1991: 40).

37 Although he too emphasizes the distinction between "traditional" swiddeners and forest squatters, Tirtosudarmo (1993) makes a persuasive argument for treating the latter as determined and ambitious farmers: having left their previous locations in search of better opportunities in the forest fringe, they are unlikely to remain in transmigration schemes offering poor returns.

38 Poffenberger (1983) comments on the low visibility of dryland farming as a result of dichotomous models.

39 See also Li (1996b).

40 It seems that environmental transgressions by large scale forestry and agricultural concerns are less surprising, and therefore less offensive.

41 Enters (1995) describes farmers in Thailand limiting their participation to a "token line" and in the Philippines Brown (1994: 56) describes the vigor with which NGOs and government agencies have promoted "sloping agricultural land technology" (SALT) and the reluctance of uplanders to adopt it, presumably because it does not benefit them.

References

Abrams, Philip, 1988, Notes on the Difficulty of the Studying the State [1977]. *Journal of Historical Sociology*, 1(1), 58–89.
Agrawal, Arun and K. Sivaramakrishan, forthcoming, "Introduction" in *Agrarian Environments*. Durham NC: Duke University Press.
Alatas, Syed Hussein, 1972, *The Myth of the Lazy Native*. London: Frank Cass.
Alexander, P., P. Boomgaard and B. White, 1991, "Introduction", *In the Shadow of Agriculture: Non-farm Activities in the Javanese Economy, Past and Present*, pp. 1–13. Amsterdam, KITLV Press.
Allen, Bryant, 1993, "The Problems of Upland Land Management" in *South-East Asia's Environmental Future*, edited by Harold Brookfield and Yvonne Byron, pp. 225–237. Tokyo: United Nations University Press; Kuala Lumpur: Oxford University Press.
Andaya, Barbara, 1993, "Cash Cropping and Upstream-Downstream Tensions: The Case of Jambi in the Seventeenth and Eighteenth Centuries" in *Southeast Asia in the Early Modern Era*, edited by Anthony Reid, pp. 91–122. Ithaca: Cornell University Press.
Andaya, Leonard, 1993, "Cultural State Formation in Eastern Indonesia" in *Southeast Asia in the Early Modern Era*, edited by Anthony Reid, pp. 23–41. Ithaca: Cornell University Press.
Angelsen, Arild, 1995, "Shifting Cultivation and "Deforestation": A Study from Indonesia. *World Development*, 23(10), 1713–1729.
Banuri, Tariq and F.A. Marglin, 1993, *Who will Save the Forest: Knowledge, Power and Environmental Destruction*. London: Zed Books.
Bappenas, 1993, *Biodiversity Action Plan for Indonesia*. Jakarta.
Barber, Charles, 1989, *The State, the Environment and Development: The Genesis and Transformation of Social Forestry Policy in New Order Indonesia*. Ph.D. Dissertation, University of California, Berkeley.
Barber, Charles, Suraya Afiff, Agus Purnomo, 1995, *Tiger By the Tail? Reorienting Biodiversity Conservation and Development in Indonesia*. Washington: World Resources Institute.
Barbier, Edward, 1989, "Cash Crops, Food Crops and Sustainability: The Case of Indonesia". *World Development*, 17(6), 879–895.
Barlow, Colin, 1996, "Introduction" in *Indonesia Assessment 1995*, edited by Colin Barlow and Joan Hardjono, pp. 1–16. Singapore: Institute of Southeast Asian Studies.
Bastaman, Henri, 1995, "The Emergence of Non-Farm Activities in the Upland Citanduy River Basin, West-Java, Indonesia". Paper presented at the conference on Agrarian Transformation in the Indonesian Uplands, 13–14 May, Dalhousie University, Halifax.
Bentley, G. Carter, 1986, "Indigenous States of Southeast Asia". *Annual Review of Anthropology*, 15, 275–305.
Bernstein, Henry, 1990, "Taking the Part of the Peasants?" in *The Food Question*, edited by Henry Bernstein *et al.*, pp. 69–79. New York: Monthly Review Press.
Berry, Sara, 1988, "Concentration Without Privatization? Some Consequences of Changing Patterns of Rural Land Control in Africa" in *Land and Society in Contemporary Africa*, edited by R.E. Downs and S.P. Reyna, pp. 53–75. Hanover: University Press of New England.
Bigalke, T, 1983, "Dynamics of the Torajan Slave Trade in South Sulawesi" in *Slavery, Bondage and Dependency in Southeast Asia*, edited by Anthony Reid, pp. 341–363. St Lucia: University of Queensland Press.
Blaikie, Piers and Harold Brookfield, 1987, *Land Degradation and Society*. London: Methuen.
Boomgaard, Peter, 1991, "The Javanese Village as a Cheshire Cat: The Java Debate Against a European and Latin American Background". *The Journal of Peasant Studies*, 18(2), 288–304.
Booth, Ann, 1991, "Regional Aspects of Indonesian Agricultural Growth" in *Indonesia: Resources, Ecology & Environment*, edited by J. Hardjono, pp. 36–60. Singapore: Oxford University Press.

Bowen, John R., 1991, *Sumatran Politics and Poetics: Gayo History 1900–1989*. New Haven: Yale University Press.

Breman, Jan, 1980, *The Village on Java and the Early-Colonial State*. Rotterdam: Comparative Asian Studies Programme.

1988, *The Shattered Image: Construction and Deconstruction of the Village in Colonial Asia*. Dordrecht: Foris Publications.

Bronson, Bennet, 1978, "Exchange at the Upstream and Downstream Ends: Notes Toward a Functional Model of the Coastal State in Southeast Asia" in *Economic Exchange and Social Interaction in Southeast Asia: Perspectives from Prehistory, History and Ethnography*, edited by Karl Hutterer, pp. 39–54. Ann Arbor: University of Michigan.

Brookfield, Harold, Lesley Potter and Yvonne Byron, 1995, *In Place of the Forest: Environmental and Socio-economic Transformation in Borneo and the Eastern Malay Peninsula*. Tokyo: United Nations University Press.

Brown, Elaine, 1994, "Grounds at Stake in Ancestral Domains" in *Patterns of Power and Politics in the Philippines*, edited by James Eder and Robert Youngblood, pp. 43–76. Arizona: Arizona State University Press.

Bryant, Raymond, 1992, "Political Ecology: An Emerging Research Agenda in Third-World Studies" in *Political Geography*, 11(1), 12–36.

Burling, Robbins, 1965, *Hill Farms and Padi Fields*. New Jersey: Prentice Hall.

Chang, Te-Tzu, 1984–85, "The Ethnobotany of Rice in Island Southeast Asia". *Asian Perspectives*, 26(1), 69–76.

1989, "Domestication and Spread of the Cultivated Rices" in *Foraging and Farming*, edited by D.R. Harris and G.C. Hillman, pp. 408–17. London: Unwin Hyman.

Colchester, Marcus, 1994, "Sustaining the Forests: The Community-based Approach in South and South-East Asia". *Development and Change*, 25(1), 69–100.

Colombijn, Freek, 1995, "The Javanese Model as a Basis for Nineteenth Century Colonial Policy in West Sumatra". *Ekonesia*, 3, 25–42.

Colombijn, Freek, David Henley, Bernice de Jong Boers and Hans Knapen, 1996, "Man and Environment in Indonesia, 1500–1950: An international workshop KITLV, Leiden, 27–29 June 1996". *Indonesian Environmental History Newsletter*, 7, 1–9.

Comaroff, John and Jean Comaroff, 1992, *Ethnography and the Historical Imagination*. Boulder: Westview Press.

Corrigan, Philip, 1994, "State Formation" in *Everday Forms of State Formation*, edited by Joseph Gilbert and Daniel Nugent pp. xvii-xix. Durham: Duke University Press.

Davies, Susanna, 1994, "Information, Knowledge and Power". *IDS Bulletin*, 25(2), 1–13.

De Koninck, Rodolphe, 1996, "The Peasantry as the Territorial Spearhead of the State in Southeast Asia: The Case of Vietnam". *Sojourn*, 11(2), 231–58.

De Koninck, Rudolphe and W. Donald McTaggart, 1987, "Land Settlement Process in Southeast Asia: Historical Foundations, Discontinuities and Problems". *Asian Profile*, 15(4), 341–356.

Departmen Kehutanan, 1994, *Pentujuk Teknis Inventarisasi dan Indentifikasi Peladang Berpindah dan Perambah Hutan*. Jakarta: Direktorat Reboisasi.

Departmen Sosial, 1994a, *Pembinaan Pemukiman Sosial Masyarakat Terasing*. Jakarta: Directorat Bina Masyarakat Terasing.

1994b, *Pedoman Kerja Petugas Lapangan Pembinaan Masyarakat Terasing*. Jakarta: Directorat Bina Masyarakat Terasing.

Dove, Michael R., 1985a, "The Kantu' System of Land Tenure: The Evolution of Tribal Rights in Borneo" in *Modernization and the Emergence of a Landless Peasantry*, edited by G.N. Appell, pp. 159–82. Williamsburg: Studies in Third World Societies, Department of Anthropology, College of William and Mary.

1985b, "The Agroecological Mythology of the Javanese and the Political Economy of Indonesia". *Indonesia*, 39, 1–36. Reprinted in East-West Center Environment and Policy Institute Reprint No. 84.

1987, "The Perception of Peasant Land Rights in Indonesian Development: Causes and Implications" in *Land, Trees and Tenure*, edited by John B. Raintree, pp. 265–272. Nairobi and Madison: ICRAF and Land Tenure Center.

1993, "A Revisionist View of Tropical Deforestation and Development". *Environmental Conservation*, 20(1), 17–24,56.

1996, "So Far from Power, So Near to the Forest: A Structural Analysis of Gain and Blame in Tropical Forest Development" in *Borneo in Transition: People, Forests, Conservation and Development*, edited by Christine Padoch and Nancy Peluso, pp. 41–58. Kuala Lumpur: Oxford University Press.

Enters, Thomas, 1995, "The Token Line: Adoption and Non-Adoption of Soil Conservation Practices in the Highlands of Northern Thailand". Paper presented at the International Workshop on Soil Conservation Extension. 4–11 June, Chiang Mai, Thailand.

Ellen, Roy, 1976, "The Development of Anthropology and Colonial Policy in the Netherlands: 1800–1960". *Journal of the History of the Behavioural Sciences*, 12, 303–24.

1979, "Sago Subsistence and the Trade in Spices: A Provisional Model of Ecological Succession and Imbalance in Moluccan History" in *Social and Ecological Systems*, edited by P.C. Burnham and R.F. Ellen. London: Academic Press.

Ferguson, James, 1994, *The Anti-Politics Machine*. Minneapolis: University of Minnesota Press.

Fox, James, 1992, "The Heritage of Traditional Agriculture in Eastern Indonesia: Lexical Evidence and the Indications of Rituals from the Outer Arc of the Lesser Sundas" in *The Heritage of Traditional Agriculture among the Western Austronesians*, edited by James Fox, pp. 67–88. Canberra: Australian National University.

Geertz, Clifford, 1963, *Agricultural Involution*. Berkeley: University of California Press.

1980, *Negara: The Theatre State in Nineteenth-Century Bali*. Princeton NJ: Princeton University Press.

Geertz, Hildred, 1963, *Indonesian Cultures and Communities*. New Haven: HRAF Press.

George, Kenneth, 1991, "Headhunting, History and Exchange in Upland Sulawesi". *Journal of Asian Studies*, 50(3), 536–564.

Ghai, Dharam, 1994, "Environment, Livelihood and Empowerment". *Development and Change*, 25(1), 1–11.

Gibson, Thomas, 1994, "Concluding Reflections on Units of Analysis in the Study of the Official and the Popular". *Social Analysis*, 35, 157–164.

Gupta, Akhil and James Ferguson, 1992, "Beyond "Culture": Space, Identity and the Politics of Difference". *Cultural Anthropology*, 7(1), 6–23.

Hall, Kenneth, 1978, "The Coming of Islam to the Archipelago: A Re-Assessment" in *Economic Exchange and Social Interaction in Southeast Asia: Perspectives from Prehistory, History and Ethnography*, edited by Karl Hutterer, pp. 213–231. Ann Arbor: University of Michigan.

Hardjono, Joan, 1994, "Resource Utilization and the Environment" in *Indonesia's New Order: The Dynamics of Socio-economic Transition*, edited by Hal Hill, pp. 179–215. New South Wales: Allen and Unwin.

Hart, Gillian, Andrew Turton and Ben White (eds.), 1989, *Agrarian Transformations: Local Processes and the State in Southeast Asia*. Berkeley: University of California Press.

Hart, Gillian, 1989, "Agrarian Change in the Context of State Patronage" in *Agrarian Transformations: Local Processes and the State in Southeast Asia*, edited by Gillian Hart *et al.* pp. 31–52. Berkeley: University of California Press.

Healey, Christopher, 1985, "Tribes and States in "Pre-Colonial" Borneo: Structural Contradictions and the Generation of Piracy". *Social Analysis*, 18, 3–39.

Hefner, Robert W., 1990, *The Political Economy of Mountain Java: An Interpretive History*. Berkeley and Los Angeles: University of California Press.

Helliwell, Christine, 1992, "Evolution and Ethnicity: A Note on Rice Cultivation Practices in Borneo" in *The Heritage of Traditional Agriculture among the Western Austronesians*, edited by James Fox, pp. 7–20. Canberra: Australian National University.

Henley, David, 1989, "The Idea of the Celebes in History" Working Paper 59, Centre of Southeast Asian Studies, Monash University.

1994, Population and Environment in Precolonial North Sulawesi, Paper for the 13th Biennial ASAA Conference, Perth, June 1994.

Henley, David and Freek Colombijn, 1995, "Environmental History Research in the Arsip Nasional". *Indonesia Environmental History Newletter* 6.

Heyzer, Noeleen (ed.), 1987, *Women Farmers and Rural Change in Asia*. Kuala Lumpur: Asia Pacific Development Centre.

Hidayati, Deny, 1991, "Effects of Development on the Expansion of Agricultural Land in East Kalimantan". *Borneo Review*, 2(1), 28–50.

——— 1994, "Adoption of Indigenous Practice: Survival Strategies of Javanese Transmigrants in South-East Kalimantan". *ASSESS Journal*, 1, 56–69.

Hoben, Allan, 1995, "Paradigms and Politics: The Cultural Construction of Environmental Policy in Ethiopia". *World Development*, 23(6), 1007–1021.

Hoffman, Carl, 1988, "The 'Wild Punan' of Borneo" in *The Real and Imagined Role of Culture in Development*, edited by Michael Dove, pp. 89–118. Honolulu: University of Hawaii Press.

Hooker, M.B., 1978, *Adat Law in Modern Indonesia*. Kuala Lumpur: Oxford University Press.

——— 1983, *Islam in Southeast Asia*. Leiden: Brill.

Joseph, Gilbert and Daniel Nugent (eds.), 1994, *Everday Forms of State Formation*. Durham: Duke University Press.

Kahn, Joel, 1993, *Constituting the Minangkabau: Peasants, Culture and Modernity in Colonial Indonesia*. Providence: Berg.

Kartawinata, Kuswata and Andrew Vayda, 1984, "Forest Conversion in East Kalimantan: The Activities and Impacts of Timber Companies, Shifting Cultivators, Migrant Pepper Farmers and Others" in *Ecology in Practice 1. Ecosystem Management*, edited by F. di Castri, F.W.G. Baker and M. Hadley, pp. 98–126. Paris: UNESCO.

Keane, Webb, 1997, "Knowing One's Place: National Language and the Idea of the Local in Eastern Indonesia". *Cultural Anthropology*, 12(1), 37–63.

Kepas, 1985, *The Critical Uplands of Eastern Java: An Agroecosystem Analysis*. Jakarta: Kepas.

King, Victor and Michael Parnwell (eds.), 1990, *Margins and Minorities: The Peripheral Areas and Peoples of Malaysia*. Hull: Hull University Press.

Koentjaraningrat (ed.), 1993, *Masyarakat Terasing di Indonesia*. Jakarta: Gramedia with Departemen Sosial.

Lee, Justin, 1995, *Participation and Pressure in the Mist Kingdom of Sumba*. Ph.D. dissertation. Adelaide: University of Adelaide.

Li, Tania Murray, 1996a, "Images of Community: Discourse and Strategy in Property Relations". *Development and Change*, 27(3), 501–27.

——— 1996b, "Household Formation, Private Property and the State". *Sojourn*, 11(2), 259–287.

——— 1997, "Producing Agrarian Transformation at the Indonesian Periphery" in *Economic Analysis Beyond the Local System*, edited by Richard Blanton *et al.*, pp. 125–146. Lanham: University Press of America.

——— nd, "Constituting Tribal Space: Indigenous Identity and Resource Politics in Indonesia".

Lian, Francis, 1993, "On Threatened Peoples" in *South-East Asia's Environmental Future*, edited by Harold Brookfield and Yvonne Byron, pp. 322–337. Tokyo: United Nations University Press; Kuala Lumpur: Oxford University Press.

Lim Tech Ghee and Alberto Gomes, 1990, "Introduction" in "Tribal Peoples and Development in Southeast Asia" *Manusia dan Masyarakat* (special issue), pp. 1–11. Kuala Lumpur: University of Malaya.

Lynch, Owen J. and Kirk Talbott, 1995, *Balancing Acts: Community-Based Forest Management and National Law in Asia and the Pacific*. Washington: World Resources Institute.

Mitchell, Timothy, 1991, "The Limits of the State: Beyond Statist Approaches and their critics". *American Political Science Review*, 85(1), 77–96.

Moniaga, Sandra, 1993, "Toward Community-Based Forestry and Recognition of Adat Property Rights in the Outer Islands of Indonesia" in *Legal Frameworks for Forest Management in Asia: Case Studies of Community-State Relations*, edited by Jefferson Fox, pp. 131–150. Honolulu: Environment and Policy Institute, East-West Center.

Moore, Donald, 1993, "Contesting Terrain in Zimbabwe's Eastern Highlands: Political Ecology, Ethnography and Peasant Resource Struggles". *Economic Geography*, 69(4), 380–401.

Nugent, David, 1994, "Building the State, Making the Nation: The Bases and Limits of State Centralization in Modern Peru". *American Anthropologist*, 96(2), 333–369.

O'Brien, Jay and William Roseberry (eds.), 1991, *Golden Ages, Dark Ages*. Berkeley: University of California Press.

Onghokham, 1975, *The Residency by Madiun: Priyagi and Peasant during the nineteenth century*. Ph.D. Dissertation, Yale University.

Padoch, Christine and Nancy Peluso (eds.), 1996, *Borneo in Transition: People, Forests, Conservation and Development*. Kuala Lumpur: Oxford University Press.

Palte, Jan, 1989, *Upland Farming on Java, Indonesia*. Amsterdam/Utrecht: Netherlands Geographical Studies.

Peluso, Nancy, 1990, "A History of State Forest Management in Java" in *Keepers of the Forest: Land Management Alternatives in Southeast Asia*, edited by Mark Poffenber, pp. 27–55. West Hartford: Kumarian Press.

——— 1992, *Rich Forests, Poor People: Resource Control and Resistance in Java*. Berkeley: University of California Press.

——— 1995, "Whose Woods are These? Counter-Mapping Forest Territories in Kalimantan, Indonesia". *Antipode*, 27(4), 383–406.

Peluso, Nancy, Peter Vandergeest and Lesley Potter, 1995. "Social Aspects of Forestry in Southeast Asia: A Review of Postwar Trends in the Scholarly Literature". *Journal of Southeast Asian Studies*, 26(1), 196–218.

Pemberton, John, 1994, *On the Subject of "Java"*. Ithaca: Cornell University Press.

Pieterse, Jan Nederveen, 1992, "Dilemmas of Development Discourse: The Crisis of Developmentalism and the Comparative Method". *Development and Change*, 22(1), 5–29.

Pigg, Stacy Leigh, 1992, "Inventing Social Categories Through Place: Social Representations and Development in Nepal". *Comparative Studies in Society and History*, 34, 491–513.

Poffenberger, Mark, 1983, "Changing Dryland Agriculture in Eastern Bali". *Human Ecology*, 11(2), 123–144.

——— 1990, "Introduction: The Forest Management Crisis" pp. xix–xxv and "The Evolution of Forest Management Systems in Southeast Asia" in *Keepers of the Forest: Land Management Alternatives in Southeast Asia*, edited by Mark Poffenber, pp. 7–26. West Hartford: Kumarian Press.

Potter, Lesley, 1993, "The Onslaught on the Forests in South-east Asia" in *South-East Asia's Environmental Future*, edited by Harold Brookfield and Yvonne Byron, pp. 103–123. Tokyo: United Nations University Press; Kuala Lumpur: Oxford University Press.

——— 1996, "Forest Degradation, Deforestation and Reforestation in Kalimantan: Towards a Sustainable Land Use" in *Borneo in Transition: People, Forests, Conservation and Development*, edited by Christine Padoch and Nancy Peluso, pp. 13–40. Kuala Lumpur: Oxford University Press.

Pred, Alan and Michael Watts, 1992, *Reworking Modernity: Capitalism and Symbolic Discontent*. New Brunswick: Rutgers University Press.

Rangan, Haripriya, 1993, "Romancing the Environment: Popular Environmental Action in the Garhwal Himalayas" in *In Defense of Livelihood: Comparative Studies on Environmental Action*, edited by John Friedmann and Haripriya Rangan, pp. 155–181. West Hartford: Kumarian Press.

Reid, Anthony, 1988, *Southeast Asia in the Age of Commerce 1460–1680, The Lands Below the Winds*. New Haven: Yale University Press.

Robinson, Kathryn, 1986, *Stepchildren of Progress: The Political Economy of Development in an Indonesian Mining Town*. Albany: State University of New York Press.

Roseberry, William, 1994, "Hegemony and the Language of Contention" in *Everday Forms of State Formation*, edited by Gilbert Joseph and Daniel Nugent, pp. 355–366. Durham: Duke University Press.

Rousseau, Jerome, 1989, "Central Borneo and its Relations with Coastal Malay Sultanates" in *Outwitting the State*, edited by Peter Skalnik, pp. 41–50. New Brunswick: Transaction Publishers.

Schuurman, Frans, 1993, "Introduction: Development Theory in the 1990s" in *Beyond the Impasse*, pp. 1–48. London: Zed Press.

Sellato, Bernard, 1994, *Nomads of the Borneo Rainforest: The Economics, Politics and Ideology of*

Settling Down. Honolulu: University of Hawaii Press.

Sherman, D. George, 1990, *Rice, Rupees and Rituals*. Stanford: Stanford University Press.

Shields, Rob, 1991, *Places on the Margin: Alternative Geographies of Modernity*. London: Routledge.

Short, John R., 1991, *Imagined Country: Environment, Culture and Society*. London: Routledge.

Skehpi and Rachel Kiddell-Monroe, 1993, "Indonesia: Land Rights and Development" in *The Struggle for Land and the Fate of the Forests*, edited by Marcus Colchester and Larry Lohmann, pp. 228–263. Penang: World Rainforest Movement.

Soebardio, Haryati, 1993, "Sambutan Menteri Sosial" in *Masyarakat Terasing di Indonesia*, edited by Koentjaraningrat, pp. vii-viii. Jakarta: Gramedia and Departemen Sosial.

Sondakh, Lucky, 1996, "Agricultural Development in Eastern Indonesia: Performance, Issues and Policy Options" in *Indonesia Assessment 1995*, edited by Colin Barlow and Joan Hardjono, pp. 141–162. Singapore: Institute of Southeast Asian Studies.

State Ministry of Population and the Environment, 1992, Act of the Republic of Indonesia Number 10 of 1992 concerning Population Development and Development of Prosperous Family.

Stoler, Ann, 1977, "Class Structure and Female Autonomy in Rural Java". *Signs*, 3(1), 74–89.

——— 1985, *Capitalsim and Confrontation in Sumatra's Plantation Belt*. New Haven: Yale University Press.

Suryohadikusumo, Djamaludin, 1995, "Pengarahan Menteri Kehutanan, Diskusi Panel: Partisipasi Masyarakat dalam Penelolaan Hutan". *Ekonesia*, 3, 63–66.

Tomich, Thomas and Meine van Noordwijk, 1995, "What Drives Deforestation in Sumatra". Paper presented at Regional Symposium on Montane Mainland Southeast Asia in Transition Chiang Mai, Thailand, 13–16 November 1995.

Thomas, Nicholas, 1991, "Against Ethnography". *Cultural Anthropology*, 6(3), 306–322.

Tirtosudarmo, Riwanto, 1993, "Dimensi Sosio-ekonomi dan Implikasi Kebijaksanaan Pemukiman Perambah Hutan". *Populasi*, 4(2), 1–12.

Tjitradjaja, Iwan, 1993, "Differential Access to Resources and Conflict Resolution in a Forest Concession in Irian Jaya". *Ekonesia*, 1(1), 58–69.

Tsing, Anna Lowenhaupt, 1993, *In the Realm of the Diamond Queen*. Princeton: Princeton University Press.

Vandergeest, Peter and Nancy Peluso, 1995, "Territorialization and State Power in Thailand". *Theory and Society*, 24, 385–426.

Volkmann, Toby, 1985, *Feasts of Honour: Ritual and Change in the Toraja Highlands*. Urbana: University of Illinois Press.

Watts, Michael, 1993, "Development 1: Power, Knowledge, Discursive Practice". *Progress in Human Geography*, 17(2), 257–272.

White, Benjamin, 1989, "Problems in the Empirical Analysis of Agrarian Differentiation" in *Agrarian Transformations: Local Processes and the State in Southeast Asia*, edited by Gillian Hart *et al.*, pp. 15–30. Berkeley and Los Angeles: University of California Press.

——— 1983, ""Agricultural Involution" and its Critics: Twenty Years After". *Bulletin of Concerned Asian Scholars*, 15(2), 18–31.

Wilk, Richard, 1997, "Emerging Linkages in the World System and the Challenge to Economic Anthropology" in *Economic Analysis Beyond the Local System*, edited by Richard Blanton *et al.*, pp. 97–108. Lanham: University Press of America.

Wolf, Eric, 1982, *Europe and the People without History*. Berkeley and Los Angeles: University of California Press.

Chapter 2

MAIZE AND TOBACCO IN UPLAND INDONESIA, 1600–1940

Peter Boomgaard

INTRODUCTION

With the arrival of the Portuguese and the Spaniards in the Indonesian Archipelago shortly after 1500, a large number of new crops, mostly originating in the Americas, were added to the existing repertoire. In this article, I deal with one subsistence crop, maize (corn), and one predominantly commercial crop, tobacco. Both crops spread rather rapidly and widely, and in many areas they came to play an important, sometimes even dominant, role in agriculture. By now, it would be as difficult to imagine every-day life in Indonesia without maize and tobacco as trying to think of a Europe without potatoes, or an Africa without cassava. Today, maize and tobacco are regarded as traditional Indonesian crops, and in some areas, such as Bengkulu [Sumatra] and Central Sulawesi, maize and/or tobacco have been incorporated in local myths of origin.[1] Claims once made for the Indonesian or Asian origin of both crops[2] have by now been refuted, but they clearly indicate that many people — Europeans and Indonesians alike — found it hard to believe that there had ever been an Indonesia without maize and tobacco.

Both crops are annuals that can tolerate a wide range of environmental conditions. In Indonesia, they are grown from sea level up to high altitudes of 1,500 m. and over. It may not come as a surprise, therefore, that they are often to be found in the same areas, and sometimes even on the same fields, albeit consecutively. They are and were grown on *sawah* (as second crop), on *tegal*, on *ladang/gaga*, and on *pekarangan*.[3] They were found from the high rainfall areas of Western Indonesia (Sumatra, West Java) to the extremely dry regions of Eastern Indonesia (Moluccas, Nusa Tenggara). Both were smallholder crops throughout the whole period, although after 1830 and even more so after 1860, tobacco became an estate crop as well. There are also differences: tobacco is more sensitive than maize to weather anomalies and pests; maize was and is often intercropped with other plants, tobacco is not; labour requirements of maize are low and those of tobacco are high. Although maize was exported in some quantities after 1900, it used to be a subsistence crop in most areas, whereas tobacco was produced for local, supralocal, and even supraregional markets. This means that the social and economic implications of the expansion of these crops may be expected to have been different.

This article is in the first place an attempt to set the record straight. Maize as an object of historical research has always remained in the shadow of rice, and smallholder tobacco has been overshadowed by tobacco grown on plantations. I trace the expansion of both crops throughout Indonesia, in so far as the sources will let me. On some areas, such as Kalimantan and Irian, information presented in this article is meagre, either because it does not exist, or because I did not look hard enough. In the second place, I try to explain the rather rapid expansion of both crops, not only to the uplands of Java, but even to (the uplands of) the most isolated parts of the Archipelago. It seems likely that maize and tobacco were instrumental in creating upland societies with quite distinct identities which have persisted to the present. Finally, I deal briefly with the environmental, economic and social consequences of this expansion.

Almost all information collected here is based on European, mostly Dutch sources. This means that areas where the Dutch came late or where they had few establishments may be underrepresented. This bias aside, I do not expect other (systematic) misrepresentations in the European sources. They are certainly much more detailed in regard to all things agricultural, which in indigenous sources are often dealt with in a rather cavalier fashion. The Dutch did have some prejudices against maize: it was regarded as a crop for lazy people, and civil servants were always trying to make the indigenous population grow rice on sawahs. Nevertheless, my impression is that these notions did not lead to an underrepresentation of maize cultivation in the sources.

MAIZE BEFORE 1800

According to a number of authors, maize could already be found in the Moluccas or on the neighbouring island of Siau (Sangihe and Talaud Islands) before the end of the sixteenth century. However, it is not inconceivable that the term for maize used in the sources (*milho*) referred to sorghum.[4] If it was really maize, it may have been introduced directly from Mexico by the Spaniards, or via the Philippines. However that may be, by the 1670s, maize had firmly established itself in the Moluccas (Amboina, Ternate, Tidore), the island of Timor (Lesser Sunda Islands or Nusa Tenggara), southwest Sulawesi (to the south of Makasar), the island of Butung [Buton], off the coast of southeast Sulawesi, and in northeast

Sulawesi (Minahasa, Gorontalo), where it may have arrived much earlier if the reports of maize on Siau are to be trusted.[5] At the headquarters of the *Vereenigde Oostindische Compagnie* (United [Dutch] East India Company, VOC for short) in Batavia, the higher echelons were apparently so impressed with the crop — according to them it could be grown everywhere and it had a yield ratio of 1,000 — that they ordered their officials in Banda in 1663 to stimulate its cultivation.[6]

Therefore, maize had found its way to the most important commercial centres of Eastern Indonesia before 1700, and was by then firmly rooted as an important crop, sometimes even the dominant staple crop. The term used for maize in the Dutch sources of that period is *Turkse tarwe* (literally Turkish wheat), in one of its many spelling variants, which excludes the possibility of a mistaken identity. Occasionally these sources give locally used equivalents, such as *milie* or *milje*, obviously their rendering of milho, and finally also an indigenous Indonesian word for maize, namely *jagung*, although the Dutch spelling is *djagon*, *jagon*, *jagong*, or even *sjagon* or *jagum*. A British source on Timor used the term Indian corn, which is of course much better than Turkish wheat. I did not encounter a spelling variant of the word maize [Dutch: *maïs*] in a Dutch source before 1682.

The first time that — to my knowledge — the word jagung was used in a Dutch source, it was not in reference to Eastern Indonesia, but to Java. In 1648, the *Daghregister* (the official diary kept at VOC headquarters in Batavia) mentioned the import of maize from Java's northeast coast at least once and possibly twice. If the second reference is also to maize, the harbour from which it came was Gresik, near Surabaya, and in that case the first shipment may have come from the same place (Gembong near Pasuruan is another possibility).[7] Eastern Java is, indeed, a likely candidate as the source of Java's early maize, given the fact that the next reference, dated 1681, explicitly names the upland area of Pasuruan as the place where *djagon* was cultivated. Around the same time, maize was also planted on the island of Bali. In the early eighteenth century, maize was mentioned again in East Java, namely in Pasuruan (1709), in Panarukan, to the east of Pasuruan (1709), and in Sumenep, part of the island of Madura, off the coast of Java's eastern salient (1737).[8] The fact that eastern Java had been exporting maize as early as 1648 certainly suggests that the crop was well established there at that time. In the 1730s, maize was reported to be grown in the coastal areas of Semarang, northern Central Java, and in the 1770s it was found in the upland teak forests of the same region. Around the latter date it had also reached the hinterland of Batavia, in West Java, where, circa 1790, it was grown in upland areas. By that time, therefore, the cultivation of maize had expanded throughout Java, including Madura, and Java's eastern neighbour Bali. In a period of slightly more than a century it had spread from east to west. A source dated 1781 dealing with food crops in Java stated that maize was such a well-known crop that there was no need to elaborate.[9]

Dobbin states that maize was introduced in Sumatra "in about the sixteenth century", but she does not present any evidence for such an early start. In the sources at my disposal, maize is not mentioned in Sumatra prior to 1780. A recent study on southeast Sumatra in the seventeenth century reports failures of the rice crop, when people either died or resorted to sago or *ubi*, but there is no mention of maize. At the end of the eighteenth century it could be found in the Batak highlands (North Sumatra) and in and around the Minangkabau upland areas (western Central Sumatra). In the Batak area it seems to have been a staple food crop at least in some places, which implies an introduction some decades earlier.[10]

The pattern that emerges from this data indicates that maize was introduced and spread over most of Eastern Indonesia, Bali and East Java (including Madura), beginning perhaps as early as c. 1550, but certainly far advanced before c. 1675. A century later, or about 1775, Central and West Java and Sumatra had been reached as well. Even out-of-the-way places often regarded as isolated and highly traditional, such as the Batak area and, in the east, Sumba, Savu and the island of Selayar near southeast Sulawesi,[11] had by the late eighteenth century not only incorporated maize into their agricultural routine, but also turned it into a staple crop. By 1790, the crop had also reached Kalimantan, where it was planted and eaten by the Bukit Dayak (as 'famine food') and the Dusun Dayak (Leupe 1864: 380, 396). In general, both within Java and the Indonesian Archipelago as a whole, the spread of maize was from the dry, low rainfall areas of the east to the much wetter regions of the west, with upland areas being favoured over lowlands.

MAIZE AFTER 1800

Between 1800 and 1810, Java went through a period of high and rising rice prices, which was caused partly by a number of dry years, partly by attempts of the VOC to produce more commercial crops such as coffee, cotton and sugar, and also by disturbed transportation links due to the Napoleonic wars (cf. Boomgaard 1989a: 99–100). In order to counter food shortages, the VOC repeatedly ordered its local representatives in Java to stimulate the cultivation of jagung. This was not done, however, without misgivings. The VOC officials regarded maize as less healthy food, and as a crop that exhausted the soil, particularly as a second crop on sawahs.[12] We cannot be sure that the injunctions to plant maize really worked, but it is clear that it was a very important crop at the beginning of the nineteenth century. When T.S. Raffles, the British Lieutenant-Governor of Java (1811–1815), formulated the Revenue Instructions for his Land Rent System in 1814, he used the potential maize crop as his measure for the taxation of tegal fields. Around 1815, it was mentioned as a crop to be found in every district, and still increasing in importance. Of the arable land under smallholder non-rice annuals (collectively called *palawija* in Java), first place went, *ex aequo*, to maize and peanuts, each with a share of about 20% (c. 1820).[13]

Around this time, we start to find more particulars on the cultivation of maize in the sources. It was found as a second crop on sawahs in all Residencies,[14] but authors emphasised that it was especially to be found in the hilly upland areas, on tegal, where rice could not be cultivated. The German botanist Junghuhn, who travelled all over Java in the 1830s and 40s, found maize everywhere up to the 4,000 feet [1,250 m.] line and over, whereas rice disappeared above 3,000 feet [1,000 m.]. Some varieties were reported to ripen incredibly quickly (40 days!), and several crops could, therefore, be harvested annually from the same field. On the other hand, maize was reputed to impoverish the soil, and in some areas pulses were planted after maize, no doubt because of their nitrogen fixating properties. In upland areas maize was sometimes the only staple crop. Its growing importance was attributed to population growth and the limited possibilities for the laying out of more sawahs. In Madura and various upland areas of Java proper maize was the dominant food crop.[15]

In 1830, on the eve of the introduction of the so-called Cultivation System (*Cultuurstelsel*), the distribution of maize over Java and Madura, both geographically and according to type of fields, was essentially the same as it would be c. 1875, c. 1920, at the end of the colonial period, and up to a point, even the same as today.[16] As regards geographical distribution, it was present everywhere, but the highest concentrations were to be found in East Java (Pasuruan, Probolinggo, Besuki) and Madura. Some areas with a high maize density were also encountered in Central Java, namely in the Residency of Rembang (bordering on East Java) and around the Dieng highlands (parts of the Residencies Banyumas, Bagelen and Kedu).[17] Regarding the types of arable land where maize could be found around 1830, the available information does not permit us to draw firm conclusions. However, the impression conveyed by the sources dating from this and earlier years is that more maize was grown on tegal, as first and second crop, than on sawah, as second crop. This was certainly the case in 1875, 1920 and 1940.

For the period 1830 to 1900, the sources are sufficiently detailed to give us some idea of the variation between regions in the actual cultivation of maize. For instance, in the upperslope areas of the Merapi and Merbabu volcanoes in Central Java, one maize crop was followed and preceded by long periods of fallow (up to six years). However, in most upland areas of Kedu and Madura tegals could produce two or even three maize harvests per year, partly because some varieties had an extremely short growing season. Normally, a second maize crop on tegal would yield 1/4th less than the first crop. In the southern districts of the regency Malang (Pasuruan), however, maize was still being grown in shifting cultivation, and plots were left after seven to eight months.[18] In some areas, therefore, maize had almost turned into a monoculture. However, it was often doublecropped or intercropped with rice, pulses or tobacco among others.[19]

During the nineteenth century, the share of maize in the fields planted with *palawija* crops gradually increased from about 20% in 1820 to more than 35% in 1880. It was not until after the turn of the century that this (relative) growth came to an end (though not the absolute growth of the area under maize), largely or perhaps even entirely owing to the increasing popularity of cassava. In the late 1930s, the shares of the three leading food crops in the total area cultivated with smallholder crops (annuals) were 45% for rice, 23% for maize, and 11% for cassava.[20]

Personally, I find the almost continuous growth of the share of maize between 1650 and 1900 the most interesting outcome of the foregoing analysis. Expansion into Java's upland areas was a major feature of this growth from the very beginning. This notion runs counter to a number of recent statements regarding the historical development of upland Java. Thus, Palte argued that the clearing of Java's uplands for peasant agriculture did not start until the 1780s. My findings also seem to be at odds with conceptions regarding the role of the Cultivation System (1830–1870) in the peopling of Java's uplands. While Palte argues that the peopling of the upland areas came to a halt during this period, two other scholars, Donner and Hefner, regard the same period as the real start of the upland clearings.[21] It cannot be denied that we find local peaks and troughs in the clearing process, but the overall picture is one of amazing continuity.

Leaving Java but not the Javanese sphere of influence, we find maize mentioned around 1830 as one of Bali's main crops. In the eastern part of Buleleng, Bali, maize had already become the staple food at the end of the nineteenth century. In the late 1930s, a relatively high proportion of arable lands were planted with maize in Bali and to a lesser extent in Lombok. It was grown on tegal and also, as a second crop, on sawahs.[22]

In Eastern Indonesia, we find the cultivation of maize on many islands mentioned for the first time in sources dating from the period 1825–1925: Makian and Muti in the North Moluccas, Seram in the Ambon group, Sumbawa, Flores, Solor, Lomblen, Pantar, and Alor, belonging to Nusa Tenggara, and most islands of the South Moluccas (Wetar, the Leti, Damar, Babar, Tanimbar, and Kai Islands), between Nusa Tenggara and Irian Jaya.[23] It should not be assumed that maize did not get here earlier. I may have missed some references, and many of these islands were hardly visited by Europeans prior to the nineteenth century. The listing of these places only shows that by the mid or late nineteenth century at the latest, maize had reached almost every nook and cranny of Eastern Indonesia. On some of these islands, maize was as important as, or even more important than rice. Most of the maize here was grown in slash-and-burn agriculture, often intercropped with rice or pulses.[24] In the mean time, Sulawesi had turned into an exporter of maize, and in some areas, such as Gorontalo, maize had become the staple food crop by the 1850s. At that time, Seram exported maize as well.[25]

Turning to Sumatra, we can conclude that maize, present at the beginning of the nineteenth century in the Batak and Minangkabau highlands, had spread to most other regions between 1800 and 1900, including some lowland areas. It could now be found in Aceh, Siak, Palembang, Lampung, Bengkulu, and the Padang lowlands as well.[26] Maize was usually planted on ladangs, often in rotation or intercropped with rice, and seldom — for instance in Aceh — as a second crop on (rainfed) sawahs. In the hill areas it was also grown in permanent gardens.[27] Agricultural statistics (arable, harvest) being largely absent for Sumatra prior to 1950, it is difficult to judge the importance of maize here. The impression is, however, that its importance was limited. Only in the upstream region of the southern Batak area is it mentioned as the main staple food crop. For Central Sumatra it is explicitly stated that maize is a "snack", and that it is only eaten on some scale in times of dearth, as a "famine food". It is likely that an unfavourable distribution of rainfall over the year had restricted the expansion of maize, as seems to have been the case in West Java. It is possible that the soils were too acidic.[28] It is also possible that cultural preferences have had some influence on the limited role of maize. From the 1880s onward, some references explicitly mentioned that maize-eating groups, such as the Batak and the Lubu, were increasing their rice consumption. One source argued that the Batak were doing this because they were (getting) ashamed of being maize-eaters, as this was regarded as a sign of poverty.[29]

Finally, turning very briefly to Kalimantan, it must be assumed that maize was by now of some importance. Rice remained the staple grain between the 1830s and 1940s, but Schwaner, travelling through southeast Kalimantan in the 1840s, mentioned maize several times, predominantly as a home-yard crop eaten by the poor and as "famine food". Bock, travelling through south and east Kalimantan in 1879, mentioned maize as an important crop among the Bukit Dayak, and as a snack for the people of the area in general, mostly grown as a home-yard crop.[30]

Summing up, it can be said that the distribution of maize over the Indonesian Archipelago at the eve of the Second World War was not all that different from its spread around the beginning of the nineteenth century. It was very important in Eastern Indonesia, Eastern Java, and to a somewhat lesser extent in Central Java and Bali. Lower densities of maize could be found in Kalimantan, Sumatra and West Java.

TOBACCO BEFORE 1800

Tobacco smoking by Javanese people is supposed to have started in 1601 according to a Javanese chronicle. The first European source on this topic mentioned Javanese smokers in Banten, West Java in 1603. By 1624 at the latest, the habit had spread to the Central Javanese court of Mataram, and in 1644 a German observer remarked that the Javanese (of Banten and Batavia presumably) were as fond of smoking as were the Germans.[31] Most of this tobacco was probably imported from China, as we know from the farming out of the tax on import and sale of Chinese tobacco in Batavia in 1643 and 1644. Some of it, however, may have come from the Philippines, where it had been introduced from Mexico in the 1570s. The tax-farm of imported tobacco was already mentioned in 1626 and 1637, but the region of origin was not specified.[32] What was specified in the Daghregister of 1637, however, was the existence of locally grown ("*alhier wassende*") tobacco, which I take to mean tobacco cultivated in the countryside around Batavia. Tobacco grown by Javanese was again referred to in 1644, the same year for which we have another source, specifying that it was cultivated by the Chinese in the Environs of Batavia.[33] In 1656, the VOC started to stimulate the cultivation of tobacco around Batavia by increasing the tariff on imported tobacco from 10 to 20% (P.10.11.1656; D.10.11.1656).

Evidence for Javanese tobacco cultivated at some distance from Batavia dates from 1648. In that year, tobacco came to Batavia in indigenous vessels from Cirebon, in West Java, and from the north-coast of Central and/or East Java.[34] The dates of the arrivals suggest that there were already then a sawah and a tegal harvest, the former taking place in October, the latter in August, as was still the practice almost 300 years later (Vleming 1925: 70). The shipments recorded in March and April do not fit this pattern so well, but the local Javanese or Chinese merchants may have saved their tobacco for a higher market, which was bound to come a few months after the harvest. A similar distribution over the year is shown in 1657, another year for which monthly lists of arriving indigenous vessels are available.[35]

Information on other production centres in Java is scarce. In 1746, the tax-farm of the later well-known Kedu tobacco was mentioned, to my knowledge the first reference to this tobacco growing area in European sources. All tobacco from Kedu being transported to the northcoast was taxed when it passed the *gedhong tembakau*, the tobacco tollgate, farmed out to the Chinese by the rulers of Mataram. In 1798, W.H. van IJsseldijk, sent out to inspect the administration of Java, mentioned the tobacco cultivated in the eastern salient of Java, for which he proposed the establishment of a similar tollgate. This implies that the tobacco from East Java must have been of some importance.[36] Finally, according to a decree from Batavia, dated 1797, Cirebon was still a tobacco exporting region at the end of the eighteenth century.[37] So prior to 1800, Java tobacco was being cultivated in East Java, Kedu, Cirebon, and — by Chinese — in the Environs of Batavia, though that may have disappeared in 1740 with the "Chinese massacre". Dirk van Hogendorp, writing just before the turn of the century, describes tobacco in Java as a crop of growing importance, in which there was a flourishing trade. It was exported to Sumatra, Kalimantan and to almost the entire eastern Archipelago (Hogendorp 1800: 110)

More surprisingly perhaps, tobacco was also cultivated in Eastern Indonesia both for consumption and for export as early as the seventeenth century. The population of Siau (Sangihe and Talaud Islands) was seen chewing tobacco in 1631. In 1661, a VOC ship arrived in Batavia from Ternate with tobacco. It is not entirely clear whether this was locally grown or imported from Mindanao, in the Philippines. By 1671, however, Ternate tobacco must have been an important product, because it is one of the four main categories of tobacco noted in the Batavia tariff of that year (the other three being Javanese, Chinese, and Japanese tobacco). The botanist Rumphius, writing between 1670 and 1680, also mentioned Ternate as a major producer.[38] During the last quarter of the eighteenth century, tobacco was reported to grow on the island of Timor. It also seems to have been cultivated by the Papuas of Irian Jaya, and at a much earlier date, 1678. As there were regular contacts between Ternate and West Irian, this is not as strange as it may seem. In Kalimantan too, tobacco was produced and exported as early as the 1660s (that is the first decade that the Daghregister recorded imports in Batavia of tobacco from places located on the coast of Kalimantan). Further imports are occasionally mentioned after those years, but they are few and far between, perhaps because tobacco had been eclipsed by pepper.[39] That the Malay aristocracy in the ports-of-trade of Kalimantan had adopted the habit of smoking is implied by a 1667 report relating the death of Pangeran Purbanegara of Banjarmasin, caused by his being hit on the head with a tobacco-pipe.[40]

In Sumatra in 1603, the ruler of Aceh was reported to be using tobacco. Dobbin mentions tobacco cultivation in eighteenth century Agam, in the Minangkabau highlands, as a byproduct of the consumption of opium, which was often smoked mixed with tobacco (Dobbin 1983: 89; Reid 1988: 44). Tobacco and opium were indeed often taken together, not only in Sumatra but also in Java. When in 1671 the sultan of Banten forbade the use of opium to his subjects, he also banned tobacco smoking. Even the VOC, although it was earning good money from the sale of opium and the tobacco farm, forbade the use of opium mixed with tobacco in the same year, because its users became "raving lunatics" (*krankzinnig en dol*). When it transpired that this decree turned all the Javanese fishermen away, they were exempted from it.[41] It is highly unlikely that this decree was ever strictly enforced.

It seems that, at the end of the eighteenth century, tobacco growing was largely restricted to the Minangkabau highlands and the surrounding areas (Marsden 1811: 323; Dobbin 1983: 33–4). We are left to speculate as to why it had not expanded over a wider area. Given the enormous success of Deli tobacco in the late nineteenth century, it is unlikely that tobacco growing was restricted by soil or weather. The fact that it was easily available from Java may form part of the explanation. An additional factor may have been the availability of another herb that could be smoked, namely *bang* or *ganja* (*Cannabis sativa* or hashish/marihuana). Ganja was (and is) cultivated in Sumatra, and although it may have been grown elsewhere in the Archipelago as well, Sumatra was always mentioned by foreign visitors as the region where it was used and grown. Areas to be found most often in these reports are Aceh and the Batak highlands, which suggests an introduction from India. In the late eighteenth century, Aceh was even exporting ganja. Rumphius stated that it was propagated in the Archipelago with Javanese seed, but he was the only European to mention Java in this context.[42]

TOBACCO IN JAVA AFTER 1800

Between 1800 and 1830, tobacco was found growing in many parts of Java. In West Java, Buitenzorg (the old Batavia Uplands), Priangan and Cirebon were often named as tobacco producing areas. The Environs of Batavia, one of the oldest tobacco producing areas, was no longer mentioned.[43] In Central Java, Kedu held pride of place, but its tobacco had spread to the surrounding areas: the regencies of Banjarnegara (in Banyumas), Ledok, later named Wonosobo (Bagelen), Batang (Pekalongan), and Kendal (Semarang).[44] In East Java, most tobacco was to be found in the regencies of Malang (in Pasuruan), Lumajang (Probolinggo), and Puger, later part of Bondowoso, and later still of Jember (Besuki).[45] Most tobacco was apparently being cultivated in upland areas, on tegal and gaga, with the exception of Kedu, where it could be found both on sawah — as a second crop — and on tegal.

The tobacco from Kedu was exported to the Malay peninsula, Sumatra, Kalimantan, Sulawesi and the Spice Islands, whereas Puger exported its tobacco only regionally, namely to the other parts of East Java and Madura. Tobacco was, after rice, the most important smallholder export commodity. In 1819, its rapid spread in and around Kedu was attributed to increased taxation, owing to the Land Rent System, introduced in 1813–4 in most Residencies of Java. It is not inconceivable that this factor was also operative in East Java.[46] In the years between 1830 and 1836, the first year in which a number of tobacco contracts were concluded under the Cultivation System, tobacco as a smallholder crop was reported to exist in four Residencies hitherto not mentioned: Rembang, Surabaya, Madiun, and Kediri. In Blitar, a Kediri regency, tobacco would become the major peasant crop within 20 years.[47]

Between 1836 and 1845, it looked as if tobacco would become an important — and compulsory — crop under the Cultivation System, in addition to the 'big three': sugar, coffee, and indigo. During these years, 37 tobacco-contracts were concluded with private entrepreneurs, 18 of which were in Rembang and 8 in Semarang. Typically, such a contract entailed that the government promised an entrepreneur, usually a European, that a specified number of villages would produce tobacco, to be bought by him at a (low) set price. By 1850, 23 of the 37 contracts had expired, and the remaining entrepreneurs were saddled with formidable losses. Therefore, tobacco under the Cultivation System had turned out to be a failure. In 1860, compulsory cultivation of tobacco was abolished in principle, and the last contract expired in 1864 (Vleming 1925: 5). Given what we know about the success of tobacco as a peasant crop in Java, it is clear that the failure of tobacco under the Cultivation System had nothing to do with soil or climate. Future developments would show that European management of peasant-grown tobacco was not to blame either. It must have been the combination of compulsory cultivation and low returns for their labour that made tobacco — as a Cultivation System crop - unpalatable to Java's peasantry.

Non-governmental tobacco production, supervised by European entrepreneurs under various contractual relations, started in Java (and Sumatra) between 1855 and 1865, producing the still well-known trade names of *Vorstenlanden, Besuki* or *Jember*, and — in Sumatra — *Deli*. In the Vorstenlanden (or Principalities, the areas still under indigenous rulers after 1830), the first small shipments of tobacco for the European market appeared in the late 1840s, but it was not until the early 1860s that production really took off. Around 1855, Europeans started to invest in tobacco in Kediri, Lumajang, and Jember. The Kediri adventure would turn into a failure within 20 years, and Lumajang was not all that successful either, but Jember was an important production centre up to the end of the colonial period and beyond. All three areas were major producers of smallholder tobacco before the Europeans moved in.[48]

Smallholder tobacco, which is my focus here, was largely produced in finely shredded form *(kerftabak)* for the local, regional and Asian market. A smaller proportion was produced for European engrossers, with a view toward European consumption. This was the so-called *krosok* (dried tobacco leaves), largely to be used as pipe-tobacco, which came predominantly from the Lumajang/Jember/Bondowoso area. Between 1900 and 1940, smallholder tobacco was cultivated in a great many districts, but it was heavily concentrated in a few areas only, namely the Dieng highlands and surrounding area (regencies of Banjarnegara, Wonosobo, Batang, Kendal, Salatiga, and the Residency Kedu), the Residency of Rembang, and the Residencies of Probolinggo and Besuki (regencies Lumajang, Jember, Bondowoso). Smaller centres were to be found in Kediri and Madura.[49] Perhaps the most surprising finding is that the list of tobacco areas is very similar to the list given above for concentrations of maize cultivation. On the relationship between tobacco and maize, more will be said below. One difference, however, between the developments of tobacco and maize should be mentioned here. Whereas the cultivation of maize expanded considerably between 1800/30 and 1940, this was basically expansion in areas where maize was already of some importance around 1830. In the case of tobacco, some production centres disappeared after 1830, while new ones arrived. Rembang and Kediri, major centres of tobacco cultivation in the twentieth century, were not mentioned as such prior to 1830. Madura, also important in 1940, was mentioned for the first time in 1845 as a tobacco producer.[50] On the other hand, Buitenzorg, a tobacco producing Residency of some note around 1815, no longer produced any tobacco around 1920. Priangan and Cirebon, both mentioned at the beginning of the nineteenth century (and Cirebon also before 1800), still produced some tobacco in the twentieth century, but had not developed into important centres. It seems safe to assume that tobacco, a high value commodity which was traded in considerable quantities, was more subject to competition, specialisation and opportunity costs than was maize. Differences in quality may have played an important role in this respect.

If we study the details of tobacco cultivation, to be found in the post-1830 literature, we find confirmation of a trend that was already visible prior to 1830: more than half of all smallholder tobacco was grown in upland areas. Junghuhn, travelling over Java in the 1830s and 40s, found it up to 4,500 ft. [1,500 m.]. He even suggested that on the slopes of two volcanos, the Sumbing and the Sundoro, both flanking the Kedu plain, tobacco was grown up to 7,000 ft. [2,275 m.]. Much lower figures of a later date (to be quoted later) indicate this was unlikely, unless, of course, these higher soils had been subject to degradation in the interim.[51] De Bie, writing around 1900, found tobacco up to 5,000 ft. [1,625 m.] and over. He stated that it was grown more on tegal and gaga than as a second crop on sawah. This was still the case in the 1920s, when tobacco was said to be grown up to 6,000 ft. [2,000 m.], and in the 1940s.[52]

Tobacco was in many areas a commercial crop, grown explicitly for local and supralocal markets. Returns per hectare in terms of money could be very high, which warranted the use of wage-labour (mentioned for the first time in 1819) and other costly inputs such as manure and seedlings bought from specialised farmers. Production of tobacco seedlings in the district of Kalibeber (Ledok, Bagelen) for the tobacco cultivators in Kedu and Banyumas, was already observed prior to 1815. In 1825, such specialists were also found in the village of Dolok (Banjarnegara, Banyumas). The trade was largely, but not entirely, in the hands of Chinese merchants, both large ones from Pekalongan and Semarang, and smaller ones living in large numbers in villages like Batur, Serojo and Parakan, around the Dieng highlands. In all probability, this situation dated back at least to the 1740s. There they prepared the tobacco — often, quite confusingly, called Chinese tobacco — and transported it to Pekalongan and Semarang for export. Chinese and Javanese engrossers bought tobacco in Kediri, to be exported from Surabaya. In the 1840s, the Chinese in Malang (Pasuruan) had even started the production of cigars for local consumption, no doubt as a sideline of their tobacco trade. Cigars and cigarettes were also produced in nineteenth century Kedu, partly with Chinese and European capital, probably under putting-out arrangements. In the twentieth century, Java witnessed the growth of many small and large-scale (mechanised) establishments producing cigars and cigarettes, using Javanese smallholder tobacco.[53]

The existence of a trade in agricultural products of a high commercial value more or less presupposes the existence of credit in any society, and certainly in a peasant economy such as the Javanese in the nineteenth and twentieth centuries. The Chinese merchants were in an excellent position to extend credit (*voorschot*) to the producers. This was reported around 1845, 1870, 1900, and in the 1920s and 40s. Sometimes the line of credit was more complicated, when Chinese merchants gave credit to Javanese *bakul*, who in turn gave advances to the producers. In the 1920s, the Government Credit Service (*Volkscredietwezen*) started to corner part of this market. In the tegal areas long-term indebtedness was no exception, and many peasants were, therefore, more or less obliged to continue tobacco cultivation whether they liked it or not. It is, however, also possible to see these arrangements in a more favourable light, stressing the mutual benefits that could be obtained in a long-term credit relationship (cf Hefner 1990: 68–9). This might explain why in some areas — such as Kedu — tobacco survived the ups and downs of the market.[54]

Besides credit and wage-labour, manure was a prerequisite for the cultivation of tobacco. It was a generally held belief among Dutch civil servants in nineteenth century Java that the peasantry did not use manure. As sawahs generally get most of the required nutrients from the annual inundation by irrigation water, this observation was accurate for many sawah areas. It did not apply, however, to the tobacco growing regions, where not only tegals but also sawahs (with tobacco) needed manuring on a regular basis. The application of manure to tobacco fields can be traced in the historical records as far back as the 1640s, when the Chinese tobacco farmers around Batavia were reported to use the city's nightsoil (human manure) for their fields (Saar 1930: 33). In China, the use of human manure as fertiliser was and is standard agricultural practice, but in Java it was only occasionally found. A reference from Cirebon in 1834 mentions the use of *bungkil*, the "cake" that remained after pressing peanuts and *jarak* (the castor-oil plant) for oil. In 1852, bungkil and farmyard-manure from cattle and buffalo were reported to be applied to tegals in Kedu. Sollewijn Gelpke, writing in the early 1880s, mentioned fertilising for tobacco, onions, and vegetables in Banten, the Dieng area, Kedu, the regencies Magetan (Madiun) and Ciputri (Cianjur, Priangan), and the island of Madura. Of these areas only Banten had no tobacco at all. He listed farmyard-manure in Kedu, a mixture of 'village dirt', manure and bungkil in Madura, and *kratok* (a pulse, *Phaseolus lunatus*), as 'green manure', in Besuki. According to De Bie, writing some 20 years later, manuring tobacco fields was common in Java. In West Java, chicken droppings, sometimes mixed with livestock manure, were used: in Central Java manure of cattle [including buffaloes?] and horses: and in East Java green manure (kratok). Around the same time manuring was reported from Wonosobo (Bagelen), Kedu, and Madura. The use of human manure was explicitly mentioned in Wonosobo in 1906. Cattle and horses, kept in stables for the production of manure, were mentioned in Wonosobo in 1906 and in Salatiga in 1912. Around 1920, animal manure, compost, and 'village dirt' were being used in Kedu, and green manure from kratok in Bondowoso (Besuki). Fruin specified that tegals in the Dieng/Kedu area were fertilised with goat and horse dung. The horses — the rather famous Kedu breed — were also kept for sale, the goats mainly for their manure. In these upland areas the fields were hoed not ploughed so buffalo and cattle manure was not available. Fruin also reported the use of artificial fertiliser (sulphate of ammonia). Vleming, writing about the same area around the same time (1925), referred to the increasing use of farmyard-manure on the better sawahs. On tegals, the use of dung was necessary, even if green manure crops had been planted. From Madura around that time, the use of farmyard-dung (from the famous Madura cattle) and bungkil were reported. Finally, in the 1940s, farmyard-manure, often from animals kept explicitly for that purpose (such as horses in Getasan, Salatiga), and increasingly, artificial fertiliser, were said to be used in some of the tobacco-growing areas. Most cultivators needed credit to buy artificial fertiliser.[55]

What makes this whole manure story so interesting is that there seems to be a connection between tobacco growing and livestock rearing. At least two famous Java breeds — Kedu horses and Madura cattle — are linked to typical tobacco areas. Equally interesting is the connection between livestock rearing and maize growing, as suggested by a number of sources. The earliest reference to maize being grown as fodder comes from Rumphius, who specifically mentioned that in Bali and Ambon maize was normally used as chicken-feed. The first reference from Java to maize as fodder (for livestock), dates, as far as I could establish, from 1847 and there are others between the 1910s and the 1940s.[56] Sparce references seem to indicate that the cultivation of maize as fodder for livestock was a relatively rare phenomenon in nineteenth century Java. A more thorough search might yield more "early" references, although one should not count on a wealth of data. Livestock rearing in Java was not a popular topic among colonial bureaucrats, who usually argued that there was no real livestock rearing to be found. According to most civil servants, the Javanese only had animals for transport, traction and for the plough, and they did not lavish much care on their buffaloes, cattle and horses. Among historians livestock rearing in Indonesia is an equally neglected topic. But a relatively late introduction of the cultivation of maize as fodder in Java cannot be ruled out. Cattle, buffalo, and horses can be fed on leaves and grasses acquired from forests and other non-cultivated areas, or even from arable lands lying fallow. Only when such areas become scarce or inaccessible is there is a need for the cultivation of fodder crops. It is plausible to assume that such a situation did not arise in most of upland Java prior to the late nineteenth century.

Be that as it may, we have now established a livestock-tobacco link and a livestock-maize link. The missing link (maize-tobacco) is supplied by the *Landbouwatlas*, which not only pointed out that maize was often grown as fodder for cattle, but also that there was a connection in a number of areas between maize growing, cattle rearing, and tobacco cultivation. This, of course, tallies nicely with the findings presented above — that in Java the areas in which maize cultivation was concentrated were more or less the same as the tobacco growing centres. I would like to argue that we may generalise the maize-cattle-tobacco complex, as postulated by the Landbouwatlas, to maize-livestock-tobacco. It thus seems that livestock rearing is the causal link in the remarkable overlap between maize and tobacco areas.

Finally, there may have been a fourth link: pulses. Due to their nitrogen-fixing properties, pulses are and were often used as green manure. Aside from improving the soil and providing good ground cover, the beans could be eaten by humans, and the leaves could be used as fodder. In some areas, such as Pasuruan, Probolinggo, and Besuki, in the early nineteenth century, several species were already specifically cultivated for the feeding of livestock. The ones most often named were *kacang iris, kedele, komak, kratok*, and peanuts.[57] Moreover, as maize could be easily intercropped or doublecropped with pulses, as I demonstrated earlier, pulses, with their 'dual function', fitted perfectly in the maize-livestock-tobacco complex, thus forming the four element complex maize-pulses-livestock-tobacco.

The complex I have identified might go a long way in explaining why both maize and tobacco growing centres, though often located in ecologically vulnerable upland areas, show an amazing continuity of at least one century of seldom-interrupted expansion, and sometimes go back two or even three centuries. Recent studies have emphasised the deterioration of Java's uplands over the last decades, and their findings seem to be well documented. Perhaps the complex, if it still exists, is finally cracking. It is possible that local deterioration set in during the period under discussion in this article. It might account for the problems and failures of some tobacco producing areas, reported as early as the 1880s. Future researchers should perhaps focus their attention on some of these regions for in-depth studies. Factors might include deforestation and the related problems of water shortages, which were already mentioned in the sources around 1850. Firewood consumption for drying the tobacco leaves at high altitudes (where sun-drying was no longer possible) must have added to the loss of forest cover. If this loss went too far, tobacco production itself may have been endangered. Locally, the peasantry of the more elevated areas such as the higher reaches of the Dieng actually planted (or rather sowed the seeds of) trees such as the indigenous *kemlandingan gunung* (*Albizzia montana*), and even various *Eucalyptus* species, introduced from outside and provided by the Forest Service. Such "plantations" were mentioned several times in Wonosobo between the 1870s and the 1920s, but the literature seems to suggest that, at least in the twentieth century, constant civil service supervision was necessary, and that shortages of firewood persisted even then.[58] Data on terracing in tobacco-growing upland areas is fairly rare, and one wonders whether this state of affairs reflects a lack of interest from the colonial bureaucracy or an actual deficiency of terraced fields. It is often assumed that continuously cultivated un-terraced upland fields are highly susceptible to erosion, but data from Wonosobo, dating from 1906 and 1923, seems to suggest otherwise.[59] The overall picture of the tobacco-growing areas is one of almost continuous expansion.

A short note on Bali and Lombok will round off the description and analysis of Java and its sphere of influence after 1800. In the 1820s and early 1830s, Bali was exporting large amounts of tobacco. This implies that its cultivation had been going on for some time. A missionary, visiting the island in 1830, said that it "produces much tobacco, of an excellent quality", listing it as one of Bali's four main crops. Bali was named as a major exporter of tobacco in the 1850s, and again in the 1880s, when reference was made to exports from Karangasem and Buleleng, and to cultivation in Badung. The people from the island of Selayar took it to Makasar and Singapore. On the neighbouring island of Lombok in the same period, tobacco was the second most important export after rice. Just after the First World War, tobacco from Bali and Lombok was exported to Europe, but it failed to make the grade there. Tobacco was grown both on sawah and on tegal, with a preference for the latter, because of the higher quality tobacco it produced.

Bali is and was — among other things — known for its cattle rearing, Bali cattle being such a famous Indonesian breed that it has elicited at least two learned monographs all of its own (Aalfs 1934; Meijer 1962). In some areas, the foliage of maize and pulses was used as fodder, and both were often cultivated for the sole purpose of feeding cattle. Particularly on tegal, the cultivation of tobacco required sufficient amounts of cattle dung. As I also found references to the use of pulses as green manure, I am inclined to regard the case of Bali as confirmation of the maize-pulses-livestock-tobacco complex hypothesis.[60]

TOBACCO IN THE OUTER ISLANDS AFTER 1800

Moving on to tobacco in Eastern Indonesia after 1800, I begin in the North Moluccas with the Ternate tobacco, which by this time appears to be cultivated mainly on the island of Makian. In the 1850s and 60s, it was still a well-known export commodity, but then disappeared from the sources. The nearby island of Muti also produced tobacco in this period. It could be found in many of the Central Moluccan islands as well, namely Buru and Seram (both exporters to regional destinations in the 1850s), and Ambon, Hila and Saparua. Particularly in Seram, an otherwise not very "commercial" island, much care was given by the tribal Alifuru to the cultivation of tobacco, including the use of separate seed-beds. Cultivation of tobacco in Irian Jaya was still mentioned in the nineteenth and twentieth centuries, though by the latter period the population had begun to prefer imported tobacco that had been prepared in the Netherlands. By the 1830s, at the latest, tobacco had also spread to the nearby Kai islands.[61] In Nusa Tenggara, tobacco had by now spread to most of the islands, but only Timor, Flores and Savu were exporting some of it to regional markets. As Timor was also a major maize producer and had a fair amount of livestock, this could be another example of the maize-livestock-tobacco complex. In Bima, Sumbawa there was the combination of maize, tobacco (though not for export), the well-known Bima and Sumbawa horses, and *kacang ijo*, cultivated as fodder.[62] Sulawesi and Selayar were also mentioned as tobacco growing areas after 1800. Most of it was grown for local consumption, but southwest Sulawesi was producing some tobacco for regional trade. In Minahasa and Gorontalo (northeast Sulawesi) we find tobacco and also maize being fed to livestock.[63]

At the western end of the Indonesian Archipelago, we witness the phenomenal expansion of plantation tobacco in the Deli area in Sumatra, already mentioned above. Smallholder tobacco, my focus here, also showed some expansion. At the end of the nineteenth century the commercial cultivation of smallholder tobacco was concentrated in the upland areas of south and west Sumatra: the Residencies of Sumatra's Westcoast (Padang Uplands, the Minangkabau area), Bengkulu (Rejang, Krui), Palembang (Ranau), and to a somewhat lesser extent Lampung. Payakumbuh (from the Padang Uplands), Ranau and Rejang tobacco were being exported to Singapore and Europe. It was often grown as a ladang crop.[64] It seems that the maize-livestock-tobacco complex did not operate here, except perhaps in the Batak highlands where maize and livestock (the well-known Batak ponies) were important and maize was used as fodder. In the nineteenth century, tobacco was grown in small quantities in most Batak regions, although it was absent in Silindung (in 1824) and was not doing well in the Toba area (1887). After 1900, it was frequently mentioned, but it does not seem to have developed into an important (export) commodity.[65]

Finally, in Kalimantan it seems likely that in the nineteenth century, its cultivation spread to more areas than before. It was, however, only of some importance in the Ulu Sungei and Dusun areas. Between 1890 and 1905, tobacco growing under European supervision was attempted without success.[66]

CAUSES AND CONSEQUENCES

This survey has documented the introduction and expansion of maize and tobacco over most of the Archipelago between 1600 and 1940. What were the reasons for the success of these crops? Starting with maize, it seems plausible to assume that its high productivity per unit of land and per unit of labour was its main attraction. Compared with "dry" rice it had a much higher yield ratio, its harvests were more "secure" (not many harvest failures), and its labour requirements were lower. Its productivity was also higher than that of a number of "archaic" cereals which in some areas — Java, Nusa Tenggara, Central Sulawesi — were probably largely replaced by maize, namely foxtail millet, sorghum, and Job's tears. The substitution of these cereals by maize has been a familiar phenomenon in many tropical countries.[67] The cultivation of maize had many other advantages. It could be grown in areas that were too high, too steep, too dry, or too infertile for "dry" rice. It was also highly adaptable to local needs, and various groups succeeded, at an early stage, in importing or developing varieties with a short to very short growing season, tailored to local cropping patterns. Thus, it could be doublecropped easily with traditional or other newly introduced crops. One of the crops with which it was often rotated was "dry" rice, which partly required different nutrients from the soil, and which also had different moisture requirements. This made it possible to shorten fallow periods on tegal and to use ladangs for a longer period. Maize could also be intercropped with various other plants, including rice. In both cases — intercropping and doublecropping — the risk of a complete harvest failure was considerably reduced. Apparently it was also possible to grow maize as a monoculture, year in year out, often two or even three crops a year. It would keep well, and it had some nutritional advantages over rice, namely its protein and vitamin A content. Compared with other high-yielding starchy competitors — roots and tubers — its nutritional value was also higher. Finally, it could be used, partly or entirely, as fodder.[68]

If the high productivity, adaptability and versatility of maize explains why people took to it so quickly and almost universally, we should now turn to the consequences of its rapid expansion. Important among these was the expansion of the population into upland areas and higher population densities in lowland areas. This seems to have been the case in Java, where growth-rates were fairly high (by Southeast Asian standards) after about 1750, and even higher in the nineteenth century. In other areas, growth rates seem to have been very low at best prior to 1850 or even 1900, which makes it rather difficult to argue that the introduction of maize was a growth inducing factor. One could argue, however, that maize enhanced the potential carrying capacity of all areas were it took root, thereby facilitating an increase in population when other obstacles to growth had been removed (cf. Henley 1993: 17/8).

Even in the Outer Islands where the increased cultivation of maize may not have influenced population growth-rates, it may still have facilitated the expansion of settlement into the mountainous uplands, the lowlands and the upland valleys losing what was gained by the mountain slopes. This is a possibility that future researchers could examine. Here it should be pointed out that the term 'upland' may be confusing: upland valleys or plateau, often quite suitable for rice cultivation, had been populated long before maize arrived, as in the Minangkabau and Batak areas. The uplands that were invaded by maize were the mountain slopes surrounding these valleys (Sumatra) and the middle and upper-slopes of mountains in general (Java and elsewhere). It was in these mountainous areas, often not suited to rice, and either sparsely populated or uninhabited, that maize made a difference.[69] Maize, then, may have enabled groups or individuals, who, for political, religious, economic, or health reasons, wanted to leave the population centres in the lowlands or the upland valleys, to survive and even flourish in hitherto sparsely populated mountain areas. According to Hefner (1990: 57), this may have applied to the Hindu groups who in the sixteenth century moved into the Tengger uplands in East Java. Something similar may have occurred in the case of the Lubu, who probably became mountain-dwellers in or before the eighteenth century (Ophuijsen 1884: 96). It is possible that the introduction of maize enabled similar migratory movements elsewhere to successfully establish themselves as 'mountaineers', thus constituting 'heathen' upland 'tribes' (masyarakat terasing in modern Indonesian parlance), who preferred to live at a distance from the coastal Muslim-Malay sultanates, the European and Chinese trading settlements, and the risk of epidemics. This, by the way, was much to the regret of the Dutch who always attempted to get the mountain-dwellers to come and live in the coastal areas, and to make them grow "wet" rice.

The introduction and expansion of maize also may have reinforced the social stratification of local societies, creating a division between a rice-eating aristocracy and maize-eating commoners and slaves or debt-bondsmen. Commoners would only eat rice on ceremonial occasions. This division is often emphasised in the sources, as is the fact that rice, in addition to being the food for the better-off, was being exported in rather large quantities. In these cases, therefore, the cultivation of maize facilitated the export of rice (e.g. Menado, Makasar, Bima). Rice tributes required by the Europeans may have been instrumental in this transformation. A comparable mechanism might also have operated in eighteenth century Java, where population growth in non-sawah areas did not lead to appreciably lower rice exports.[70] Finally, the reverse side of all the positive characteristics of maize must also be noted. By facilitating the movement of large numbers of people into the upland areas, the introduction and expansion of maize was also partly responsible, at least potentially, for the removal of the forest cover of these ecologically fragile areas.

Tobacco was in many respects a very different crop. It was from the outset almost entirely grown for the market, be it the local, regional or even supra-regional. It required constant vigilance and labour, and non-labour inputs such as manure. The necessary labour input was so high that the tobacco cultivating peasant often had to rely at least partially on hired hands. In Java, it was also regarded as a rather risky crop, sensitive to drought and flooding. Nevertheless, it seems to have been grown successfully both in semi-arid areas in Eastern Indonesia, and in the very wet uplands of Sumatra. This seems to imply that it could be adapted to climatic extremes, which might explain its Indonesia-wide expansion.

It is also significant that tobacco cultivation was not limited to the richer peasants. Relatively poor upland farmers were apparently able to grow some tobacco as well. One is tempted to assume that, though maize made it possible to live in the upland areas, it was tobacco — often combined with livestock rearing and the cultivation of pulses — that made it a rewarding alternative to do so. Nevertheless, with tobacco being the risky crop it is, many upland tobacco cultivators in Java could not have survived without a system of patronage involving Chinese merchants-cum-moneylenders. It is not unlikely that these traders kept the tobacco cultivators from accumulating capital, but that they also sheltered them from the more negative influences exerted by participation in the world market. Detailed information on the fortunes of tobacco cultivators outside Java is lacking, but it seems likely that in these regions middlemen-traders played a similar role, sheltering Papuans, Alifuru and various other groups quite effectively from the fluctuations of the world market. Credit, therefore, seems to have been the fifth element that was needed to turn the four-element complex maize-pulses-livestock-tobacco into a sustainable system. Recall also that increased taxation was mentioned as a stimulating factor in early nineteenth century Java, and it may have been an incentive to tobacco growing in other times and at other places.

CONCLUSIONS

Both maize and tobacco were introduced successfully in many areas of the Indonesian Archipelago between 1600 (or earlier) and 1850. Although maize has the reputation that it can grow anywhere, it seems to have been less successful in the wetter areas of Western Indonesia. Tobacco, with a more "difficult" reputation, was at the end of the nineteenth century grown commercially in both wet and dry areas. It seems, however, that there are more similarities than differences between the crops. They were both grown on sawah, tegal, and ladang/gaga. Throughout the period under consideration the cultivation on 'dry' upland fields seems to have predominated. They were often to be found in the same regions.

In a number of areas, the two crops seem to have been part of a remarkably sustainable "complex". Maize was grown here partly as fodder, as were pulses, which, moreover, improved the tobacco and maize fields through their nitrogen fixating properties (green manure). Livestock, reared with this fodder, produced the manure, needed by the tobacco fields. Most of the evidence for this complex comes from post-1850 Java. This is in keeping with the hypothesis that one does not expect to find this complex in areas where fodder and land are not scarce.

The high productivity, adaptability and versatility of maize must have greatly contributed to its almost ubiquitous introduction and fairly rapid expansion. The availability of maize may have contributed to increasing population growth rates, to the survival of "mountaineers" with their 'deviant' identity, to a more marked division between aristocracy and commoners, and to rice-exports from non-sawah areas. On the other hand, the expansion of maize into upland areas was also a potential threat to these ecologically vulnerable areas.

In the case of tobacco, it is likely that its adaptability to climatic extremes contributed to the Indonesia-wide adoption of this crop. Although a risky commercial crop, it was nevertheless grown by poor upland cultivators, often as part of the above-mentioned four-element complex, to which credit has to be added as the fifth element. In the twentieth century, the presence of smallholder tobacco triggered the rise of an important local branch of industry.

The role of the state in promoting these crops was probably restricted to indirect measures, namely rice tributes in the case of maize and taxation in the case of tobacco. These "incentives" apart, the introduction and expansion of both crops seem to have been spontaneous.

ACKNOWLEDGEMENTS

I owe a debt of gratitude to Christiaan Heersink, David Henley, Bernice de Jong Boers, and Han Knapen, members of the EDEN-team, KITLV, who provided me with data on their respective research areas, and who gave their comments on an earlier version of this article. I am also indebted to Tania Li and an anonymous referee for their suggestions.

NOTES

1. Westenenk (1962: 12); Fox (1977: 73); Hefner (1990: 8); Henley (1993: 4/5).
2. Rumphius (1741/55, V: 225/6); Wigboldus (1979: 20/3).
3. The term *sawah* is used for irrigated or rainfed bunded fields, usually planted with rice (as first crop). *Tegal* is the usual term for more or less permanently cultivated 'dry' (not inundated) fields, *gaga* (Java) or *ladang* for fields under shifting cultivation, and *pekarangan* is the Javanese term for home-gardens.

4   Wigboldus (1979: 23); Reid (1988: 19); Henley (1993: 2/3).
5   *Daghregister* 1663, 15 June (further references to the *Daghregister* daily register, kept by the VOC at Batavia will have the following form: D.15.6.1663); KITLV, H 802 (2): Notitie Speelman over Makassar, 1669, p. 117; *Generale Missiven* (GM for short) III, 782 (31.1.1672); Rumphius (1741/55, V: 202); Padtbrugge (1866, 324); Stapel (1927/54, II/1: 78 and 97); Skinner (1963: 105).
6   D.15.6.1663. Fox (1977: 76/7) suggested that the VOC also attempted to stimulate maize cultivation in Timor, but that is based on an erroneous interpretation of GM III, 782 (31.1.1672). Actually, the VOC was trying to get more rice planted.
7   D.18.4.1648; D.12.6.1648. The first shipment came from Gresik, Jepara, Gembong near Pasuruan, or "Chillemaijo" (Indramayu, Cimanuk?) The second shipment consisted of 300 pieces of dried *tagons*, of which I can only make sense if it is assumed that it was a clerical or a printing error for a j. The *tagons* came from Gresik.
8   D.13.12.1681; Rumphius (1741/55, V: 202); Nagtegaal (1988: 37).
9   Algemeen Rijks Archief (General State Archives, The Hague, ARA for short), Collection van Alphen/Engelhard (Coll.A/E for short), 1900, 166: Resolutie Semarang (14 April 1777); Hawkesworth (1773, III: 732/3); Aanmerkingen (1781: 177); Teisseire (1792: 83); Nagtegaal (1988: 37).
10 Marsden (1811: 72, 323, 371, 380); Raffles (1830b: 349); Dobbin (1983: 28); Andaya (1993: 66, 228).
11 GM IX, 264 (28.12.1731); Hogendorp (1779: 297); Wharton (1893: 344); Roo van Alderwerelt (1906: 221). Dampier (1939: 170); Heersink (1995: 84). According to Fox (1977: 117) the "Indian corn" mentioned for Savu was a mistake (for sorghum or foxtail millet), but I am not convinced by his arguments.
12 ARA, Coll.A/E (1900: 208): Semarang to Batavia, 31 Dec. 1804, par. 420; ARA, Coll.A/E (1916: 112): Extract Secreete Notulen Geresolveerde in Raade van Indiën (26 Nov. 1805, par. 5 and 10); *Plakaat* 23.1.1800 (in *Plakaatboek*, to be cited as P.23.1.1800); P.27.9.1800; P.21/4.4.1807; P.14.7.1809; Rijstcultuur (1854: 12). The VOC functionaries had a similarly low opinion of roots and tubers, such as *ketela*, *tales*, and *ubi*.
13 Raffles (1814: 199); Crawfurd (1820, I: 367); Raffles (1830a, I: 134/5); Boomgaard (1989b: 88-96).
14 In the nineteenth century, the highest administrative level in Java was the residency (*residentie*), of which there were 20 to 25, depending on the period. Residencies were subdivided into regencies (*regentschap*), the *kabupaten*. A regency consisted of a number of districts. I will use the names of the mid-nineteenth century residencies. There were frequent changes in the arrangement of the residencies in the twentieth century. There was much less change on the level of the regencies.
15 ARA, Collection Baud, 452: "onderzoek wegens de betrekkelijke voordeelen der Kultuur van Rijst, met die van suiker en indigo", t.g.v. circulaire 4 sept. 1834: 474; Crawfurd (1820, I: 367); Raffles (1830a, I: 134/5); Domis (1830: 366); Domis (1832: 328); P. (1845: 115/6); Steyn (1851: 107); Junghuhn (1853/4, I: 227); Deventer (1865/6, II: 18/9).
16 On the geographic distribution of maize in 1875, *Koloniaal Verslag* 1876. appendix NN; for c. 1920, *Landbouwatlas* 1926, I: map 14, and II, 107/8; for c. 1940, Veer (1948: 115/7); for c. 1975, Kano (1987: 98).
17 On the Dieng area c. 1830, Crawfurd (1820, I: 369); Domis (1830: 363/81); Junghuhn (1853/4, II: 268/414); Deventer (1865/6, II: 18/9); Residentie (1871: 94). On East Java and Madura c. 1830, ARA, Coll. Baud, 452: "onderzoek", 1834, report Besuki; Raffles (1830a, I: 134/5); Domis (1832: 328); Domis (1835: 105/7, 125); Kussendrager (1861: 199).
18 Arsip Nasional (National Archives, Jakarta, ARNAS for short), Arsip Daerah Madura 15/1: Verslag zending naar Pamakasan, 1854; ARNAS, Appendix to Geheim Besluit 6.5.1857, La.E(I): Rapport Ass.Rest. Pamakasan; Teijsmann (1855: 86/7); Residentie (1871: 94); Veth (1875/82, III: 985/8); Sollewijn Gelpke (1885: 30/4).
19 Sollewijn Gelpke (1885: 34); Sollewijn Gelpke (1901: 117/9); Bie (1901/2, I: 99).
20 Boomgaard (1989b: 88); Boomgaard and van Zanden (1990: 44).

21 Donner (1987: 69/70); Palte (1989: 40/2); Hefner (1990: 10).
22 Journal (1830: 38); Bloemen Waanders (1859: 188/91); *Indisch Verslag* (1941, II: 284/6); Liefrinck (1969, 61: 64/5).
23 Zollinger (1850: 9/10, 73, 102); Bosscher (1854: 424); Francis (1856, II: 130, 163); Doren (1863: 257); Doren (1864: 86); Ludeking (1868: 90, 117); Ternate (1980: 190, 196); Coolhaas (1985: 97).
24 Wetering (1926: 552); Ormeling (1956: 107). It is not clear how Fox can argue that "maize is less successfully intercropped with other plants" (Fox 1977: 76/7). Actual practice in Java and Eastern Indonesia suggests otherwise.
25 Ludeking (1868: 117); Paerels (1918: 835); Henley (1993: 3).
26 Müller (1855: 50); Nieuwenhuijzen (1858: 415); Bogaardt (1859: 37); Steck (1862: 111); Mohnike (1874: 137/8); Raedt (1888: 434/6); Helfrich (1889: 569/77); Snouck (1893/4, I: 276/82, 308); Jacobs (1894, II: 128).
27 Hagen (1886: 330, 344, 369); Snouck (1893/4, I: 276/82); Jacobs (1894, II: 128); Dobbin (1983: 28).
28 Hasselt (1882: 347); Neumann (1886/7: 263); Veer (1948: 117); Purseglove (1972: 310).
29 Neumann (1886/7: 236); Kerckhoff (1890: 576); Kreemer (1912: 307).
30 Schwaner (1853/4, I); Hageman (1855: 95); Bock (1882); Lindblad (1988: 22/6).
31 Jonge (1862/95, V: 42); Behr (1930: 31); Foster (1967: 173); Reid (1988: 44).
32 P.31.8.1626; D.7.4.1637; P.3.6.1643; D.1.1.1644; Reid (1988: 44).
33 D.7.4.1637; Behr (1930: 31); Saar (1930: 33).
34 D.21.1.1648; D.22.1.1648; D.15/6.3.1648; D.10.4.1648; D.2.9.1648; D.22.9.1648; D.23.9.1648; D.29.9.1648; D.6.11.1648; D.19.12.1648.
35 D.31.1.1657; D.30.4.1657; D.31.7.1657; D.30.11.1657; D.31.12.1657.
36 Tollgates in VOC territory were, however, against the official policy of Batavia, and his suggestion was never acted upon.
37 ARA, Coll.vA/E (1900: 191): Memoir of Transfer, Java's Northeast-Coast, J.F. van Reede tot de Parkeler to N. Engelhard, August (1801, par. 390; P.12.12.1797); Reis (1853: 417); Jonge (1862/95, X: 65); Carey (1981: XLII).
38 D.31.5.1661; P.27.2.1671; Rumphius (1741/55, V: 225/6); Hogendorp (1779: 297); Leupe (1875: 146); Jacobs (1974/84, III: 477). It is possible that the Ternate mentioned in the sources is not the island but the wider area of the sultanate.
39 Evidence for the substitution of pepper for tobacco on ladang can be found in the Lampung area, south Sumatra, at the end of the nineteenth century (Broersma 1916: 112). Tobacco and pepper are in many respects similar; they are both commercial, labour intensive crops, which tax the soil rather heavily.
40 D.14.5.1661; D.30.9.1661; D.30.6.1664; D.31.8.1664; D.30.11.1664; GM III, 527 (25.1.1667); D.31.7.1667; D.30.4.1686; D.31.7.1690. The last two entries are taken from microfilms of the unpublished Daghregister manuscripts kept at the ARNAS.
41 D.12.11.1671; D.22.12.1671; Haan (1910/2, I: 247).
42 Lockyer (1711: 61/2); Salmon (1729/33, II: 280, 295); Rumphius (1741/55, V: 208ff); Forrest (1792: 42); Burton and Ward (1856: 286, 303); Dampier (1931: 88).
43 ARA, Coll.vA/E (1900: 118): Dagregister Pieter Engelhard Inspectie der Bovenlanden (23 June 1802); ARA, Ministerie van Koophandel en Koloniën, 147: Generaal Rapport der Commissie tot Inspectie der gezamentlijke Jaccatrasche Regentschappen en Preanger Bovenlanden, 29.1.1808, par. 168; ARA, Collection Schneither, 72: N.Engelhard, Kort verhaal..., 17.6.1816; Wilde (1830: 100); Haan (1910/2, IV: 366).
44 ARA, Coll.vA/E (1900: 210): Aparte Missive Governor Java's Northeast Coast to GG (30 June 1804, par. 90); Report Agricultural Commission 1827, in *Javasche Courant*, (24 April 1828); Raffles (1830a, I: 149/50); Domis (1830: 363/81); Deventer (1865/6, I: 289); II: 13/4, 21, 654/5); Residentie (1871: 95).
45 ARA, Coll.vA/E (1896: 30): Berigten nopens den opneem der onderscheidene cultures op Javas noordoost kust, Ao. 1802/3; ARA, Coll. Schneither, 90: Statistiek Pekalongan (1822/3); ARA, Archief Ministerie van Koloniën (AMK for short), Verbaal 10.4.1830, 29: Report Agricultural Commission (1828); Raffles (1814: 141); Raffles (1830a, I: 149/50).

46 Crawfurd (1820, III: 416/7); Raffles (1830a, I: 149/50); Steyn (1851: 22); Deventer (1865/6, II: 13/4); Residentie (1871: 96).
47 ARA, Coll. Baud, 452: "onderzoek" (1834); Teenstra (1852, I: 137–240); Teijsmann (1856: 152); Kussendrager (1861: 253–301).
48 Staverman (1868: 39); Vleming (1925: 6/8, 93); Houben (1994: 266/7, 301/3).
49 Bijlert (1919: 21/32); Landbouwatlas (1926, I: map 20; II: 140/3); Fruin (1923a: 269); Fruin (1923b: 353); Broek (1949: 525/6).
50 Temminck (1846, I: 338); Teenstra (1852, I: 240); Jonge (1988: 94).
51 Sollewijn Gelpke (1885: 90) mentioned a decline in tobacco cultivation in some areas of Kedu and Probolinggo, but he may have been referring to temporary effects of the trade-cycle.
52 Junghuhn (1853/4, I: 408); II: 316/22; Bie (1901/2, II: 24); Vleming (1925: 67); Landbouwatlas (1926, II: 108); Broek (1949: 526).
53 Crawfurd (1820, III: 416/7); Raffles (1830a, I: 149/50); Domis (1830: 363/80); Jukes (1847, II: 100); Deventer (1865/6, II: 13/4, 654/5); Staverman (1868: 39); Residentie (1871: 96/7); Beschrijving (1894: 50); Bijlert (1919: 23/4).
54 Epp (1852: 407); Residentie (1871: 97); Fruin (1923a: 294/9); Vleming (1925: 67); Broek (1949: 525/6); Alexander and Alexander (1991: 83/4).
55 ARA, Coll. Baud, 452: "onderzoek" (1834); ARNAS, Arsip Daerah Kedu, 48: *Algemeen Verslag* Kedu (1852); Sollewijn Gelpke (1885: 32); Bie (1901/2, II: 27/8); Breda de Haan (1906: 516); Horst (1912: 272); Hasselman (1914: 30); Bijlert (1919: 25); Fruin (1923a: 273); Vleming (1925: 70); Wijers (1928: 127/8); Vink (1941: 152); Broek (1949: 529); Alexander and Alexander (1991: 83/4).
56 Rumphius (1741/55, V: 202); Hoëvell (1849/51, II: 150); Weidegang (1913: 7); Hoen (1919: 57, 97/9); Landbouwatlas (1926, II: 107); Vink (1941: 152); Veer (1948: 145).
57 ARA, Coll. Baud, 452: "onderzoek" (1834 (Besuki/Banyuwangi)); ARA, AMK, Commission Umbgrove (1854/6: 67 (Probolinggo)); Hoëvell (1849/51, II: 150); Bie (1901/2, II: 27/8); Weidegang (1913: 6/7); Hoen (1919: 57, 97/9); Bijlert (1919: 25); Vink (1941: 151).
58 Teijsmann (1855: 34); *Koloniaal Verslag* (1883: 194/7); Tobi (1894: 150); Breda de Haan (1906: 511); *Verslag Dienst Boschwezen* (1907: 3, and 1923: 3); Horst (1912: 275); Fruin (1923a: 268, 273); Fruin (1923b: 363/5); Zwart (1928: 25, 35); Fokkinga (1934: 168/9).
59 Teijsmann (1855: 86/7); Breda de Haan (1906: 512); *Onderzoek Mindere Welvaart*, Va (1908: 227/30); Horst (1912: 273); Fruin (1923a: 268); Schuitemaker (1949: 165); Hefner (1990).
60 Journal (1830: 24, 38); Broek (1835: 228, table facing p. 236); Crawfurd (1856: 436, 537); Bloemen Waanders (1859: 192, 198); Jacobs (1883: 118, 146, 199); Bijlert (1919: 35); Aalfs (1934: 51/2, 141); Vink (1941: 151/2); Broek (1949: 523); Meijer (1962: 52); Liefrinck (1969: 61/5); Kraan (1980: 12).
61 Ceram (1856: 84/8); Doren (1863: 257); Ludeking (1868: 90, 98, 117); AStWb (1869, III: 811); Bijlert (1919: 31); Ternate (1980: 190/6).
62 Zollinger (1850: 73, 102); Francis (1856, II: 215); AStWb (1896, III: 811); Bijlert (1919: 34).
63 Zollinger (1850: 9/11); AStWb (1869, III: 810/1); Broek (1949: 536); Henley (1993: 15).
64 Köhler (1856: 144); Bois (1857: 27, 37); AStWb (1869, III: 808/10); Hasselt (1882: 176/7, 344); Bijlert (1919: 29/34); Vleming (1925: 65, 175/6); Broek (1949: 536); Dobbin (1983: 33/4).
65 Junghuhn (1847, II: 202/4); Burton and Ward (1856: 303); Brenner (1894: 353); Volz (1909: 264); Bijlert (1919: 21).
66 Schwaner (1853/4, I: 148, 223); (II: 137); AStWb (1869, III: 810); Bijlert (1919: 30/4); Lindblad (1988: 27/30).
67 On the substitution of 'archaic' cereals by maize (and rice), see Purseglove (1975: 134, 262); Fox (1977: 76/7); Wigboldus (1979: 26); Boomgaard (1989a: 92); Fox (1991); Henley (1993: 10).
68 D.15.6.1663; Rumphius (1741/55, V: 202); Aanmerkingen (1781: 177); Crawfurd (1820, I: 367); Raffles (1830a, I: 134/5); Junghuhn (1853/4, I: 227); Veer (1948: 141/6); Wigboldus (1978: 22/4); Henley (1993: 11/4).

69 E.g. Raffles (1830b: 349); Junghuhn (1847, II: 190, 201/2); Müller (1855: 135, 145/6, 156/7, 179/80); Köhler (1856: 144); Kerckhoff (1890: 576); Ormeling (1956: 107/8); Dobbin (1983: 28).

70 Hogendorp (1779: 297); Zollinger (1850: 102); Rijstcultuur (1854: 110); Francis (1856, II: 134, 201); Wigboldus (1978: 22); Boomgaard (1989c: 332/3, 338).

References

Aalfs, H.G., 1934, *De rundveeteelt op het eiland Bali*, PhD. dissertation. Utrecht.
Aanmerkingen, 1781, 'Aanmerkingen op de vraag, welk zijn de beste en spoedigst voortkomende wortelen, om de behoeftigen, bij misgewas van graanen te spijzigen?' *Verhandelingen Bataviaasch Genootschap*, 3, 175–186.
Alexander, J. and P. Alexander, 1991, 'Trade and petty commodity production in early twentieth century Kebumen', in *In the Shadow of Agriculture. Non-farm Activities in the Javanese Economy, Past and Present*, edited by P. Alexander, P. Boomgaard and B. White, pp. 70–91. Amsterdam: Royal Tropical Institute.
Andaya, B. Watson, 1993, *To Live as Brothers: Southeast Sumatra in the Seventeenth and Eighteenth Centuries*. Honolulu: University of Hawaii Press.
AStWb, 1869, *Aardrijkskundig en Statistisch Woordenboek van Nederlandsch-Indië*, 3 Vols. Amsterdam: Van Kampen.
Behr, J. von der, 1930, *Reise nach Java, Vorder-Indien, Persien und Ceylon, 1641–1650*. Reisebeschreibungen von deutschen Beamten, 4. Den Haag: Nijhoff 1st ed. 1668.
Beschrijving, 1894, *Beschrijving der tentoonstelling te Magelang van producten van inlandsche nijverheid uit de Residentie Kedoe, 1891*. Batavia: Landsdrukkerij.
Bie, H.C.H. de, 1901/2, *De landbouw der inlandsche bevolking op Java* Mededeelingen uit's Lands Plantentuin, 45 and 58, 2 Vols. Batavia: Kolff.
Bijlert, A. van, 1919, 'Tabak', in *Dr. K.W. van Gorkom's Oost-Indische Cultures*, edited by H.C. Prinsen Geerligs, 1st ed. 1913, Vol. 3, pp. 1–163. Amsterdam: De Bussy 3 Vols.
Bloemen Waanders, P.L. van, 1859, 'Aanteekeningen omtrent de zeden en gebruiken der Balinezen, inzonderheid die van Boeleleng'. *Tijdschrift Bataviaasch Genootschap*, 8, 105–279.
Bock, C., 1882, *The Head-Hunters of Borneo: A Narrative of Travel up the Mahakkam and down the Barito; also, Journeyings in Sumatra*. London: Sampson Low, et al.
Bogaardt, T.C., 1859, 'Moko-Moko in 1840'. *Bijdragen tot de Taal-, Land- en Volkenkunde*, 6, 26–42.
Bois, J.A. du, 1857, 'De Lampongsche distrikten op het eiland Sumatra'. *Tijdschrift voor Nederlandsch-Indië*, 19(1), 1–49, 89–117.
Boomgaard, P., 1989a, *Children of the Colonial State; Population Growth and Economic Development in Java 1795–1800* CASA Monographs, 1. Amsterdam: Free University Press.
Boomgaard, P., 1989b, 'Java's agricultural production, 1775–1875', in *Economic Growth in Indonesia 1820–1940*, edited by A. Maddison and G. Prince, pp. 97–132, Verhandelingen KITLV, 137. Dordrecht/Providence: Foris.
Boomgaard, P., 1989c, 'The Javanese rice economy 800–1800', in *Economic and Demographic Development in Rice Producing Societies. Some Aspects of East Asian Economic History, 1500–1900*, edited by A. Hayami and Y. Tsubouchi, pp. 317–344. Tokyo: Keio University.
Boomgaard, P. and J.L. van Zanden, 1990, *Food Crops and Arable Lands, Java 1815–1940*, Changing Economy in Indonesia, Vol. 10. Amsterdam: Royal Tropical Institute.
Bosscher, C., 1854, 'Statistieke schets der Zuidwester-Eilanden'. *Tijdschrift Bataviaasch Genootschap*, 2, 419–458.
Breda de Haan, J. van, 1906, 'Verslag over de tabakscultuur in het district Garoeng der Residentie Kedoe'. *Teysmannia*, 17, 579–591.
Brenner, J. Freiherr von, 1894, *Besuch bei den Kannibalen Sumatras. Erste Durchquerung der unabhängigen Batak-Lande*. Würzburg: Woerl.
Broek, P.J. van den, 1949, 'Bevolkingstabak', in C.J.J. van Hall en C. van de Koppel (eds.), *De landbouw in de Indische Archipel*, 3 Vols. in 4 parts, Vol. 2b, pp. 522–558. 's-Gravenhage: van Hoeve
Broek, H.A. van der, 1835, 'Verslag nopens het eiland Bali'. *De Oosterling*, 1, 158–236.

Broersma, R., 1916, *De Lampongsche districten*. Batavia/Rijswijk: Javasche Boekhandel en Drukkerij.
Burton and Ward, 1856, 'Verslag van eene reis in het land der Bataks, in het binnenland van Sumatra, ondernomen in het jaar 1824'. *Bijdragen tot de Taal-, Land- en Volkenkunde*, 5, 270–308.
Carey, P.B.R., 1981, *Babad Dipanagara; An Account of the Outbreak of the Java War (1825–1830)*. Kuala Lumpur: Malaysian Branch Royal Asiatic Society.
Ceram, 1856, 'Iets over Ceram en de Alfoeren'. *Bijdragen tot de Taal-, Land- en Volkenkunde*, 5, 72–88.
Coolhaas, W.Ph., 1985, *Controleur B.B. Herinneringen van een jong bestuursambtenaar in Nederlands-Indië*. Utrecht: Hes.
Crawfurd, J., 1820, *History of the Indian Archipelago*, 3 Vols. Edinburgh: Constable.
Crawfurd, J., 1856, *A Descriptive Dictionary of the Indian Islands and Adjacent Countries*. London: Bradbury and Evans.
Daghregister, 1888/1931, *Daghregister gehouden int Casteel Batavia vant passerende daer ter plaetse als over geheel Nederlandts-India*. Batavia/'s-Gravenhage: Landsdrukkerij/Nijhoff 30 volumes covering selected years between 1624 and 1682.
Dampier, W., 1939, *A Voyage to New Holland*, edited by J.A. Williamson. London: The Argonaut Press.
Deventer Jzn, S. van, 1865/6, *Bijdragen tot de kennis van het landelijk stelsel op Java (...)*, 3 Vols. Zalt-Bommel: Noman.
Dobbin, C., 1983, *Islamic Revivalism in a Changing Peasant Economy. Central Sumatra, 1784–1847*. London/Malmö: Curzon Press.
Domis, H.J., 1830, 'Journaal eener reis van Welerie naar het gebergte Praauw'. *Verhandelingen Bataviaasch Genootschap*, 12, 359–383.
Domis, H.J., 1832, 'Aanteekeningen over het gebergte Tinger'. *Tijdschrift Bataviaasch Genootschap*, 13, 325–356.
Domis, H.J., 1835, 'Byzonderheden betreffende Sourabaya en Madura'. *De Oosterling*, 2(2), 87–125.
Donner, W., 1987, *Land Use and Environment in Indonesia*. London: Hurst.
Doren, J.B.J. van, 1863, 'De Keij-eilanden, ten N.W. van de Arroë-eilanden'. *Bijdragen tot de Taal-, Land- en Volkenkunde*, 10, 238–259.
Doren, J.B.J. van, 1864, 'De Tenimber-eilanden, ten Zuid-Westen van de Keij-eilanden'. *Bijdragen tot de Taal-, Land- en Volkenkunde*, 11, 67–101.
Epp, F., 1852, *Schilderungen aus Holländisch-Ostindien*. Heidelberg: Winter.
Fokkinga, J., 1934, 'Boschreserveering en inlandsche landbouw van overjarige gewassen op Java en Madoera'. *Tectona*, 27, 142–189, 338–385.
Forrest, T., 1792, *A Voyage from Calcutta to the Mergui Archipelago (...)*. London: Robson.
Foster, W. (ed.), 1967, *The Voyage of Thomas Best to the East Indies 1612–14*, Works Hakluyt Society, 2nd series, 75, 1st ed. 1934. Nendeln/Liechtenstein: Kraus Reprint.
Fox, J.J., 1977, *Harvest of the Palm. Ecological Change in Eastern Indonesia*. Cambridge, Mass.: Harvard University Press.
Fox, J.J., 1991, 'The Heritage of Traditional Agriculture in Eastern Indonesia: Lexical Evidence and the Indications of Rituals from the Outer Arc of the Lesser Sundas', in *Indo-Pacific History 1990. Proceedings of the 14th Congress of the Indo-Pacific Prehistory Association, Yogyakarta, Indonesia, Vol. 1*, edited by P. Bellwood, pp. 248–262. Bulletin of the Indo-Pacific Prehistory Association, 10 Canberra/Jakarta.
Francis, E., 1856, *Herinneringen uit den levensloop van een' Indisch' Ambtenaar van 1815 tot 1851*, 3 Vols. Batavia: Van Dorp.
Fruin, Th.A., 1923a, 'Een en ander over de tabakscultuur voor de Inlandsche markt in de regentschappen Bandjarnegara, Wonosobo, Temanggung en Magelang'. *Blaadje voor het Volkscredietwezen*, 11(9), 267–326.
Fruin, Th.A., 1923b, 'Kerftabak op Java'. *Koloniale Studiën*, 7(4), 347–385.

Generale Missiven, 1960/88, *Generale Missiven van Gouverneurs-Generaal en Raden aan Heren XVII der Verenigde Oostindische Compagnie*, edited by W.P.Coolhaas et al. RGP, grote serie, 9 Vols. 's-Gravenhage: Nijhoff.

Haan, F. de, 1910/2, *Priangan: de Preanger Regentschappen onder het Nederlandsch Bestuur tot 1811*, 4 Vols. Batavia: Kolff.

Hageman Jcz., J., 1855, 'Aanteekeningen omtrent een gedeelte der Oostkust van Borneo'. *Tijdschrift Bataviaasch Genootschap*, 4, 71–106.

Hagen, B., 1886, 'Rapport über eine im Dezember 1883 unternommene wissenschaftliche Reise an den Toba-See (Central Sumatra)'. *Tijdschrift Bataviaasch Genootschap*, 31, 328–385.

Hasselman, C.J., 1914, *Algemeen Overzicht van de Uitkomsten van het Welvaart-onderzoek, gehouden op Java en Madoera in 1904–1905*. 's-Gravenhage: Nijhoff.

Hasselt, A.L. van, 1882, *Volksbeschrijving van Midden-Sumatra*, edited by P.J. Veth, Vol. 3, *Midden-Sumatra. Reizen en Onderzoekingen der Sumatra-Expeditie 1877–1879*. Leiden: Brill.

Hawkesworth, J. (ed.), 1773, *An Account of the Voyages undertaken (...) for making Discoveries in the Southern Hemisphere (...)*, 3 Vols. London: Strahan and Cadell.

Heersink, C.G., 1995, *The Green Gold of Selayar. A Socio-Economic History of an Indonesian Coconut Island, c. 1600–1950, Perspectives from a Periphery*, PhD. dissertation. Amsterdam: Free University.

Hefner, R.W., 1990, *The Political Economy of Mountain Java. An Interpretive History*. Berkeley: University of California Press.

Helfrich, O.L., 1889, 'Bijdrage tot de geographische, geologische en ethnographische kennis der Afdeeling Kroë'. *Bijdragen tot de Taal-, Land- en Volkenkunde*, 38, 517–632.

Henley, D., 1993, 'Maïs, mens en milieu op Noord Sulawesi', paper Studiedag Milieugeschiedenis, Utrecht, 19 Nov.

Hoen, H. 't, 1919, *Veerassen en Veeteelt in Nederlandsch-Indië*. Weltevreden: Kolff.

Hoëvell, W.R. van, 1849/51, *Reis over Java, Madura en Bali in het midden van 1847*, 3 Vols. Amsterdam: Van Kampen.

Hogendorp, D. van, 1800, *Bericht van den tegenwoordigen toestand der Bataafsche bezittingen in Oost-Indië, en den handel op dezelve*, 1st ed. 1799. Delft: Roelofswaert.

Hogendorp, W. van, 1779, 'Beschryving van het eiland Timor, voor zoo verre het tot nog toe bekend is'. *Verhandelingen Bataviaasch Genootschap*, 1, 273–306.

Horst D.W.zn, H.A., 1912, 'Indische landbouw op onbewaterde gronden (tegallans) der noordelijke hellingen van de Merbaboe (...)'. *Cultura*, 24, 271–276.

Houben, V.J.H., 1994, *Kraton and Kumpeni. Surakarta and Yogyakarta 1830–1870*. Verhandelingen KITLV, 164. Leiden: KITLV Press.

Indisch Verslag, 1931–1941, Batavia: Landsdrukkerij (annual; 2 vols. per year).

Jacobs SJ, H., 1974/84, *Documenta Malucensia*, Monumenta Historica Societas Iesu, 109, 119, 126. Rome: Jesuit Historical Institute 3 Vols.

Jacobs, J., 1883, *Eenigen tijd onder de Baliërs. Eene reisbeschrijving met aanteekeningen betreffende hygiène, land- en volkenkunde van de eilanden Bali en Lombok*. Batavia: Kolff.

Jacobs, J., 1894, *Het familie- en kamponleven op Groot-Atjeh. Eene bijdrage tot de ethnographie van Noord-Sumatra*, 2 Vols. Leiden: Brill.

Javasche Courant, 1828–1942, Batavia (weekly).

Jonge, H. de, 1988, *Handelaren en handlangers. Ondernemerschap, economische ontwikkeling en islam op Madura*, Verhandelingen KITLV, 132. Dordrecht/Providence: Foris.

Jonge, J.K.J. de and M.L. van Deventer, 1862/95, *De opkomst van het Nederlandsch gezag in Oost-Indië. Verzameling van onuitgegeven stukken uit het oud-koloniaal archief (1595–1814)*, 17 Vols. 's-Gravenhage/Amsterdam: Nijhoff/Muller.

Journal, 1830, *Journal of a Tour along the Coast of Java and Bali etc., with a Short Account of the Island of Bali, particularly of Bali Baliling*. Singapore: The Mission Press.

Jukes, J.B., 1847, *Narrative of the Surveying Voyage of H.M.S. Fly (...) during the Years 1842–1846; together with an Excursion into the Interior of the Eastern Part of Java*. London: Boone.

Junghuhn, F.W., 1847, *Die Battaländer auf Sumatra*. Berlin: Reimer 2 Vols.
Junghuhn, F.W., 1853/4, *Java, zijne gedaante, zijn plantentooi en inwendige bouw*, 3 Vols. 's-Gravenhage: Mieling.
Kano, H., 1987, 'The Long-Term Development of Farm Agriculture in Java: A Rice Economy?', in *Human Ecology of Health and Survival in Asia and the South Pacific*, edited by T. Suzuki and R. Ohtsuka, pp. 93–110. Tokyo: University of Tokyo Press.
Kerckhoff, Ch.E.P. van, 1890, 'Eenige opmerkingen betreffende de zoogenaamde "orang loeboe" op Sumatra's Westkust'. *Tijdschrift Nederlandsch Aardrijkskundig Genootschap*, 2nd series, 7(2), 576–577.
Kreemer, J., 1912, 'De Loeboes in Mandailing'. *Bijdragen tot de Taal-, Land- en Volkenkunde*, 66, 303–335.
Köhler, J.C., 1856, 'Verslag eener reis door een gedeelte der Lampongse Distrikten'. *Tijdschrift Bataviaasch Genootschap*, 5, 132–149.
Koloniaal Verslag, 1849–1930, Batavia (annual).
Kraan, A. van der, 1980, *Lombok: Conquest, Colonization and Underdevelopment, 1870–1940*, Asian Studies Association of Australia Southeast Asia Publications Series, 5. Singapore: Heinemann.
Kussendrager, R.J.L., 1861, *Natuur- en aardrijkskundige beschrijving van het eiland Java*, 1st ed. 1841. Amsterdam: Weyting and Brave.
Landbouwatlas, 1926, *Landbouwatlas van Java en Madoera* (text by C.W. Bagchus, in cooperation with A.M.P.A. Scheltema), Mededeelingen van het Centraal Kantoor voor de Statistiek, 33. Weltevreden: Departement van Landbouw, Nijverheid en Handel.
Leupe, P.A., 1864, 'Beschrijving van eenen togt naar de bovenlanden van Banjermassing enz. in het jaar 1790'. *Kronijk van het Historisch Genootschap*, 4th series, 20(5), 313–404.
Leupe, P.A., 1875, 'De reizen der Nederlanders naar Nieuw-Guinea en de Papoesche eilanden in de 17e en 18e eeuw'. *Bijdragen tot de Taal-, Land- en Volkenkunde*, 22, 1–162, 175–311.
Liefrinck, F.A., 1969, 'Rice Cultivation in Northern Bali', in *Bali. Further Studies in Life, Thought and Ritual*, Selected Studies on Indonesia 8, pp. 1–73 original version dated 1886/7. The Hague: Van Hoeve.
Lindblad, J.Th., 1988, *Between Dayak and Dutch. The economic history of Southeast Kalimantan 1880–1942*, Verhandelingen KITLV, 134. Dordrecht/Providence: Foris.
Lockyer, Ch., 1711, *An Account of the Trade in India (...)*. London: Crouch
Ludeking, E.W.A., 1868, 'Schets van de Residentie Amboina'. *Bijdragen tot de Taal-, Land- en Volkenkunde*, 15, 1–272.
Marsden, W., 1966, 1811, *The History of Sumatra* (reprint of the third edition 1811 introduced by John Bastin), 1st ed. 1783. Kuala Lumpur: Oxford University Press.
Meijer, W.Ch.P, 1962, *Das Balirind*, Neue Brehm-Bücherei, 303. Wittenberg: Ziemsen.
Mohnike, O., 1874, *Banka und Palembang nebst Mittheilungen über Sumatra im Allgemeinen*. Münster: Aschendorffschen Buchhandlung.
Müller, S., 1855, *Reizen en onderzoekingen in Sumatra, gedaan op last der Nederlandsche Indische regering, tusschen de jaren 1833 en 1838, door Dr. S. Müller en Dr. L. Horner*. 's-Gravenhage: Fuhri for KITLV.
Nagtegaal, L., 1988, *Rijden op een Hollandse tijger. De noordkust van Java en de VOC 1680–1743*, PhD. dissertation. Utrecht.
Neumann, J.B., 1886/7, 'Het Pane- en Bila-stroomgebied op het eiland Sumatra (Studiën over Batahs en Batahsche landen)'. *Tijdschrift Koninklijk Nederlandsch Aardrijkskundig Genootschap*, 2nd series, 3(2), 1–99, 215–314, 459–543.
Nieuwenhuijzen, F.N., 1858, 'Het rijk Siak Sri Indrapoera", *Tijdschrift Bataviaasch Genootschap*, 7, 388–438.
Onderzoek Mindere Welvaart, 1905–1920, *Onderzoek naar de mindere welvaart der inlandsche bevolking van Java en Madoera*, 14 Vols. Batavia: Landsdrukkerij.
Ophuijsen, C.A. van, 1884, 'De Loeboes'. *Tijdschrift Bataviaasch Genootschap*, 29, 88–100.
Ormeling, F.J., 1956, *The Timor Problem. A Geographical Interpretation of an Underdeveloped Island*. Groningen-Djakarta/'s-Gravenhage: Wolters/Nijhoff.

P., 1845, 'Indigocultuur, vooral in Pekalongan; inlandsche huishouding'. *Indisch Magazijn*, 2(7,8), 115–125.

Padtbrugge, R., 1866, 'Beschrijving der zeden en gewoonten van de bewoners der Minahassa, door den Gouverneur der Molukken –, 1679'. *Bijdragen tot de Taal-, Land- en Volkenkunde*, 13, 304–331.

Paerels, A.E., 1918, 'Tweede gewassen', in H.C. Prinsen Geerligs (ed.), *Dr. K.W. van Gorkom's Oost-Indische Cultures*, 3 Vols.; 1st ed. 1913, Vol. 2, pp. 813–862. Amsterdam: De Bussy.

Palte, J.G.L., 1989, *Upland Farming on Java, Indonesia. A socio-economic study of upland agriculture and subsistence under population pressure*, PhD. dissertation. Utrecht.

Plakaatboek, 1885/1900, *Nederlandsch-Indisch Plakaatboek 1602–1811*, edited by J.A. van der Chijs, 16 Vols. Batavia: Landsdrukkerij.

Purseglove, J.W., 1974/5, *Tropical Crops*, 2 Vols. London: Longman.

Raedt van Oldenbarnevelt, H.J.A., 1888, 'Tochten in het stroomgebied der Beneden-Ketaun, en een vierdaagsch uitstapje in de Lebong'. *Tijdschrift Koninklijk Nederlandsch Aardrijkskundig Genootschap*, 2nd series, 5, 178–211, 417–440.

Raffles, T.S., 1814, *Substance of a Minute*. London: Black and Parry.

Raffles, T.S., 1830a, *The History of Java*, 2 Vols. 1st ed. 1817. London: Murray.

Raffles, S. (ed.), 1830b, *Memoir of the Life and Public Services of Sir Thomas Stamford Raffles (...)*. London: Murray.

Reid, A., 1988, *Southeast Asia in the Age of Commerce 1450–1680. Vol. 1. The Lands below the Winds*. New Haven/London: Yale University Press.

Reis, 1853, 'Reis van den Gouverneur-Generaal van Imhoff, over Java, in het jaar 1746'. *Bijdragen tot de Taal-, Land- en Volkenkunde*, 1, 291–440.

Residentie, 1871, *De Residentie Kadoe naar de uitkomsten der statistieke opname en andere officiële bescheiden, bewerkt door de Afdeeling Statistiek ter Algemeene Secretarie*. Batavia: Landsdrukkerij.

Rijstkultuur, 1854, 'De rijstkultuur op Java vijftig jaren geleden'. *Bijdragen tot de Taal-, Land- en Volkenkunde*, 2, 1–117.

Roo van Alderwerelt, J. de, 1906, 'Historische aanteekeningen over Soemba (residentie Timor en Onderhoorigheden)'. *Tijdschrift Bataviaasch Genootschap*, 48, 185–316.

Rumphius, 1741/55, *Het Amboinsch Kruid-boeck*, 5 vols. Amsterdam/Den Haag: Changuion et al.

Saar, J.J., 1930, *Reise nach Java, Banda, Ceylon und Persien, 1644–1660*, Reisebeschreibungen von deutschen Beamten, 6, 1st ed. 1662. Den Haag: Nijhoff.

Salmon, Th., 1729/33, *Hedendaegsche historie, of tegenwoordige staet van alle volkeren* transl. from English original, 5 Vols. Amsterdam: Tirion.

Schuitemaker, B., 1949, 'Maatregelen tot het behoud van de bodem in het inheemse landbouwbedrijf op Java'. *Landbouw*, 21(4,5), 153–176.

Schwaner, C.A.L.M., 1853/4, *Borneo. Beschrijving van het stroomgebied van den Barito en reizen langs eenige voorname rivieren van het zuid-oostelijk gedeelte van dat eiland (...) 1843–1847*. Amsterdam: Van Kampen for KITLV 2 Vols.

Skinner, C. (ed.), 1963, *Sja'ir Perang Mengkasar (The Rhymed Chronicle of the Macassar War) by Entji' Amin*, Verhandelingen KITLV, 40 's-Gravenhage: Nijhoff.

Snouck Hurgronje, C., 1893/4, *De Atjèhers*, 2 Vols. Batavia/Leiden: Landsdrukkerij/Brill .

Sollewijn Gelpke, J.H.F., 1885, *Gegevens voor een nieuwe landrente-regeling; Eindresumé der onderzoekingen bevolen bij het Gouvernementsbesluit van 23 Oct. 1879, No. 3*. Batavia: Landsdrukkerij.

Sollewijn Gelpke, J.H.F., 1901, *Naar aanleiding van Staatsblad 1878, No. 110*. Batavia: Landsdrukkerij.

Stapel, F.W. (ed.), 1927/54, *Pieter van Dam: Beschryvinge van de Oostindische Compagnie* Rijks Geschiedkundige Publicatiën, 5 Vols. in 7 parts. 's-Gravenhage: Nijhoff.

Staverman, H.J., 1868, 'Beschrijving der wijze, waarop tegenwoordig de tabaks-kultuur in de Residentie Kedirie gedreven wordt (...)'. *Tijdschrift voor Nijverheid en Landbouw in Nederlandsch Indië*, 13, 39–47.

Steck, F.G., 1862, 'Topographische en geografische beschrijving der Lampongsche Distrikten'. *Bijdragen tot de Taal-, Land- en Volkenkunde*, 8, 69–113, 123–126.
Steyn Parvé, D.C., 1851, *Het koloniaal monopoliestelsel getoetst aan geschiedenis en staathuishoudkunde; nader toegelicht door den schrijver*. Zaltbommel: Noman.
Teenstra, M.D., 1852, *Beknopte beschrijving van de Nederlandsche overzeesche bezittingen*, 3 Vols. Groningen: Oomkens.
Teijsmann, J.E., 1855, *Uittreksel uit het dagverhaal eener reis door Midden-Java*. Batavia: Lange and Co.
Teijsmann, J.E., 1856. 'Uittreksel uit het dagverhaal eener reis door Oost-Java, Karimon Java en Bali Boleling'. *Natuurkundig Tijdschrift Nederlandsch-Indië*, 11, 111–206.
Teisseire, A., 1792, 'Beschrijving van een gedeelte der Omme- en Bovenlanden dezer hoofdstad'. *Verhandelingen Bataviaasch Genootschap*, 6, 23–94.
Temminck, C.J., 1846/9, *Coup-d'oeil général sur les possessions Neerlandaises dans l'Inde Archipélagique*, 3 Vols. Leiden: Arnz.
Ternate, 1980, *Ternate. Memorie van Overgave, J.H. Tobias (1857). Memorie van Overgave, C. Bosscher (1859)* Penerbitan sumber-sumber sejarah, 11 Jakarta: Arsip Nasional.
Tobi, E., 1894, 'Reboiseering van den Goenoeng Sendoro en Goenoeng Soembing'. *Tijdschrift Nijverheid and Landbouw Nederlandsch-Indië*, 48, 147–151.
Veer, K. van der, 1948, 'Maïs', in C.J.J. van Hall and C. van de Koppel (eds.), *De landbouw in de Indische Archipel*, 3 Vols. in 4 parts, Vol. 2a, pp. 111–156. 's-Gravenhage: van Hoeve.
Verslag Dienst Boschwezen, 1901–1947, Batavia (annual).
Veth, P.J., 1875/82, *Java, geographisch, ethnologisch, historisch*, 3 Vols. Haarlem: Boon.
Vink, G.J., 1941, *De grondslagen van het Indonesische landbouwbedrijf*. Wageningen: Veenman PhD diss.
Vleming Jr., J.L., 1925, *Tabak. Tabakscultuur en tabaksproducten van Nederlandsch-Indië*. Weltevreden: Dienst der Belastingen in Nederlandsch-Indië.
Volz, W., 1909, *Nord-Sumatra. Bericht über eine im Auftrag der Humboldt-Stiftung der Königlich Preussischen Akademie der Wissenschaften zu Berlin in den Jahren 1904–1906 ausgeführte Forschungsreise. Band I: Die Batakländer*. Berlin: Reimer.
Weidegang, 1913, *Weidegang en stalverpleging van vee*. Buitenzorg: Departement van Landbouw, Nijverheid en Handel.
Westenenk, L.C., 1962, *Waar mens en tijger buren zijn*, 1st ed. 1927/1932. Den Haag: Leopold.
Wetering, F.H. van de, 1926, 'De Savoeneezen'. *Bijdragen tot de Taal-, Land- en Volkenkunde*, 82, 485–575.
Wharton, W.J.L. (ed.), 1893, *Captain Cook's Journal during his First Voyage round the World made in H.M. Bark "Endeavour" 1768–71*. London: Stock.
Wigboldus, J.S., 1978, 'A promising land. Rural history of the Minahasa, about 1615–1680', paper for the second Indonesian-Dutch Historical Conference, Ujung Pandang, June 26–30.
Wigboldus, J.S., 1979, 'De oudste Indonesische maïscultuur', in F. van Anrooij et al. (eds.), *Between People and Statistics. Essays on Modern Indonesian History Presented to P. Creutzberg*, pp. 19–32. The Hague: Nijhoff.
Wijers, E.W., 1928, 'De bevolkings-tabakscultuur op Madoera'. *Landbouw*, 4, 122–135.
Wilde, A. de, 1830, *De Preanger Regentschappen op Java gelegen*. Amsterdam: Westerman.
Zollinger, H., 1850, 'Verslag van eene reis naar Bima en Soembawa en naar eenige plaatsen op Celebes, Saleyer, en Flores gedurende de maanden mei tot december 1847'. *Verhandelingen Bataviaasch Genootschap*, 23, 1–224.
Zwart, W.G.J., 1928, *Herbosschingswerk in Bagelen 1875–1925*, Mededeelingen Proefstation Boschwezen, 17. Weltevreden: Landsdrukkerij.

Chapter 3

CULTURALISING THE INDONESIAN UPLANDS
Joel S. Kahn

In her introduction to this volume, Tania Li points to significant differences between lowland and upland agrarian regimes in Indonesia, drawing our attention among other things to the dangers inherent in generalising from lowland to upland contexts. But of course diversity has long been thematised by observers of the Indonesian economic, social and cultural landscape — it is even enshrined in the nation's official motto. The lowland:upland distinction is an ancient one within Indonesia itself, finding expression directly in the myths and rituals of probably the majority of Indonesia's precolonial peoples, not just as a means of distinguishing one group from another, but also of underpinning status differentials within groups. What makes Li's comments particularly salient, therefore, is not so much that they speak of diversity and the dangers of generalisation *per se*, but that they draw our attention to the flaws in existing systems for classifying that diversity.

We are now, for example, all too aware of the flawed evolutionary assumptions inherent in representing uplands peoples in Indonesia as somehow more "traditional" than those in the lowlands. The notions of timeless (upland) traditions on which such arguments are based are, as Li so cogently argues, extremely problematic in the case of Indonesia. The more general problems with such evolutionary or primitivist narratives are twofold. On the one hand there are the intellectual shortcomings of arguments that project contemporary spatial diversity onto an historical axis, with otherness consequently explained away as being more or less "primitive", according to the laws of history.[1] On the other hand as Edward Said and others have been pointing out for some time now, certain civilisational discourses were directly implicated in the imperial projects of Britain, France and, one might add, the Netherlands at least in the nineteenth century (Said 1978; Jan Mohammed 1985).

But while primitivist images of upland cultures still circulate in Indonesia, particularly in "frontier" areas like Borneo and West Irian, another view of the differences between upland and lowland peoples has become increasingly prevalent. This is the view that what we are dealing with is a case of cultural divergence, such that upland and lowland peoples can be distinguished on *cultural* grounds. It is precisely this concept of the cultural diversity of Indonesia that informs the national motto which captures neatly what has long been taken to be the "problem" for Indonesian nationalism, namely the forging of some sort of imagined or artificial unity among a culturally disparate people. To quote from the oft-cited *Imagined Communities*:

> The case of Indonesia affords a fascinatingly intricate illustration of this process [the promotion of colonial nationalisms], not least because of its enormous size, huge population, geographical fragmentation, religious variegation and ethnolinguistic diversity .... Furthermore ... its stretch does not remotely correspond to any precolonial domain ... its boundaries have been those left behind by the last Dutch conquests. (Anderson 1983: 110)

In this view the project of constructing or imagining an Indonesian nation is conceived in terms of overcoming or transcending pre-existing primordial group loyalties which are in some sense more real than the nation. The latter, by contrast, was comprised at its birth of people linked only by the accident of colonial boundaries. Thus Indonesian nationalism has come to be seen as the establishment of *intercultural* relations.

It is the culturalisation of the diversity that exists within the Indonesian nation — the view that relations between uplands and lowlands, core and periphery, inner and outer Indonesia, rich and poor, powerful and marginal are *intercultural relations* — that seems to have come increasingly to characterise the discourse of intellectuals, government officials, and economic and cultural élites in the New Order period. An excellent example of this way of thinking is manifest in the phenomenon of what John Pemberton (1994) has called "Mini-ization" after Jakarta's Taman Mini theme park, reputedly conceived by Madam Soeharto after a visit to Disneyland. By fitting Indonesia's peoples into a multicultural grid, Taman Mini gives concrete form to a multicultural discourse that is widely manifest in the media, the arts, museums, cultural "festivals", and tourism promotions and attractions, to say nothing of similar theme parks, "traditional" houses and the like scattered throughout the archipelago and modelled on the original Taman Mini (see Schrauwers, this volume).

In this discussion of what I shall call the constitution of a particular uplands group — the Minangkabau of central western Sumatra — I shall not be concerned with documenting yet again the flaws of primitivism. Instead I shall examine the multicultural model of Indonesia's diverse population in general, and of the divergence between upland and lowland peoples in particular. In other words, I want to look critically at the view that the modern nation of Indonesia is in fact composed of a large number of more or less related but separately identifiable cultural groups — the Javanese, Balinese, Minangkabau, Toraja, Buginese, Dayak — so that what might otherwise be understood as inequalities in access to resources and/or power are read instead as problems in intercultural relations.

Culture is being used here in an extended sense to refer not just to the "learned" component of human social behaviour, but to what writers like Geertz, and before him Max Weber, call the "meaningful" aspects of human existence. In this usage culture denotes that dimension of human adaptation that goes beyond utility, that which, to use Charles Taylor's term, testifies to the "expressive" dimension of human creativity (see Taylor 1976; Kahn 1995). Humans might, therefore, use a language (a human creation) to communicate, or a tool to construct, but words and tools are more than utilities with which to communicate or build. They are also meaningful, and hence express something deeper about human endeavour. Once this is recognised, so the argument goes, some notion of multiculturalism follows almost immediately, since it is easy to see that the meanings of things that are socially-created (e.g. language) are, unlike their utilities, specific to the social groups that produce them. In the view of Herder, one of the first to advance such an expressivist critique of Enlightenment utilitarianism, French, German and English are not merely accidentally-differing means of communicating the same basic human instincts, they are expressions of the quite different "spirits" (*volksgeisten*) of the French, German and English "peoples".

We might begin by stating the obvious — that the discourse of Indonesian multiculturalism is as much a *construction* as is any other way of representing Indonesian diversity in the sense that it requires the creation of boundaries where none in fact exist. By this I mean simply that defining, for example, a discrete and somehow unchanging Minangkabau culture, or Balinese culture or whatever cannot be done by reference to pre-existing cultural differentiation *on its own*, since cultural variation across the archipelago tends to be continuous. Were one to travel from village to village from, say, Bukit Tinggi in the "cultural heartland" of Minangkabau through eastern and northern Sumatra, it is unlikely that one would observe any sharp break in cultural practices, rituals or world views. The transition from one "culture" to another could be marked instead only by reference to administrative boundaries created by the Indonesian state, or by eliciting statements from locals about cultural self-identification. In undertaking such a journey we would pass through "borderlands" or "margins" in which the sense of cultural hybridity and ambiguity would be particularly strong, making the drawing of boundaries seem especially arbitrary. But to describe such regions as special would be misleading to the extent that this preserves the sense of a purer cultural core that lies elsewhere. Defining one point as centre and the other as margin or "borderland" is as much an exercise in arbitrary identification as is the drawing of the boundary itself.

Just as the boundaries around cultures are in some sense arbitrary, so too does the definition of cultural purity at some imagined centre or core require processes of exclusion. All Minangkabau villages have, just as one suspects do all Javanese villages, people who are adjudged "outsiders" or "marginals", people variously assumed to be bearers of "foreign" culture, or "insiders" who are "uncultured" or "decultured" through poverty, powerlessness, ignorance, youth, sexuality, mental instability and so on. Moreover all Minangkabau villagers, to varying degrees, have beliefs and engage in practices that are considered to be non-Minangkabau, such as going to the cinema to watch a Hindi weepie or listening to "western" popular music; attending (excessively) to religious ritual at the expense of one's cultural commitments; riding around on Japanese-made motorcycles; or neglecting one's social obligations.

Frederick Barth's lament on the problems of defining "real" Balinese culture is relevant here: "Observe the litany of authorities within the tradition that make conflicting claims to be heard in Bali-Hinduism's variously instituted liturgies and priesthoods...", among which he lists the Sanskrit manuscripts, the highest ranking priests, the main body of temple priests, the family and descent group priests, the deceased ancestors and the gods. "To approach such a raucous cacophony of authoritative voices with expectations that their messages and their teaching will be coherent" would, says Barth, require a very "dogmatic anthropologist" (1989: 127f).

But to speak of the "constructedness" of cultural classifications, and the "dogmatism" of anthropologists and others who believe in them is no grounds for complacency, since as Pemberton (1994:13) points out, anthropologists may themselves be implicated in something closely approximating Mini-ization:

> At issue, then, is not the force of *one* tradition that happens to be dominant, one reigning representation among a diversity of representations, but the representational force of the idiom of "diversity" itself. In contemporary Java, the distinction between what may be the remnant particulars of a now unsuitable past ... and New Order displays of an elaborate cultural diversity is more often than not highly ambiguous. This ambiguity is unsettling for anthropologists not just for practical ethnographic reasons but because the ethnographic enterprise itself is called into question; the search for conventions that might inform Javanese culture, or any culture, parallels the process of Mini-ization in many

respects. It is as if the ethnographic move to document and interpret customary practices winds up cataloguing "diversity" in all its myriad forms, as anthropological and New Order disciplinary interests in culture coincide, yet again.

Rather than simply abandon culturalist images in favour of universalist notions like class, which are equally products of discursive construction, it would seem fruitful to examine the context within which discourses of intercultural relations between uplands and lowlands peoples in Indonesia emerge. Only in this way can such a discourse be properly evaluated. In the central part of this chapter I shall look at a particular case of an upland people (the Minangkabau of central western Sumatra) who, precisely at a moment of intense economic and political modernisation, came to be seen, and often to see themselves, as a culturally distinctive group.

CONSTITUTING THE MINANGKABAU

> In this world each people (*bangsa*) has its own *adat* organisation in its own country. This adat is used to facilitate social interaction in these countries. With this adat organisation people are able to love and sympathise with each other. With this adat people are able to establish social ties with each other. With this adat people are able to honour and raise the level of their "race" (bangsa). With this adat people are able to cause their race and their countries to progress. And with this adat people are able to achieve all their noble objectives — of their welfare, the progress of their nation, and the like.
>
> The adats of the different peoples of this world are not the same. Just as with the variety of customs, so it is with adat. There are so many different varieties that even the experts cannot tell us how many there are.
>
> From the above, we must draw the conclusion that there are many meanings when it comes to the use of adat. Such was also the case with our ancestors, the people of the Minangkabau world. In earlier times, when the nagari [villages] began to be formed, when the wells were dug, our ancestors worked to collect and construct a favourable adat in the regulation of our interaction as Minangkabau, so that we could live in peace and harmony together in the villages. As a result we could honour our subdistricts and districts, along with our race, our land and our world. The adat was used for the welfare of the whole nagari and its inhabitants.[2]

The above is an extract from an article by one Datuk Sanggoeno di Radjo, the president of Perkoempoelen Minangkabau. This was an organisation established in the mid-1920s by conservatives, mainly clan chiefs from the Bukit Tinggi area, with the professed aim of contributing to the "search for goodness and welfare following the road of Minangkabau adat." Perkoempoelen Minangkabau and other conservative organisations were concerned to counter the inroads made by the newly established Indonesian Communist Party (PKI) on Sumatra's Westcoast in the years leading up to the outbreak of armed rebellion against the Dutch on January 1, 1927 in the village of Silungkang (see Abdullah 1971:41).

What is notable about this extract is the way in which terms like *bangsa*, usually translated as race, and adat, glossed by terms like custom or customary law, take on meanings typically associated with the anthropological concept of culture. As the article demonstrates, for at least some Indonesian intellectuals in the 1920s, the peoples or races (bangsa) of Indonesia and of the world, can each be seen to have its own distinctive customs (adat), created by the ancestors and handed down to the present generation. Minangkabau is taken to be such a group, with a discrete and distinctive culture that must be preserved and continue to guide social behaviour. Thus in the 1920s adat, which could in the past have been taken merely to mean the cultural practices of social groups, has become "crystallised" (to use Abdullah's term), at least in the language of its pre-eminent spokespersons the *Panghulu*, heads of the matrilineal kin groups or *suku*.

In the above instance the discourse of culture and culturalisation was embedded in a conservative political project to combat the spread of radical nationalism, communism and Islamic modernism, and to defend the overarching rule of the colonial power. But it was not always so. Other organisations engaged in constituting the Minangkabau as a distinctive cultural group were engaged in a politics both populist and radical. One such was an association first known as the Boedi Tjaniago Association founded in the village of Bukit Sarungan near Padang Panjang in 1919, later renamed the Organisation of Villagers of Padang Panjang, which dedicated itself to "Learning, Knowledge of Adat and Skill". "We live in a world," writes a contributor to the organisation's newspaper, "organised according to three sets of regulations: religious, legal and those deriving from adat. Religious rules are oriented to the afterlife (*achirat*), adat regulations are for the organisation of our daily lives, while laws function to ensure our security." "The Minangkabau religious rules come from Islam, and that is right and proper since for historical reasons the religion in West Sumatra is Islam." But what is adat? According to our anonymous writers, adat is no more than custom (*kebiasaan*). All countries have their own custom, indeed in the Indies itself there are a variety of adats — Minangkabau, Batak, Acehnese, etc. Each people should live according to its own individual custom.[3]

Although embedded in a very different political project, the discourse of culture and multiculturalism in the pages of *Boedi Tjaniago* is remarkably similar to that of the far more conservative members of the

Perkoempoelan Minangkabau. Human diversity in general, and diversity within the Netherlands East Indies in particular, is understood as a diversity of more or less discrete "cultures". The implication is that political movements, whatever their hue, should aim for some form of cultural autonomy — a society free of Dutch rule, and based on the principles of a specifically Minangkabau adat for the "nationalists" of the Padang Panjang grouping, or the cultural autonomy guaranteed through rule by clan elders (backed up by a culturally-sympathetic colonial power) for the more conservative Perkoempoelan Minangkabau.

Thus the language of adat, bangsa, custom and tradition served to construct diversity within Indonesia (indeed within the Dutch empire as a whole, to the extent that the Dutch were also seen as a distinctive European group) as cultural diversity, and the relations among people primarily as intercultural relations. Taking a lead from Taufik Abdullah, I have suggested that this was a relatively new way of understanding the relationship between the Minangkabau, other Indonesian peoples, and their Dutch rulers. This was perhaps part of the first wave of the culturalist mode of understanding differential relations of wealth and power (including relations between upland and lowland peoples) that we have seen re-emerge in the New Order period.

While it is tempting to see in this language of cultural diversity a simple traditionalist response to colonial rule, it is important to recognise that it makes sense only in the context of changes that were themselves distinctly modern. For this reason, primordialism does not adequately explain this phenomenon any more than it explains the subsequent rise of Indonesian nationalism. How adat ideologues in the uplands of western Sumatra in the 1920s came to cast the social landscape in culturalist idioms is understandable only in the context of the sweeping transformations of economy, society and culture that began towards the end of the nineteenth century. More specifically, an understanding of the new discourse of cultural differentiation requires attention to the debates generated by Islamic modernism, the parallel Dutch critique of colonial liberalism, the processes of peasantisation that followed on from massive alienation of land from Minangkabau communities and, finally, the ways in which the "modernisation" of the colonial state generated new relations of power and authority between colonial rulers, local élites and peasant villagers.[4]

LIBERALISM, MODERNISM, ISLAM

> While facing the challenge of successive religious reform movements... adat changed from a collection of commonly accepted forms and traditions into a

statement of regulations and philosophy. In response to the activities of the Islamic modernists, however, adat began to assume the status of an ideology. Its ideas and institutions were being crystallized into a universal system. (Abdullah 1971:15)

The arguments about the integrity of Minangkabau cultural traditions sketched above can be understood first and foremost as a contribution to the debate over modernism, and particularly to the arguments and activities of Islamic modernists in West Sumatra. If the label "modernist" is used to describe those seeking a rapprochement with contemporary Dutch liberalism, then doubtless the first modernists on Sumatra's Westcoast were those "coastal aristocrats" and "native officials" who, in the last years of the nineteenth century formed a number of Dutch-style clubs and began to publish European-style newspapers. These early Sumatran modernists took an interest in the world outside the colony, and began propagating a notion of *kemajuan* (progress) as both a yardstick against which the achievements of the Indies could be measured, and a desirable path that they should follow.

While the modernist movement seems to have had its origins in an initially secular debate over the desirability of progress, there is no doubt that religious developments were responsible for expanding its scope, as well as generating conflicts that revolved around the proper role in a future society of Minangkabau adat. From the mid-1910s, the banner of modernism changed hands and, at least in Minangkabau, was carried increasingly by the advocates of Muslim reform. Characteristic of Minangkabau's Muslim reformers of the 1910s and 1920s was a combination of fundamentalist theology and a firm commitment to kemajuan that was not altogether different from their more secular forebears. The important logic here was that returning to the fundamentals would serve to clear the ground of tradition, making way for the full exercise of humanity's capacity for reason (*akal*). [5]

As a movement committed to progress and the triumph of reason over uncritical acceptance of doctrine and the teachings of established religious leaders, Islamic modernism in West Sumatra resulted in open conflicts over religious practices between supporters of the so-called *Kaum Muda* (Young Generation) and the *Kaum Tua* (Old Generation) of religious scholars and teachers. But in challenging what they took to be un-Islamic practices, Kaum Muda supporters, not surprisingly, touched off reactions on the part of others who saw themselves as supporters of Minangkabau tradition in opposition to the modernists whom they took to be hostile to it. It is in this context that the development of the kind of culturalist discourse on the uniqueness of Minangkabau, and the multicultural nature of the peoples of the Netherlands Indies is best understood. By most accounts such disagreements were fought out on the local village level with, among other things, conflicts over the holding of adat ceremonies between fellow villagers.

It might be the case that in more recent times, Minangkabau villagers in general see no problem in reconciling their religious beliefs and cultural practices. The presumption that began to emerge toward the end of the second decade of the twentieth century that the two were incompatible was therefore a consequence of particular historical circumstances that need further investigation.

INDONESIA IN COLONIAL DISCOURSE: FROM EVOLUTIONISM TO HISTORICISM

Before jumping to the conclusion that the culturalist response to modernism outlined above was a purely indigenous phenomenon, we should note that some Dutch intellectuals and colonial officials were discovering the uniqueness of Indonesian cultures at much the same time as were indigenous critics of modernism. In other words, contrary to the assumption that civilisational narratives always and everywhere characterised the discourse of European colonial rule in Asia, a counter- or anti-evolutionism began increasingly to inform the way the Dutch represented their colonial subjects in the first few decades of this century. It led them, like their indigenous counterparts, to a view of relations between themselves and their subjects as intercultural relations first and foremost.

The shift in Dutch colonial discourse from a more or less liberal evolutionism to cultural relativism or historicism is discussed in some detail in another publication (Kahn 1993: 68–109). The so-called Leiden school of jurists and ethnologists perhaps represented this new attitude most clearly. Their leading member, Cornelius van Vollenhoven, according to a recent commentator "believed in the possibility of merging existing traditions with western modernism by fostering respect for traditional Indonesian laws and customs (the so-called adat) and wished to leave tradition respectable and intact, particularly on the local or regional level, so that society could cope with modernising influences in its own specific Indonesian ways"(Schöffer 1978:90)". Hooker has characterised the school as a whole by their attitude towards westernisation:

> Members of the school disapproved strongly of rapid Westernisation of Indonesia, especially where this was to be accompanied by the introduction of a codified Western legal system. They warned against a forced pace of Westerni-

zation and advocated a gradual social evolution through the growth of stable adat communities especially in the Outer Provinces. This resulted in the formation of a school of jurisprudence whose whole philosophy came to rest upon a distinction between the laws of various races. (Hooker 1978:15)

The Leiden School produced an exhaustive classification of cultural legal groups in Indonesia and was responsible, perhaps more than any other group of scholars in this period, for a picture of the colony as "multicultural", in which intergroup relations were construed as relations between related but separate and discrete groups, each with its own system of *adatrecht* (see Dutch East Indies, Commissie voor het adatrecht, 1911–1955).

The emergence of the Leiden School in the early decades of this century, with its particular brand of legal historicism, marks a sharp break with the evolutionist ideals of the 1860s and 1870s. The ideas of the School had an important effect on colonial policies, not least because Leiden became the training ground for a whole generation of colonial officials who took up their posts in the East Indies strongly influenced by images of the strength and integrity of indigenous cultural traditions, and convinced of the inappropriateness of 'western' notions of progress and modernity in the Indonesian context.

A similar shift towards a more relativistic and hence culturalist account of Indonesian society is found in the field of "colonial economics", the best-known advocate of which was Julius Herman Boeke who argued strenuously against the imposition of "western" economic theory in colonial situations because of the distinctively non-western nature of the latter. Boeke is perhaps best known for his often criticised notion of a "dual economy", but what is frequently not appreciated is the extent to which his arguments represented an important attack on grand evolutionary narratives and an imperial project dedicated in this case to the economic civilisation of the backward native. While he talked of different stages of economic development in West and East, as the following extract from his dissertation shows, Boeke was not in support of policies that would have the effect of turning Indonesians into westerners. Of the distinctive nature of colonial societies, Boeke wrote:

> ... there is not one homogeneous society but a Native society side by side with a society of foreigners, not one people but a multiplicity of peoples, not one course of development but a clash between two heterogeneous stages of development, not a sense of solidarity but one of ruling and being ruled. And finally, there is ... a group which is interested primarily in the products of the soil and asks no more of its people than a certain amount of labour of a quality that does not necessarily entail their advancement (1910, quoted in Boeke *et al.*, 1966: 10f).

Thus Boeke, from the perspective of economics, did what Van Vollenhoven had done from the standpoint of legal studies: contribute to the picture of Indonesian society as a system with its own internal cultural logic that would not necessarily (and should not be forced to) develop along the evolutionary path laid down by the West. Along with a growing number of lesser-known Dutch jurists, ethnologists, agricultural specialists, and government advisers, these two colonial scholars contributed to an emerging culturalist imaginary, one which in this case led to the view that Europeans and "natives" differed fundamentally in their cultures rather than, say, in their access to power and resources.

It is difficult to say whether there were direct relationships between the emerging culturalist discourse of Indonesians and Europeans, or whether they discovered the language of cultural differentiation more or less independently. Certainly this language was put to different uses by Europeans and Indonesians. Indeed, as we have seen, Indonesians themselves had both radical and conservative variants of the multicultural model of local society. More significant, perhaps, is the fact that this discursive transformation was taking place in the context of sweeping economic and political changes in colonial society, particularly in upland regions like those inhabited by the Minangkabau. Far from leading to the kinds of autonomous development favoured by scholars like Van Vollenhoven and Boeke, these developments had the effect of integrating uplands peoples much more closely, if in unexpected ways, into colonial regimes of accumulation and the matrices of power associated with colonial rule.

LAND AND POWER

A fascinating example of the interaction between the discursive, political and economic dimensions of colonial modernisation is provided by the case of the resistance to land alienation from Minangkabau villages by the then Governor of Sumatra's Westcoast, J. Ballot. At his own expense, Ballot published a slim volume in 1911 entitled *Ontwerp Agrarische Regeling voor Sumatra's Westkust* (Draft Agrarian Regulation for Sumatra's Westcoast). In the book the Governor expressed himself very critically on the subject of colonial land policy, at the time a very sensitive issue indeed. Partly because he expressed open criticism of the governments in Batavia and The Hague, and also because he chose to air dirty laundry in such a public way, Ballot was severely chastised and eventually eased out of the Governorship of Sumatra's Westcoast, to be replaced by one of his underlings, Lefebvre, who was prepared to toe the official line.[6]

In his actions Ballot was casting his lot with those Dutch opponents of liberalism whose critique of colonial policies had come to be called "ethical". The ethical position emphasised a paternalistic concern for the welfare of Indonesia's "native" population, presumed to be threatened by untrammelled commerce and westernisation. Those associated with the ethical turn in colonial thought included well-known scholars like Van Vollenhoven, and they opposed, as we have seen, colonial policies which were designed to move colonial economy and society along paths already trodden by the nations of Europe.

To the extent that it was able to overcome the "ethical" resistance of people like Ballot on Sumatra's Westcoast, the colonial state was empowered to open the gates of liberal reform. In the matter of land policy, to which Ballot's criticisms were directed, Ballot's defeat gave the state increased freedom to distribute land in the province as it saw fit, without having to concern itself unduly with the rights of the Minangkabau people. Combined with other significant changes in colonial policy in the region — the imposition of money taxation, the dissolution of the system of forced coffee deliveries, the reorganisation of nagari government and so on — the activities of the colonial state contributed to very substantial changes in the social, political and economic environment within which Minangkabau villagers were henceforth forced to operate. This is not to say that after 1915 the colonial state was free to remake West Sumatran society after the liberal image. "Traditional" property rights were maintained to the extent that land in permanent use for cultivation was rarely alienated. The result was a hemmed in "peasant" economy in the interstices of the colonial economy of plantations and mines, albeit one that differed greatly from the precolonial situation. Furthermore, there was a good deal of indigenous resistance to colonial land alienation which prevented outright dispossession.

There is, however, a second feature of Ballot's criticisms of government policy which is significant here. It concerns the extent to which, in the context of an attack on colonial liberalism, Ballot rests his argument on a particular construction of "traditional Minangkabau society". His piece provides an excellent example of the way in which concepts of Minangkabau tradition become implicated in discourses other than those of scholars. It clearly demonstrates that anthropological models are as much part of the reality studied by anthropologists as they are external reflections on it, a point made by Pemberton in a different context.

This is not the place for a detailed exegesis of Ballot's argument. Suffice it to point out that in his submission to the government in Batavia[7], Ballot challenged the validity of the so-called Domain Declaration of 1874 whereby all land within the boundaries of the Government of Sumatra's Westcoast was declared the domain of the state. Ballot raised several objections to this, but a significant one was that it was in direct conflict with Minangkabau adat, according to which all land in the province was already the property of extant Minangkabau village communities (the nagari). While the Domain Declaration made certain provisions to protect at least the use rights of Minangkabau villagers to land already in cultivation (mainly wet rice fields), the main point of contention was the status of uncultivated land, or land which had been previously cultivated but on which cultivation had lapsed. According to the Domain Declaration this land, classified as waste land, was declared to be the 'free domain' of the colonial state, and its distribution was to be entirely controlled by the colonial government. Ballot argued that the land classified as free domain was, in fact, subject to the communal property rights of Minangkabau nagari, rights known in Minangkabau as *hak ulayat*.

Objections notwithstanding, between the late 1870s and the mid-1920s more than 110,00 hectares of so-called waste land was alienated from Minangkabau villages and leased mainly to European-owned companies for periods of up to 75 years. In the same period considerably more land was taken out of circulation and handed over to mining companies for exploration or locked away in "forest reserves" by the colonial Forestry Service. Not all the land alienated from Minangkabau villages was actually used for mining or plantation agriculture. In fact at any given time only a very small proportion of it was in productive use. But what we are here concerned with are the effects on the village economy of these "enclosures".

By the middle to late 1920s the organisation of village economic activities in one part of Sumatra's Westcoast, a region I have called the Southern Frontier (see Kahn 1993: 224–260), had taken on a character substantially different from that prevailing in the years before the transformation of colonial society. In the nineteenth century the cultivation of rice, particularly on wet fields, was organised through an overarching set of relationships of clanship, while the production and distribution of other commodities was embedded, in most cases, within a set of relations defined by the structure of nagari communities. While the organisation of work in nineteenth century village communities was overwhelmingly individualised or household-based, individuals and households were themselves constituted according to the principles of clanship, territory and gender as the nagari and the suku (matrilineal kin groups) became dominant principles of social and economic organisation in the first period of colonial rule.

The development of more clearly autonomous small-scale and household-based economic units, pursuing that combination of capitalist and non-capitalist economic strategies we commonly associate with the notion of peasantry (minimising risk, maximising the naturalised component of productive investment) emerged as a consequence of developments after 1874. Important factors were the abolition of forced deliveries, the imposition of a money tax, the reform of nagari government, and the heightened pace of land alienation, all of which took place roughly in the years 1908–1920.

Villagers responded in different ways to threats to their modes of livelihood. Some abandoned village life, at least on a temporary basis. Those fortunate enough to gain basic educational qualifications were able to find jobs in the expanding bureaucracy associated with the reorganisation of the colonial state. Others, even more fortunate according to prevailing values, were able to set up relatively large-scale trading or productive enterprises, employing fellow Minangkabau as wage labourers. Many were forced, often unhappily, to work as full time wage labourers on European-owned mines, factories and plantations.

But the largest number of Minangkabau villagers, in this period and subsequently, followed a different path. Seeking to achieve a degree of economic independence, and unable to build up successful large scale enterprises, individuals and domestic groups sought to exploit what opportunities there were for economic survival. In most cases this meant engaging in a range of economic activities, including the cultivation of rice on ancestral land or dry fields to supply households with at least a proportion of their needs, and earning cash to supply the rest. In order to earn cash it made sense to engage in forms of production and distribution which required minimal cash outlays. Thus an important part of the economic strategy of individuals and households involved the exploitation of productive inputs with little or no (monetary) cost. If additional labour was required, attempts were made either to use family labour or, more frequently, to join together in small working groups to share proceeds. If land was required for agriculture, then individuals struggled to gain access to land without having to pay cash for it — by sharecropping, or planting a cash crop on rice fields in the off-season, or occupying land illegally which had been granted as leasehold or on concessions or which had been reserved by the Forestry Service. If none of this was possible, then peasantisation required migration to areas where land was more plentiful. If cash was required for setting up an enterprise, then the strategy was to minimise fixed investments even if raw materials had to be purchased at regular intervals.

In short, processes of land alienation, modified both by "ethical" policies and the ever-present threat of resistance, greatly affected the economic lives of villagers in the upland regions of West Sumatra in the early decades of this century, resulting in new forms of economic organisation neither genuinely "traditional", nor modern, if the latter is characterised by large-scale capitalist enterprises employing wage labour. Far from being external to these processes of economic modernisation, upland Minangkabau cultivators and craftspeople were consigned to its interstices as marginal producers and merchants squeezing out a tenuous existence on the margins of the colonial economy. In the context of these agrarian transformations, community forms and gender relations were significantly re-shaped, as villagers sought to maximise the naturalised component of their economic activities through the "self-exploitation" of family labour and ancestral land.

At the same time, therefore, as many observers, both indigenous and European, were casting the relations between upland peoples and the outside world in intercultural terms, economic relationships were being established between upland "peasants", and the managers of plantations and mines that might be termed relations of marginality (following Tania Li's discussion in the opening chapter), emphasising always that margins "are an essential part of the whole", that "marginality is a relational concept involving a social construction" and that "there is an obvious asymmetry between margin and centre".

Just as new economic relations were being formed between Minangkabau villagers and outsiders in the early decades of this century, so too new matrices of power emerged linking upland peoples to the colonial society of which they formed a significant part.

THE EMERGENCE OF TRADITIONAL POLITICAL RELATIONS

To characterise upland societies as peasant societies is misleading not just for the timelessness of the imagery of a "traditional peasantry", but also because it treats regions like Minangkabau, and peasant enterprises within it, as isolates. As anthropologists since Kroeber, who described the peasantry as a "part society" and a "part culture" have argued — and as the above discussion clearly demonstrates — peasantries exist only in relation to non-peasants. Moreover the economic system of peasant enterprises/ peasant households cannot be treated as though it were some pure form of premodern autarchy. To understand the nature of these relations is inevitably to shift the focus from basic units of production and distribution in upland societies to the systems of social relations within which peasants are enmeshed.

At one level, of course, market relations link peasant enterprises to each other and to non-peasant producers and consumers. But here I wish to focus on the emergence of a particular institutional or socio-political structure associated with this mode of accumulation, one that is often characterised by the term "patronage". Many analysts have pointed to the prevalence of patron-client ties in "traditional" Southeast Asian systems of social stratification, drawing attention to social hierarchies characterised ideologically by reciprocal flows of labour, political support, and favours. It is as much the supposed cultural legitimacy of such status hierarchies as their on-the-ground forms that has led some to write of a particularly Asian or Southeast Asian form of social stratification, which is characterised by a shared vision of a natural status hierarchy spatialised in terms of relative closeness to, or distance from, a socio-religious "centre". In an early article (1972) the political scientist Benedict Anderson, for example, outlined the "traditional Javanese idea of power", arguing both that it should be distinguished from notions of power prevalent in the West, and that it continues to influence political structures in Indonesia. O'Connor (1983) has suggested that a broadly uniform nexus between power and distance underpins most "traditional" Southeast Asian status hierarchies.

For upland groups, such analyses leave at least two questions unanswered. First, how do such overarching systems of ritual and/or political hierarchy articulate with local systems of social stratification? Second, to what extent are on-the-ground systems of stratification indeed the expression of genuinely pre-colonial status hierarchies?

Local studies of lowland villages suggest that village-based stratificational systems arise out of unequal access to property, particularly land. For example Wertheim (1969) finds that landlords play the dual role of patron and what Eric Wolf has called "broker" between the peasant village and the outside world, a role often legitimised by a belief that such traditional patron-client relations are characterised by reciprocity rather than exploitation.[8] In the upland regions of West Sumatra and East Kalimantan where I have carried out research, however, the situation is rather different, for two main reasons. First, given the distribution of rights in land, one would be hard put to draw any clear distinction between landlords and tenants. Since land distribution is *relatively* more egalitarian than that described for lowland Java (or the rice bowl regions of peninsular Malaysia) and there is no class of people who are able to command the labour of fellow villagers on any systematic basis, it is difficult to speak of a discrete class of landlords. Looked at another way, the power and status of local élites does not generally seem to arise from the ownership of land. Indeed, where powerful and/or high status villagers do own more than average amounts of land, it is largely because they have been able to translate their power and status into land ownership rather than the other way around.

Local élites in the upland villages with which I am most familiar more often than not derive their positions of power from some kind of relationship either to the Indonesian state and/or to large non-state corporations, including foreign-owned multinationals. Thus, in villages in highland West Sumatra, the village elite is made up largely of active, or more often pensioned, civil servants, soldiers and policemen, rather than large landlords or even successful local entrepreneurs. In East Kalimantan, at least in the vicinity of large multinational enterprises, it is the village-resident employees of these enterprises who have both the highest status and the most influence among their fellow villagers. The reason for this state of affairs is fairly obvious, since it is these people who can provide their fellow villagers with the kinds of "favours" they most urgently need — employment in a large private enterprise or government, corporate or government contracts, access to the plethora of licenses and letters of authority required to conduct business, obtaining or validating use rights to land for cash-cropping and residential purposes, and intervention with the army and police who both officially and unofficially exercise a good deal of local control, including permission to move around Indonesia.

Particularly in a country like Indonesia, where the powers of both the state and large enterprises (foreign-owned or local conglomerates closely linked to the governing elite) are all-pervasive, a personal contact with someone with access to the powers-that-be is essential. This power is manifest through the formal-legal reach of the state with its extensive powers to tax, grant licenses for a wide range of activities, control access to all land not in permanent use under conditions of local customary law, and so on. But, perhaps even more importantly, this power is manifested in the extraordinary and sometime extra-legal ways that state officials, acting as semi-private citizens (particularly members of the police and the armed forces) become involved in a whole gamut of local economic activities with the tacit or explicit backing of the state. There must be very few peasant villagers who do not at some point in their lives need privileged access to one or other branch of the Indonesian state in order to go about their everyday lives, more especially because the New Order bans and restrictions on political parties, trade unions and non-governmental organisations mean that channels for redress which exist in other countries are unavailable for Indonesians.

In upland areas, especially those with frontier-type economies, where the potential conflicts between a commercialised and expansive peasantry on the one hand and the state or large-scale agricultural or extractive industries on the other is perhaps greater than lowland regions, the result of all this is not so much a bipolarised class or status hierarchy, but the development of chains or matrices of power relations that extend into peasant villages linking "patrons" and "clients" into reciprocal relations involving the flow of money, services, favours and support. Since, of course, upland villagers rarely establish such relations directly with the highest-level of state officials, those who serve as patrons to villagers are themselves generally low or middle-level state or private sector employees, clients of others higher up in the pecking order. Under these conditions, to talk of "the state" as an institution that is somehow separate from the rest of society is somewhat misleading. A good deal is known about how patronage relationships operate, and how they serve to structure economic flows within the nation as well as the kinds of cultural hegemonies and resistances they imply and are implicated in at the national level. But it is important to recall that patronage relations within the formal state apparatus extend down to the bottom of the hierarchy to clients who lie outside the formal apparatus of the state.

Seen in this light, it is not surprising that most members of the elite in Minangkabau villages — in addition to the very small number of active officials — are retired civil servants and low ranking members of the armed forces and the police. Having returned to their villages upon retirement, they have the established networks which permit them to play both the role of patron to their fellow villagers on the one hand, and broker between villagers and the state on the other. A study of the impact of large foreign-owned mines in Kalimantan conducted by myself and others[9] provided the occasion to investigate these relationships in a somewhat different context. In this case, the influential village-resident "élites" were made up of lower and middle-level, largely Indonesian (rather than expatriate) employees of the company. This is the group that has the most intensive interaction with both the company and residents of the local community. This group is significant for the purpose of my analysis for the way its existence is (more or less) determined, and its role structured, by two (often contradictory) forces which it does not control. These are: the profit-making imperatives of the company on the one hand, and certain general structural principles operating in Indonesian society on the other. These two principles come together most sharply in the activities of this group of employees.

Middle-level jobs in a large Western-owned company, or elsewhere in various instrumentalities of the Indonesian state, are highly valued among educated/ qualified, middle class Indonesians for obvious reasons (relatively high salaries, good benefits, job security, etc.). At the same time, to the extent that these people are (or become), unlike most of their expatriate counterparts, members of some "local"[10] community, they also face certain pressures. In particular, regardless of their individual inclinations, most such employees experience the expectation that, because they have a highly desirable position in the company, they will help those to whom they are obligated (by ties of kinship, marriage, neighbourhood, ethnicity, religion, etc.). Such "help", of course, means a variety of things — giving money, assisting with housing, channelling contracts, finding employment, and the like. These forms of help sometimes come into conflict with what managers, particularly expatriates, think of as good business principles, and hence are often labelled "corruption". But while not wishing to deny that numerous examples of blatant corruption exist, such as individuals abusing positions of trust to enrich themselves (although I doubt they are any more frequent here than in other parts of the world), much of what is labelled corruption is in fact not outright theft, but is instead the kind of preferential treatment described above.

It should by now be clear that the kinds of social hierarchies that have emerged in upland regions of Sumatra and Kalimantan are closely linked not to any simple persistence of tradition, but instead to distinctly modern processes of state formation and foreign-investment. In other words, the appearance of traditionalistic power relations — "patrimonial", paternalistic or patriarchal in form — has been embedded in modernising processes. This seems most obvious in the Kalimantan case because the processes are so recent. But it is equally true of Minangkabau where they are related to the development of what I have elsewhere (1993, Chapter 8) termed a hypertrophic modern state.

The historian Robert Elson has singled out five major ways in which the "organisational principles and practice" of the new states which began to emerge in Southeast Asia in the last few decades of the nineteenth century differed from those of earlier states in the region. These are: the enormous growth in the size of bureaucracies as a consequence of the managerial requirements of the new order; an expansion in the scope of bureaucratic functions; an increase in the intensity of governance, with more officials doing a greater range of jobs more frequently, more regularly and more efficiently; new styles of governing, marking a move from the manipulation of personal ties and followings to administration through clearly defined, formal and impersonal institutions; and, finally, increased centralisation of state powers (Elson, 1992).

These five characteristics of modern colonial states in Southeast Asia are clearly manifest on Sumatra's Westcoast. The size of government bureaucracy on Sumatra's Westcoast, for example, increased substantially after 1870. The biggest spurt took place in the last quarter of the nineteenth century and the early years of the twentieth. The figures, moreover, confirm Elson's assertion that a major feature of the growth in bureaucracy was the increase in government employment of members of indigenous groups.[11] The scope of the bureaucracy and its penetration to regional and local levels also increased. By the late 1920s, for example, there were eight government departments in the Dutch East Indies, all of which were represented on Sumatra's Westcoast, namely the Departments of: Justice, Finance, Transport, Internal Affairs (Binnenlands Bestuur), Agriculture, Industry and Trade, Public Works, Government Enterprises and Education and Religion. The other features described by Elson, such as increased intensity in the style of governing and increased centralisation, were also present in the decades after 1870, and were manifest at the local level as well. Perhaps the clearest illustration of this is found in the reorganisation of village government which took place on Sumatra's Westcoast as a consequence of the so-called Nagari Ordinance of 1914[12]. Ironically, the Nagari Ordinance was described by the Dutch colonial government as a means of promoting greater village autonomy as part of the reforms of the Ethical Policy. But in fact, as Oki's (1977) analysis clearly shows, the subsequent reorganisation of village government increased village autonomy only by decreasing its financial dependence on the state, at the same time in fact giving the state a much more significant role in the organisation of village affairs.

It was processes such as these that served to create a new kind of village elite, more often than not educated in government schools and more closely tied to the processes of colonial state formation. The new kinds of village officials, now increasingly dependent on the backing of the colonial state for their position, together with a growing class of (usually low-level) civil servants drawn from, and hence embedded in social relations within, peasant villages increasingly performed the function of patrons to fellow villagers, and intermediaries between villagers and the state. The relations of power and prestige forged between this latter group and lower-status villagers lay outside the formal structure of the state. However, by taking a broader view of the state as a matrix of power relations that extends into and hence encompasses civil society, it becomes possible to speak of the development of a new "traditionalised" system of status and power — one that continues to define hierarchical relations within upland societies and between upland villagers and élites whose power increases the closer they are to the (new) centres of power in postcolonial Indonesia.

CONCLUSION

There are many ways in which the relationships between upland and lowland peoples in Indonesia can be categorised and analysed. In this chapter I have examined the changing nature of social relations among Indonesians and between Indonesians and the Dutch in the colonial period, focusing in particular on one upland people, the Minangkabau. Economic and political changes in the colony that began in the latter part of the nineteenth century resulted in new kinds of institutional linkages between upland Minangkabau and the outside world, leading to a process of economic marginalisation and the formation of power matrices linking peasants, village élites and Dutch rulers. At the same time, this period witnessed the development of a new language for understanding human diversity in the Netherlands Indies: a language of culture and multiculturalism. Consequently, Dutch scholars and colonial advisers as well as certain members of the local rural élite began to understand social interaction in the colony, and even in the empire as a whole, as instances of intercultural interaction.

It is impossible merely to dismiss these multicultural models of diversity in the archipelago as wrong, although clearly this is only one way in which the situation might have been understood — others focusing instead on economic marginalisation, or political hegemony. To the extent that uplanders and others came themselves to read the colonial situation through the lens of culture, so cultural differentiation became phenomenologically real. There is little doubt that the culturalist critique of modernity and modernisation captures something inherently problematic about the civilising projects of both empires and aspiring nation states. It might, however, be possible to argue that such a culturalist discourse itself goes wrong when it becomes exclusivistic. There is nothing quite so alarming as when authoritarian regimes — like the Dutch colonial government towards the end of the 1920s, or the New Order government in more recent years — take on the rhetoric of an exclusivistic culturalism, reserving for themselves the right to assign people to particular cultural categories and excluding others who do not fit. This, I believe, is the gist of Takdir Alisjahbana's criticism of the Leiden school:

> [The] partisans of [the] historical school of law looked for the permanent and official establishment in Indonesia of a traditional and largely unwritten customary law .... By using ethnological concepts and comparing the existing customary law in the various regions of Indonesia, van Vollenhoven demonstrated with great skill that there were certain common elements in all. He found, *inter alia*, a preponderance of communal over individual interests, a close relationship between man and the soil, an all-pervasive 'magical' and religious pattern of thought, and a strongly family-oriented atmosphere in which every effort was made to compose disputes through conciliation and mutual consideration ....
>
> In the broadest sense, ... van Vollenhoven ... found in Indonesia only what he, as a European reacting against the individualism and formalism of European law, ... was looking for .... In this light we can see that the exaggerations in his analysis and picture of Indonesian customary law were the results of his own personal ideals and sentiments; and that these ideals and sentiments were in their turn simply the manifestations of certain currents within one legal school flourishing in Europe at that time ....
>
> But the attempts to put this theory into practice ... revealed the paradoxical dualistic quality of colonial society .... In the general framework of colonial relationships it was inevitable that the final word in the form and context of customary law would be with van Vollenhoven and other jurists of similar views. This meant, essentially, that customary law could only be applied where it did not conflict with the interests and policies of the colonial system. (Takdir Alisjahbana 1966:72)

The abandonment of the universalistic, and hence inclusive, impulse of the nation-building project is, it seems to me, premature. Perhaps the problem with existing universalisms is not their pretence to universality, but the fact that they are not really universal, and therefore not inclusive enough.

NOTES

1. For an extremely good description of the specificities of primitivist discourse, see Jacques (1997).
2. From "*Awal Kato*" by *Datuk* Sanggoeno di Radjo in the first issue of *Berito Minangkabau* (April, 1926), a journal available in the Indonesian National Library in Jakarta.
3. From "*Hindia merdeka*" (A Free/Independent Indies) in the first issue of *Boedi Tjaniago* dated 1 January 1922. Readers at the time would not have missed the significance of the discussion of the proper function of religion, a far from subtle critique of the argument that religion, rather than adat, should organise social and political life in the here and now.
4. For a more detailed discussion of these issues see Kahn 1993.
5. For accounts of the origins of Islamic modernism in Sumatra see Abdullah (1971) and Alfian (1969).
6. See Ballot (1911). The whole affair is charted in Verbaal (V) 3 Feb 1913 A3 no.4 (this contains a longer draft of Ballot's submission together with a copy of the 1911 book and Nolst Trenit's reply) and in a subsequent exchange of correspondence, first between Ballot and Batavia, and then Lefebvre (still Resident, before officially replacing Ballot) and Batavia

(see, for example, V 17 June 1915 A3 no.16; V 6 July, 1915 A3 no.19;V 15 June, 1916 A3 no.41; and V 2 October 1916 A3 no.23) Documents located in the archive of the former Ministry of Colonies in the Hague.

7   Unless otherwise noted reference is to a paper entitled "*Over de grondrechten en het grondbezit ter Sumatra's Westkust*" (On Land Rights and Occupation on Sumatra's Westcoast) dated 19 May, 1911 by Ballot. This is contained in V 3 Feb 1913 A3 no.4 and formed the basis for the book published in the same year.

8   Probably the best known of such analyses is James Scott's (1985) discussion of the contrast between "traditional" landlord — tenant relations, and the increasingly exploitative relationships characterizing the capitalist transformation of rice farming in Malay villages. See also various chapters in Hart *et al.* eds., 1989.

9   This study involved an examination of the social impact of two large mining enterprises in Kalimantan Timur carried out by Ramanie Kunanayagam, Ken Young and myself in the early 1990s.

10   By local I do not mean necessarily that the people themselves come from the area. Some do come from the region and thus have kin and/or fellow villagers nearby. Others originate elsewhere but live in nearby communities and establish ties of neighbourhood, marriage, contract marriage, or develop a sense of community with fellow members of their ethnic or religious groups.

11   Information on the number of government employees is provided in the annual colonial reports (*Koloniale Verslagen*). For estimates for Sumatra's Westcoast see Ginkel *et al.*, (1928, v.2: 21).

12   The definitive study of these administrative changes remains Akira Oki (1977).

References

Abdullah, Taufik, 1971, *Schools and Politics: The Kaum Muda Movement in West Sumatra*, Modern Indonesia Project, Monograph Series. Ithaca, N.Y.: Cornell University.
Alfian, 1969, *Islamic Modernism in Indonesian Politics: The Muhammadijah Movement during the Dutch Colonial Period*, PhD thesis. University of Wisconsin.
Anderson, B., 1972, "The idea of power in Javanese culture". In *Culture and Politics in Indonesia*, edited by Claire Holt, pp. 1–69. Ithaca, N.Y.: Cornell University Press.
1983, *Imagined Communities: Reflections on the Origin and Spread of Nationalism*. London: Verso.
Ballot, J., 1911. *Ontwerp Agrarische Regeling voor Sumatra's Westkust*. Padang: De Volharding.
Barth, F., 1989, "The analysis of culture in complex societies." *Ethnos*, 54(3–4): 120–142.
Boeke, J.H. *et al.*, 1966, *Indonesian Economics: The Concept of Dualism in Theory and Policy*. The Hague: W. Van Hoeve.
Dutch East Indies, Commissie voor het adatrecht, 1911–1955, *Adatrechtbundels*. The Hague: Koninkilijk Instituut voor taal-, land- en volkenkunde.
Elson, Robert E., 1992, "International Commerce, the State and Society in Southeast Asia". In *The Cambridge History of Southeast Asia*, edited by Nicholas Tarling, pp. 131–195. Cambridge: Cambridge University Press.
Ginkel, F. De M. Van *et al.*, 1928, *Westkust Rapport*. Rapport van de Commissie van Onderzoek ingesteldt bij het Gouvernementsbesluit van 13 Februari 1927. Weltevreden: Landsdrukkerij.
Hart, Gillian *et al.*, 1989, *Agrarian Transformations: Local Processes and the State in Southeast Asia*. Berkeley: University of California Press.
Hooker, M.B., 1978, *Adat Law in Modern Indonesia*. Kuala Lumpur: Oxford University Press.
Jacques, T. Carlos, 1997, From Savages and Barbarians to Primitives. *History and Theory*, 36(2): 190–215.
Jan Mohammed, Abdul R., 1985, The Economy of Manichean Allegory. *Critical Inquiry*, 12(1): 59–87.
Kahn, Joel S., 1993, *Constituting the Minangkabau: Peasants, Culture and Modernity in Colonial Indonesia*. Oxford and Providence: Berg.
1995, *Culture, Multiculture, Postculture*. London: Sage.
O'Connor, R.A., 1983, *A theory of indigenous Southeast Asian urbanism*. Singapore: Institute of Southeast Asian Studies.
Oki, Akira, 1977, *Social Change in the West Sumatran Village : 1908–1945*, PhD Thesis. Australian National University.
Pemberton, John, 1994, *On the Subject of "Java"*. Ithaca and London: Cornell University Press.
Said, Edward, 1978, *Orientalism*. New York: Pantheon.
Schöffer, I., 1978, "Dutch 'Expansion' and Indonesian Reactions: Some Dilemmas of Modern Colonial Rule (1900–1942)". In *Expansion and Reaction. Comparative Studies in Overseas History*, edited by H.L. Wesseling, pp. 78–99. Leiden: Leiden University Press.
Scott, James, 1985, *Weapons of the Weak: Everyday Forms of Peasant Resistance*. New Haven: Yale University Press.
Takdir Alisjabana, S., 1966, *Indonesia: Social and Cultural Revolution*. London, Kuala Lumpur, Melbourne: Oxford University Press.
Taylor, Charles, 1976, *Hegel*. Cambridge: Cambridge University Press.
Wertheim, W.F., 1969, "From aliran towards class struggle in the Javanese Countryside", *Pacific Viewpoint*, 10(2): 1–17.

Chapter 4

"ITS NOT ECONOMICAL":
THE MARKET ROOTS OF A MORAL ECONOMY IN HIGHLAND SULAWESI, INDONESIA

Albert Schrauwers

What are we to make of a peasant who opposes an "economically rational" course of action with the words, "its not economical"? And what are we to make of a state which charts its development policies in terms of the selective preservation of "tradition"? The juxtaposition of these conundrums underscores the contingent nature of capitalist transformations of peasant agriculture in marginalized areas like the highlands of Central Sulawesi. Groups like the To Pamona have been politically and economically incorporated within a state which has reformed and reconstituted their livelihoods as "peasants". By examining how moral economies emerge out of this development process, I invert the historical presuppositions of the moral economy model and the related concept of the "persistence of tradition". A moral economy is assumed to be a natural, universal characteristic of the peasantry which existed prior to the introduction of capitalist relations. It is characterized by subsistence production, the provisioning of "subsistence insurance", and is driven by a risk averse non-market rationality. A historical analysis which relies on such models obscures or ignores the mutualistic links between moral and market economies and their co-emergence under colonial tutelage (Breman 1988). The moral economy model implies historical continuity, the "persistence of tradition", yet is usually explained in abstract terms not unlike those models of "Homo Oeconomicus" against which it argues (Peletz 1983: 732). Although the To Pamona appear to be perpetuating a "traditional moral economy" as an obstacle against capitalist development, closer scrutiny indicates that these forms arose in the context of state directed capitalist transformation and class differentiation (cf. Li 1997,[1] Attwood 1997). That is, I would like to demonstrate that the moral economy is not a survival of some "natural economy", but is an invented tradition which must be viewed through the dual processes of what Kahn (1993) calls "peasantization" and "discursive traditionalism".

The To Pamona were isolated swidden cultivators in the highlands of Central Sulawesi until the first decade of this century, when intervention by the Netherlands East Indies government led to their reconstitution as a peasantry through forced resettlement and the compulsory adoption of wet-rice agriculture (Kruyt 1924). Fundamental shifts in relations of production accompanied this change in agricultural technologies; landed property, wage and market relations were all introduced. The introduction of new technologies and relations of production flowed from Dutch liberal critiques of "native communalism" which saw an ethic of individualism and "economic rationality" as the only road to economic, political and spiritual development (Schrauwers 1995a). "Modern" social and capitalist transformation, conceived teleologically, required that multiple family longhouses be broken down into constituent nuclear family households which would own and work their own property to ensure their own subsistence, and sell their surplus production in the market. However, these liberal economic prescriptions were ironically predicated upon evolutionary models, and hence sought to recapitulate earlier forms of European development in the hinterland; distance from the center represented a measure of backwardness which could never be bridged, producing a distinctly "peasant" economy.

Despite decades of official government policy and pressure, the state appears to have been unsuccessful in its attempts to create either an "economically rational" peasantry or simple family households among the To Pamona. The area is characterized by vague household boundaries, a feasting economy supported by a gift exchange network, forms of non-waged labor exchange and patterns of petty merchandising all of which underscore an apparent economic "irrationality". However, this apparent "irrationality" in the local market economy cannot be explained as the strategic use of some preexisting, culturally insulated, "moral economy". Each of these strategies was adopted through rational economic calculation and uses commodity forms linked to the market. Since the state development models by which they were incorporated and marginalized limited their ability to fully engage the market, relations of production and social reproduction have shifted to create a Janus-faced economic strategy simultaneously dependent upon both market and non-market mechanisms.

The apparent persistence of tradition is, according to Kahn(1993:65), a "discursive traditionalism" which developed during the process of "peasantization", whereby farmers seek to "ensure or even extend the exploitation of the 'free' productive inputs available to them". These "free" inputs are not acquired through the market, although they may have a social "cost". They are enveloped in a state sanctioned "discursive traditionalism" which casts non-market relations as a "perpetuation of tradition" (cf. Roseberry 1989: 217-223). These new "traditions" depend upon, and indeed could not exist outside, a commoditized economy. Kahn thus describes a continuum of peasant economic behaviors from the "traditional" peasant whose activities are made possible only through commoditized inputs, to the "capitalist" peasant whose success is owed to maximizing the exploitation of "traditional" resources such as family land and labor. In each case, both commoditized and "free traditional" inputs are necessary for the reproduction of household production.

I will briefly describe the features of the local moral economy, followed by a critique of the dualist definitions in which it is usually couched. The availability of the commodity circuit alters the moral dynamic of subsistence production. Moral economy arguments based upon cultural notions of altruism never address the issue of how moral economies are manipulated to meet the labor needs of "altruistic" benefactors who have no need for "subsistence insurance". In impugning their "altruism" I am not asserting that they are driven by profit motives aimed at the accumulation of capital; rather, to paraphrase Wolf (1966:13-7), their aims are to maintain the viability of their households, not their businesses. They seek to maximize the "free" inputs available to them in an otherwise commodified economy where the margins are small. Central to this process is the way in which "kinship" has been discursively shaped by a series of "traditional" gift exchanges (*posintuwu*). The character of these gift exchanges establishes the official tenets of the moral economy which has practical implications in the organization of non-commodified labor.

The historical precedence of the moral economy is not a simple "chicken and egg" question. Rather, its emergence has important implications in the discussion of class formation. Moral economies are normally considered to be "threatened" when agricultural production is commodified (Attwood 1997: 149). Yet, if market and moral economies emerge simultaneously in the process of marginalization by which specifically "peasant" communities are created, then "moral economy" strategies can be seen to assume a role in defusing class tensions; the way in which subsistence insurance is provided limits rebelliousness. In this, the role of the state has been central. On the one hand, the state has set the terms of the green revolution which has accentuated class tensions and necessitated economic strategies of a "moral economy" kind. On the other hand, the state has phrased the discursive traditionalism of the moral economy in terms of nationalist discourses of consensus and communalism to dispel the revolutionary consequences of the development of underdevelopment. The state selectively preserves such traditions as part of an encompassing ideology which casts itself as the modern, paternalistic font of development, and peasants as "traditionally" communal, their ethnic rationality the cause of their own underdevelopment.

THE DISCOURSE OF KINSHIP

The district of North Pamona (*Pamona Utara*) is the ancestral homeland of a diverse set of highland groups who are now collectively known as the *To Pamona* (people of Pamona). The 28 villages in the district (total population in 1991, approx. 25,000) are the product of forcible resettlement, which also entailed the adoption of wet-rice agriculture (where possible), the introduction of a hierarchical "kingdom" through which the Dutch colonial state ruled, and conversion to Christianity. Many of these villages (including four in which I did fieldwork) lie on the shores of a large, picturesque highland lake. The Trans-Sulawesi highway links the area to *Tana Toraja*, the second most popular tourist destination in Indonesia. The hotel "*Pamona Indah*" (Beautiful Pamona), complete with satellite television reception and paddle boat swans imported from a bankrupt Javanese amusement park, overlook the "Lake Poso Festival" grounds where "traditional" houses display the arts and crafts of the area under the tutelage of the New Order's cultural policies. The hotel, like most of the stores, is owned by a member of the small ethnic Chinese community. "Beautiful Pamona" thus encapsulates the multiple processes I wish to capture: a created tradition that conforms to western preconceptions of a rural "other", which benefits a capitalist class and not the "backwards" people who are the object and subject of their discourses.

Looking for a more authentic culture less driven by state-sponsored tourism, I was eventually introduced to a second traditionalist discourse on the moral imperatives of kinship. It was this discourse, and the explicit limits it placed upon market relations, which led to my initial investigation of the moral economy of this village. Yet, as I dug deeper into Tentena's brief history (it was founded within the living memory of its oldest citizens) I was struck by the recurring remark: "Our traditions are not what they were". The specific cultural features of kinship which might have been interpreted as the persistence of a "natural economy" had developed during the colonial transformation of the local economy. The "traditionalism" of this "official" kinship discourse (Bourdieu 1977: 36) was no less subject to state revision, and is proudly summarized on a mammoth ten-meter high sign erected at the entrance to the district capital. The motto *sintuwu maroso, tuwu siwagi, tuwu malinuwu*, roots Pamonan culture in communalism, mutual aid, and shared common roots.

The significance of the motto is found in a series of gift exchanges referred to as posintuwu, the material and symbolic expression of mutual aid (tuwu siwagi). The exchange network which results defines the most encompassing kin group, the *santina*, which extends laterally to include third cousins. Because genealogical knowledge is a form of social capital husbanded by elders, younger generations are afflicted by genealogical amnesia, and cannot clearly articulate the exact relations tying their santina together. Rather, their continuing identification of kinship ties is shaped by their participation in these gift exchanges. Posintuwu was traditionally given by kin on only two occasions, marriage and funerals, as a means of making large feasts possible. Now, during any number of church sponsored feasts, members of the extended family bring their contributions to the host's house, where they are carefully recorded. Participation in these feasts constitutes the kin group as a group by bringing it physically together, as well as establishing the special character of kin relations in contrast to the commodified relations of the market within which these feasts are embedded. The special character of these ties is expressed as "*be maya mombereke*" (the calculation of mutual costs and benefits is prohibited). Posintuwu exchange thus constitutes what Parry and Bloch have called a "transactional order" able to "decontaminate" morally suspect commercial exchanges (1989: 23, cf. Carsten 1989). This transactional order has been encapsulated within state discourses on "local tradition", casting the To Pamona as "communal" peasants whose economic rationality relegates them to backwardness.

In the pre-colonial period, posintuwu exchange was marked by generalized reciprocity of specific, magically powerful items given by the groom's kin for his bridewealth (*oli mporongo*). The root word of bridewealth, *maoli*, referred to the direct exchange of magical goods for other equally powerful goods. However, over the last 85 years, as the To Pamona have converted to Christianity, maoli exchanges have lost their magical connotations. Maoli now refers to market exchange, and hence the word for bridewealth literally means "the price of the wedding." It is only as bridewealth has been redefined in terms of its cash value, that posintuwu exchanges have been ideologically contrasted with the bridewealth and market exchanges to define the special character of kin relations. Posintuwu is contrasted with oli, the gift of kin with the cash paid to non-kin.

Posintuwu exchanges have gained a great deal of ideological importance in constituting a "kinship" sphere independent of the competitive ethos of market values. None of the early Dutch missionaries in the area wrote of posintuwu exchanges. Yet by the time of my fieldwork, posintuwu had become a topic of daily conversation, the idiom in which kinship responsibilities were phrased. The occasions on which such gifts were given has broadened immensely. All of my informants emphasized that calculation, the *sin qua non* of market exchange was inimical to the free gift giving which characterized posintuwu, even though the gift given was usually cash. All of my informants emphasized that posintuwu was a free gift for which one could not demand a return payment, nor even expect a return gift of equal value. However, since there is no "natural" opposition between these two spheres of exchange (Parry & Bloch 1989: 7), we should not be surprised that there is leakage of market rationality in this ideologically constituted kinship sphere as wealthier patrons calculate the relative investment benefits of their scarce cash reserves.

Posintuwu is thus a negotiation of the limits of kinship in both a practical and an ideological sense. The ideology of the posintuwu exchange system emphasizes that it is a means of assisting poorer kin to meet the demands of expensive rituals of social reproduction like weddings. In the practical sense, posintuwu is an exchange of cash, commodities and labor which are invested to maintain specific diadic kinship ties. One's ability to give gifts influences the breadth of one's remembered family. The wealthy have bigger families, both in the sense of numbers of domestic dependents as well as the breadth of their santina. They were the only informants who had prepared genealogies. Poorer peasants on the other hand, find their family shrinking over time, and find themselves dependent upon a restricted number of patrons. Wealthier peasants can maintain larger kin groups, and thus are able to give larger feasts. Since these feasts constitute the kin group as a group, the larger the feast, the greater the status of the host. The hosts become the natural center of a kin group whose relations can be characterized as a moral economy, a sphere where calculation should not enter, ideologically contrasted with market exchange.

The ideology of posintuwu exchange should not be taken at face value. Bourdieu has argued that gift giving is based upon a fundamental "misrecognition" of the exchange "which an immediate response brutally exposes: the *interval* between gift and counter-gift is what allows a pattern of exchange that is always liable to strike the observer and also the participants as *reversible*, i.e., both forced and interested, to be experienced as irreversible" and hence as generosity (1977: 5-6). In the case of posintuwu, where most of the gifts are now in cash, it is only the interval between exchanges which allows participants to ideologically distinguish their "gift-giving" from the direct reciprocity typical of the market. Although it was emphasized by all that posintuwu exchanges were "true" gifts in which generosity was the only motivating force, contradicting exchange strategies were evident. Villagers themselves feel this tension between the balanced and generalized aspects of posintuwu exchange, a tension which they say reflects the different motives of the participants. It was common for elders to decry one element of current practice with reference to the *buku posintuwu*, the notebook in which the hosts of wedding or funeral feasts will transcribe the donors and the amount of their gifts. These elders claim that in the old days, one gave what one had, without calculation. The offer was made publicly before one's kin group, where one's generosity could be seen. Now, they claim, posintuwu (consisting of money) is slipped in envelopes and written in books so that the hosts will know exactly how much they must return. Thus, according to these elders, there has been a noticeable shift from generalized to balanced reciprocity.

The difference between past and present gift-giving strategies is less marked than these elders made out, and itself reflects state and church intervention in the practice of posintuwu exchange. Bourdieu's analysis demonstrates that all gift-giving is based upon the misrecognition of its "interestedness" and hence changes over the last century cannot be characterized in terms of a lack of "generosity" and "disinterested giving". My younger informants denied this shift in attitude, explaining that these records were an innovation tailored to a type of posintuwu which is itself an innovation introduced by the Indonesian state when it modernized traditional wedding practices as part of its general drive towards rural development. Only *posintuwu umum* (general gifts) are recorded in the gift books. Posintuwu umum refers to the numerous small (but hardly negligible) gifts made by more distant family and neighbors. The gift amounts to the daily wage for an agricultural worker (i.e. Rp. 5.000 or less, or its equivalent in rice), a value set by the village government. Poorer families, I noted, gave little posintuwu umum outside of their small circle of neighbors. According to my informants, the paradox of posintuwu umum is that it is only these smaller gifts which are recorded. One farmer bitterly pointed out that it was obvious people didn't check their gift books when making their gifts, since he had given a neighbor a sizable contribution and received little in return. Others pointed out that because the amounts were fairly standard, they gave all that they had on hand and even borrowed if they had nothing. They referred to their books only when making a return gift to someone who was a lazy gift giver (*malose mosintuwu*). Such individuals are known to be stingy, not just with their gifts, but also with their participation. Poverty was taken into account as an obviating factor, and those who had little to give often donated inordinate amounts of labor instead.

This innovative form of posintuwu exchange is contrasted with *posintuwu keluarga dekat* (gifts of close family), which is regarded as the original form, and the model for state-regulated posintuwu umum. Posintuwu keluarga dekat is pledged in an open meeting, as in the past, but now consists of cash and purchased commodities. Even those who pledge cattle, as in the past, usually pledge only an "*adat* cow", the officially set (and undervalued) price of a cow. It is this form of exchange which is played off posintuwu umum to define the boundaries of the kinship group as people slip from one form of gift giving to the other by increasing the size of their gift. "Close family", it should be emphasized, can potentially include anyone up to a third cousin (and sometimes an anthropologist). The recognition of the kin tie is dependent, however, on the continuation of the exchange relationship. The boundaries of the kin group are open to negotiation, and hence manipulation. Over the past decade, as more hosts have substantially enlarged their "close family", they have also taken to specifying a floor amount for this gift (a practice borrowed from posintuwu umum). The high status host who introduced this practice, well known as a generous gift-giver and community leader, stated that he did so in an attempt to modernize the accounting practices of the wedding. To ensure that the wedding remained within budget, it was essential to have an accurate knowledge of one's resources. The application of accounting principles has, however, led to spirited public debate about the very nature of gift-giving, as others have utilized the innovation in a calculated manner to increase contributions to their feasts.

The more people the hosts can include as close family, the larger the feast they can hold, and the greater the status they derive. Yet such a maximizing strategy carries its own risks. The benefits of specifying a floor value for posintuwu are now often hotly contested. Specifying this value is closely related to extending the boundaries of one's "close family" to maximize income. Those who have adopted the innovation have found that calculation begets calculation, not generosity. Thus, by including more people as "close family", the hosts feel obligated to provide hospitality to demonstrate their own generosity in the face of public criticism. The family meeting at which the gift is made often turns into a feast itself. After the costs of the meeting are deducted, the increase in posintuwu pledged seldom nets more than posintuwu umum, while obligating the host to give more to all those he has invited as "close family". Attempts to gain status by holding a large feast through the calculated manipulation of invitation lists and posintuwu networks inevitably backfired as gossip deflated the host's reputation and hence the size of future gifts.

The introduction of the floor value for posintuwu keluarga dekat exposed the calculation implicit in gift exchange which until then had been obscured by the non-commodified nature of the gift. As the goods contributed to and required for the feast have become marketable commodities, the calculation of those requirements was increasingly made in terms of money. This was equally true of those hosts who did not set a floor value for the gift. Each feast-giver was faced with escalating costs resulting from feasting inflation. Like the community leader noted above, wedding hosts found that the ideal of generosity is "uneconomic", that they could not budget appropriately unless they set a posintuwu floor value. They also knew that the opportunity cost of setting a floor value could outweigh its benefits. The availability of a choice of gift-giving strategies has thus served to highlight the opposition between generosity and balanced reciprocity, kin and non-kin, posintuwu and market exchanges. It is careful calculation which leads to the adoption of one strategy over the other, including the decision to be "generous" and "not calculate". In both cases, however, gift-givers find it to their advantage to emphasize the posintuwu ideal of the "free gift", officially defining kinship through an ideologically asserted refusal to calculate. To the "generous" gift givers eager to maintain their reputation, the ideal of the free gift underscores their magnanimity. To the "calculating" feast giver attempting to benefit from the generosity of others, the ideal of the free gift limits his obligation to repay an equal amount. This is, then, a case which supports Jonathan Parry's more comprehensive argument that "an elaborated ideology of the 'pure' gift is most likely to develop in state societies with an advanced division of labor and a significant commercial sector" rather than being an evolutionary precursor of the "free" market (1986: 467).

Posintuwu exchange has come to play an increasingly prominent role in the ideological definition of kin ties in an otherwise commodified market economy. This predominance is the result of state intervention, as the example of posintuwu umum shows. The colonial state consistently defined ethnic groups in terms of a local "tradition" (adat) through which they ruled indirectly; this "tradition" was distinct from the market, which was regulated by a different, national set of laws. The colonial state, influenced by nineteenth century evolutionary schema, cast this "traditional" polity in kinship terms, which they saw as the implicit contrast with their own market society (*gemeinschaft-gesellschaft*) (Parry & Bloch

1989: 7). The terms of incorporation adopted under colonial rule have been perpetuated by the New Order state. The New Order has similarly cast the To Pamona as a kin-based, communal society based upon mutual aid, this aid taking the form of posintuwu exchange. The state, through village government, has sought to modernize this "tradition" as a necessary part of rural development, but importantly, has never altered the terms in which To Pamona "tradition" were cast. Within the larger march to development, the To Pamona are consistently relegated to the position of communal "peasants", their ethnic "rationality" somehow making them inherently unfit for participation in the wider market economy.

THE MORAL ECONOMY OF KINSHIP

Just as he contrasts the "interests" which hide behind the ideology of generosity in gift-giving, Bourdieu also contrasts "practical" kinship with the "official" kinship shared by the elders and the state. The strategies by which the goals of gift-giving are disguised are similar to the strategies of "official" kinship discourse, which hide the way in which these categories are bent "towards the satisfaction of material and symbolic interests and organized by reference to a determinate set of economic and social conditions" (1977: 36). It is then, the practical application of the kinship injunction against calculation established through the "transactional order" of posintuwu exchange which constitutes the moral economy of the district. The economic strategies of the To Pamona invoke the kinship sphere in meeting their production, consumption and market activities. This moral economy is characterized by forms of non-waged labor exchange, vague household boundaries, and "irrational" patterns of petty merchandising. These features, predicated upon "kinship" and a "non-market" rationality, are held responsible by the state for the continued "backwards" economic condition of the To Pamona.

The villages of Pamona Utara were created through government fiat in 1908. In an attempt to hasten the social and cultural evolution of the To Pamona, the state has consistently sought to eliminate the multiple family longhouses typical of the area in the nineteenth century and to foster an ideal simple family household which is both unit of production and consumption. It was argued that individual production of household consumption needs would stimulate self-interest and individuality over communally, hence culturally, determined needs. The New Order state has similarly directed its development efforts at nuclear family households. Although all of my informants emphasized this as a personal ideal as well, actual household composition was exceedingly complex. This complexity

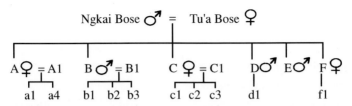

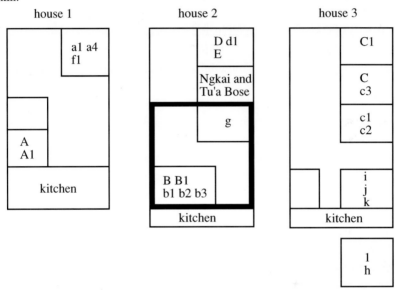

Genealogical ties of the residents of three households in Tentena, shown below. Capital letters are used for the children of Ngkai and Tu'a Bose, and lower case letters for their grandchildren. F lives elsewhere, and g through l are more distant kin.

FIGURE 1 HOUSEHOLD BOUNDARIES.

has been interpreted by the state in terms of the perpetuation of tradition, or more brutely, as "backward" resistance to state development efforts.

The majority of villagers I surveyed lived in what could be classified as nuclear family households, although a significant minority did not. Few households could be treated as units of production and consumption. Figure 1 illustrates the layout and membership of the housing complex of three related government officials — people whom one would assume would most readily adhere to the state policies they were enjoined to encourage. The extended family of Ngkai and Tu'a Bose have complied with government policy by building three separate houses approximately 3 meters apart. However, these physical structures only roughly approximate household boundaries which are purposely left vague so as to provide subsistence insurance for the extended family. There are three households resident in the three houses: Household A, living in House 1, Household B living in House 2, and Household C living in House 3. The boundaries of these three households varies, however, depending upon our criteria for defining the household.

In terms of co-residence and the accomplishment of household chores, each physical house comprises a single unit dedicated to the upkeep of that structure. Note, however, that in no case is the membership of the house that of a nuclear family. In the case of House 1 for example, a nephew is co-resident, while one of their own sons and two of their daughters are not. House 2 contains a multiple family household: Ngkai and Tu'a Bose with two of their married children and their children. House 3, like House 2, contains a number of more distant relatives, signified by the lower case letters. Interestingly, if we define the household as a unit of consumption, these household boundaries shift. There are still only three households, but House 2 is split. Ngkai and Tu'a Bose and one of their married children are a part of the household economy of House 3, which provides for all their needs. The individual labeled "E" is part of House 1. At a later point in my fieldwork, one of the married sons, "D" moved out of House 2 and established residence in his own small store. He nonetheless remained a part of House 1's unit of consumption.

These two different ways of defining the household correspond with a To Pamona distinction between the *banua*, or (long)house, and the *sombori*, defined in terms of a hearth. Here then, would appear to be evidence of the "persistence of the tradition" of the longhouse despite government policies designed to foster simple family households. Inter-household boundaries are purposely kept vague because of a married couple's explicitly stated responsibility to ensure the social reproduction of their larger kin group. Kin could be offered food and shelter either as individuals incorporated within the sombori, or as an independent sombori within the banua, or through an intermediate type of co-householding (*sanco-ncombori*). It is this responsibility to their extended family which accounts for the large number of distant kin living in all three houses. The vagueness of household boundaries is one strategy used to provide subsistence insurance, hence this "persistence of tradition" could be taken as evidence of a moral economy.

Relations of production are similarly complex, obfuscated by the vagueness of household boundaries. A wage labor market was slow in developing in the area due to the ubiquity of small plots. Wet rice fields are worked by domestic labor supplemented by labor exchanges for specific tasks such as planting and harvesting. This pattern of labor exchange, called *pesale*, was adapted from shifting cultivation. It is characterized by the exact exchange of a day's labor for a day's labor, although "kin" labor is not subject to a rigid accounting. The workers bring their own cooked rice, but the host provides fish and coffee for all. Many farmers emphasized that they hold labor exchanges to enliven the drudgery of field work.

The hiring of wage workers was introduced by Chinese merchants in the 1930s, and they were emulated by indigenous civil servants in the republican era who were unable to reciprocate labor. Caught in a network of kin relationships, it was difficult for the civil servants to hire wage labor. Rather, they were forced to hold a "labor exchange" and offer a cash replacement for the days labor they were unable to return. This entailed issuing a general call for labor and waiting to see who showed up; the hiring was out of their hands. All who came had to be paid even if their number exceeded that required for the task itself. This added immeasurably to the festive atmosphere of the labor exchange, but precluded the widespread hiring of wage labor. A hiatus thus initially appeared between the Chinese merchants who worked their sizable holdings with wage labor, and the To Pamona who continued to depend upon "traditional" labor exchanges and kin labor. This hiatus was short lived, ending with the introduction of green revolution inputs which needed to be purchased. Those peasants who did not have sufficient land to meet both their subsistence needs and production costs, had no need for labor exchanges to work their own minuscule plots. They thus began to demand wages.

Labor exchanges were still held at the time of my fieldwork, although their adoption was subject to rigorous calculation. It was emphasized to me that "labor exchanges are not economical". During interviews, my research assistant pulled me aside to express his fear that I might misinterpret these examples of the self-exploitation of domestic labor for marginal returns as irrational, and cast these peasants as "stupid". All the peasants with whom I spoke were quick to commodify the costs involved in a labor exchange: they had to both provide food for the workers and then also replace each worker day with a day's labor of their own at a potential loss of a day's wage. The labor exchange could potentially end up costing them almost twice as much as hiring wage workers if they lost an opportunity to earn wages due to labor exchange commitments. It was, nonetheless, the preferred option if they had a surplus of domestic labor because of the lower up-front cost (the cost of the meal rather than the payment of wages).

Under the labor exchange system, the cost of fore-gone wages may also be reduced as certain types of "kin labor" used in labor exchanges do not need to be repaid. For example, one large landholder with insufficient cash to hire laborers held a labor exchange instead, and partly repaid the labor with that of a dependent of his non-resident father-in-law. He also convinced a distant elderly relative to move in. She performed domestic chores, freeing the man's wife for agricultural work. In yet another case, landless cousins of a poor farmer volunteered to work for free in planting, and hence were called to help with the harvest, which is paid with a percentage of the crop. The continued use of self-acknowledged, economically-irrational labor exchanges, and the use of such exchanges to provide subsistence insurance, is the second element of the local moral economy.

It was the availability of kin labor which also explained the wide spread pattern of petty trade. During the course of my fieldwork, I saw 5 petty trade ventures established at a single corner, just a stone's throw from several large stores operated by Chinese merchants. These petty trade ventures resold commercial goods like salt, soap and candy purchased in these nearby stores to the surrounding 15 households. The small profit they made in trading at the margin rarely exceeded their bad debts, and they were frequently forced to close when a shortage of capital prevented them from buying in bulk. None of this dampened their enthusiasm for establishing a petty trade venture.

Every aspect of petty trade seemed economically "irrational", from the number and location of such ventures, the similarity of stock, the long hours spent for only marginal returns, and the necessity of selling on credit. Indeed credit is integral to their "market ethic". Most To Pamona are "too shy" to enter a Chinese merchant's store to purchase a single item like a block of salt. They prefer to purchase such small items at a nominally higher cost from a petty trader rather than display their poverty to the wealthy. These buyers frequently lack the cash for even small purchases; the larger stores will not offer credit for small purchases whereas the petty traders must, since their customers are family, and those who have must share with those that need.

The irrationality of these petty trade ventures stems, however, from viewing them as businesses. My informants emphasized that these trade ventures were an integral part of their household budgets, not capitalist ventures seeking profit. By purchasing larger amounts from Chinese merchants, they received marginal discounts, thus lowering the costs of these goods for their own households. By opening a petty trade venture, they were able to sell off the surplus of these goods which they could not consume themselves. They were thus devoting long hours to staff shops to marginally reduce the cash needs of their household economies rather than deriving a "profit". Again, such ventures depended upon having surplus domestic labor.

These irrational features of the local economy could easily be interpreted as evidence of a moral economy, and as the persistence of tradition in the face of encroaching market relations imposed from without. In particular, the definition of kinship through gift exchanges combined with the obligation to provide "subsistence insurance" for them, the dependence on "kin labor" in "traditional" labor exchanges, and obtaining credit from the petty market stalls of "kin", all underscore the importance of an ideologically distinct kinship sphere. I have emphasized, however, that the continued dependence of To Pamona peasants on their kin-based moral economy does not imply a deficit of rational planning, nor is it a repudiation of the market. Further, each of these elements of the moral economy are dependent upon commodity forms. This moral economy emerges, rather, out of strategies for the maximal utilization of cash, symbolic capital, resources and labor. As Kahn has argued, peasants are so economically marginal that they depend upon the exploitation of all the "free" (i.e. non-market) inputs available to them to produce their cash incomes. They are aware that the ultimate cost of their "free" inputs may be high, yet their low capitalization makes the "uneconomic" choice imperative.

PEASANTIZATION

These strategies have emerged because of real obstacles to capital accumulation. The differentiation of land holdings over the last 65 years has not resulted in the polarization of landholders and landless workers, but in the increasingly fine partition of ancestral plots. Rather than polarization, differentiation has produced a relatively homogeneous class of smallholders who can be characterized as "just-enoughs" and "not-quite-enoughs" (cf. Geertz 1963: 97). These smallholders have become increasingly immiserated as agricultural production has been commodified, due to their inability to accumulate capital: hence their efforts to maximize the "free" inputs provided by the kin-centered moral economy. The green revolution radically boosted yields through double cropping but did so in a way which transferred the bulk of the gains to the local merchant class. The green revolution has made these smallholders increasingly dependent upon highly capitalized inputs such as tractors, fertilizers, mechanical threshers and rice mills which can only be obtained at a premium through these merchants. The local economy thus has an apparent dualistic cast, with an ethnic Chinese merchant class, characterized by a capitalist rationality, gaining the bulk of the benefits from the agricultural development of the "communal" To Pamona, whose kin-centered ethnic rationality is said to make them unfit for participation in the wider market economy. Such a characterization is predicated, however, upon official state development and kinship discourses, and ignores the practical utilization of these discourses by the To Pamona.

Capital accumulation has first of all been limited by the restricted availability of land suitable for wet-rice cultivation, the colonial government's chosen development path (Kruyt 1912 II: 232-33). Since 1925, when all such land was converted, access to land has been through inheritance which is usually granted at or about the time of marriage. But with the limited amount of land suitable for wet-rice cultivation, parental holdings were often insufficient to provide a livelihood for all their children. Since 1925, a process of demographic differentiation has divided the local population into those with viable holdings sufficient for subsistence, and those with unviable holdings. This demographic differentiation has occurred within sibling groups. Although customary (adat) inheritance law calls for the equal subdivision of parental property, the actual distribution of property at the time of marriage frequently disenfranchises younger unmarried siblings from productive resources. Older married siblings attempt to form the ideal, self-contained unit of production and consumption and so claim a larger share on a first-come basis. Younger siblings are left to form "unviable" households dependent on the resources of their kin for their survival. The pattern which develops is not one of large landholders and landless, but of those with "just enough" and those with "not quite enough" tied by mutual claims to land.

After several generations and sixty-five years, a survey of households and their histories of resource allocation in the district revealed the long term implications of this demographic differentiation on household forms. The household development cycle was not unilinear, but followed two major pathways depending upon the viability of the peasant enterprise (Figure 2) (Schrauwers 1995b; cf. Wong 1991). In the first case, poorer families were more likely to live in simple family households denuded of dependents, or in multiple family households forced to share sparse resources. In contrast, wealthier families with viable land holdings were more likely to live in extended family households, supplemented by their unmarried poorer relatives. It is important to note that this variety of household forms is not a simple persistence of the longhouse tradition; longhouses were composed of independent nuclear family units of production and consumption. The creation of multiple family households, which

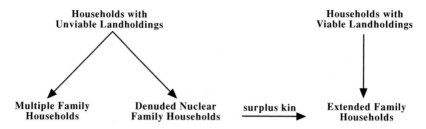

FIGURE 2 HOUSEHOLD DEVELOPMENT CYCLES.

was once a support for parents no longer able to farm swiddens, is now a parental support for married children with no wet rice fields of their own. Fosterage is a new phenomenon, which increases the amount of domestic labor available to larger landholders which can be exploited as a "free" input in agricultural production or petty trade. Differentiation, once introduced, shapes the trajectory of further household creation and perpetuates this dual household development cycle.

Viable households remain viable only by exploiting the "free" productive inputs available to them through practical kinship relations. With their larger land holdings, they need to mobilize a larger work force for planting, weeding and harvesting. This was initially managed through a combination of labor exchanges and domestic labor. Unviable households which did not require reciprocal exchanges of labor began demanding wages in the 1960s. Faced with the high cost of labor, viable households responded in two ways. The first response was to expand their sources of "free" domestic or kin labor. Participation in labor exchanges is only economically practical, I was told, if the household has a surplus of domestic labor. The members of this surplus labor pool, mainly older children and women with small children, are rarely able to obtain wage work despite being capable and productive workers. They can, however, be used in labor exchanges and wherever "free" labor can reduce the costs of either production or consumption. Viable households with a need for "free" domestic labor foster large numbers of their poorer kin. On the one hand, richer householders explicitly stated they were moved by the pitiable (*kasihan*) condition of their poor relatives. On the other hand, they generally did not transfer scarce economic resources to them if they could use them themselves; rather, they provided "subsistence insurance" in exchange for the "free" labor of another domestic dependent. It was thus in the interests of these viable peasant households to buy into and expound the official kinship ideology being perpetuated by the state; kinship was cast as a transactional order antithetical to the market dominated by Chinese merchants.

This dependence on practical kinship as a source of labor has become more important as other aspects of the means of production have been commodified. The introduction of green revolution inputs such as tractors and fertilizers, while increasing yields, has none-the-less embedded agriculture more firmly in the market. Unable to afford these necessary inputs, farmers are forced to either cease production and rent their land out, or to borrow the money from local ethnic-Chinese merchants. Since the 1970s, these peasant farmers have been forced by the state-directed "development" of agriculture to engage in yet another economically irrational practice, the sale of their unharvested crops at cut-rate prices to pay for the needed inputs. Known as *mengijon*, goods like fertilizer are bought on credit and paid for in rice at the rice mills owned by the same shopkeeper. The capital these shopkeepers accumulated through their usurious loans has been reinvested in mechanical threshers, tractors, rice mills and stock for their stores. Through the fees which must now be paid for the use of these inputs, the merchants collect up to half of the area's rice production, nullifying the benefits of double cropping. To Pamona peasants work twice as long to obtain the same net result for their domestic budget. It is the terms under which they have been "developed" and incorporated in the market which create and perpetuate their need to exploit kin labor.

The moral economy of kinship typical of the highlands thus stands at odds with the ideal type against which it is usually evaluated. Differentiation in land holdings has not created class polarization among To Pamona peasants, despite the exploitation of kin labor which takes place. The decision to use kin labor (and the household's investment in posintuwu exchange by which the kinship sphere is defined) is the result of careful calculation, despite the injunctions of the official kinship ideology against "calculating costs and benefits". This state-supported kinship ideology relegates the To Pamona to the position of economically irrational communal peasants who are culturally responsible for their own underdevelopment; yet such negative characterizations have to be accepted as the price of their dependence on "free" kin labor. Their structural position in the local economy makes the "rational" alternative "uneconomic".

ECONOMIC "IDEAL TYPES" IN HISTORICAL CONTEXT

Moral economy models are inherently dualistic, phrased in terms of polar oppositions. These models are based upon an untheorized natural economy in which moral imperatives are viewed as environmentally determined, unlike the class relations predominating in the market economy. Rebelliousness, or the lack of it, is thus tied to subsistence levels and not to social relations of production. Since this morality is assumed to be "natural", it is never examined as an "invented tradition" subject to ideological manipulation within rural capitalist relations of power and class. Starting from Parry and Bloch's position (1989), which rejects any inherent natural dualism of gift and market exchanges, I have rooted the "communal", kinship-based moral economy of the To Pamona in a state-supported traditionalist discourse which is used to "blame the victims". That is, this economically-necessary moral economy which guarantees marginal peasants the factors of production, is utilized in the development discourse of the state to demonstrate that the ethnic rationality of the To Pamona makes them unfit for participation in the wider market economy. It thus makes "culture", rooted in the past, the cause of underdevelopment, and not the development policies pursued by the state. By contrasting "practical" kinship with this "official" kinship discourse, I have attempted to show how the moral economy of the To Pamona is the product of careful calculation which runs counter to the explicit ideology said to regulate their behavior.

It is important to underscore that the moral economy model was developed within the wider context of the Vietnam War and the pressing question of "peasant resistance". Scott's *The Moral Economy of the Peasant* (1976) is subtitled "Rebellion and Subsistence in Southeast Asia". His approach was innovative in its appreciation that active rebellion may not be rooted in the absolute amounts of surplus production extracted from the peasantry, so much as the amount of their production that remained with them. Citing Tawney, he prefaced his book with the metaphor of the "rural population" being like "that of a man standing permanently up to the neck in water, so that even a ripple is sufficient to drown him" (Scott 1976: 1). Scott argues for the existence of a "subsistence ethic" which amounts to a social guarantee of survival. Given their marginal situation, peasants view only those exactions which transgress this right to survival as unjust, and as cause for rebellion. Peasant resistance is thus rooted in a traditional "moral economy" characterized by patron-client bonds, communal land-holding and other kinds of risk-sharing and social welfare institutions which underwrite this guarantee. Peasant resistance occurs when these "traditional" institutions come under attack through the introduction of market mechanisms which encourage individual profit seeking over prior socially grounded welfare rights.

The explicit conjunction of a "moral economy" with resistance to the market (now equated with armed rebellion) polarized the debate in ways which did not seriously question the presuppositions of the moral economy model. Popkin's *The Rational Peasant* (1979) was the greatest challenge to anthropologically rooted conceptions of distinctive "peasant" behaviors. Drawing on neoclassical economic theory, he argued that peasants are less tradition-bound, and more responsive to innovation than is generally credited. They are but a particular form of a universal "economic man" whose decision making is shaped by a materialistic cost-benefit calculus. Popkin, in effect, refutes an empirically grounded ethnographic tradition with deductive economic models in an effort to challenge the specific linkage of peasant resistance with the introduction of the market. He highlights the benefits of these markets for peasants. But, by denying the existence of a moral economy, he simply hardens the opposition inherent in Scott's analysis which views market and moral economies in antithetical, rather than mutualistic terms.

Popkin's assumptions lead him to ignore the historical links between the development of patron-client bonds, communal land-holding and other kinds of risk-sharing and social welfare institutions said to characterize the moral economy, and the colonial conditions which shaped the ways in which peasants gained access to both land and markets (cf. Bailey 1991: 137). This chapter has carefully documented these processes among the To Pamona. Moral economies are said to be both "traditional" (i.e. culturally specific), yet also a universal and "timeless" feature of peasantries throughout the world. Although claiming a phenomenological appreciation of the difficulties "Homo Peasanticus" faces in obtaining a livelihood, Scott's model (like the state modernization model) remains thoroughly entrenched in abstract concepts of "natural" economy (Roseberry 1989: 218).

The polarization of market and moral economies is a product of these presuppositions. Scott creates an abstract, "natural" rationality for peasant economic behavior mirroring that of market rationality, which makes it then possible to trace a unilinear transformation from a "traditional" peasantry to capitalism, and place the whole within the familiar "transition to capitalism" literature (cf. Roseberry 1989: 216). Such a treatment ignores the possibility of alternate forms of capitalist transformation (Roseberry 1989: 58-9, 222). Further, its characterization of "traditional" peasants is limited; they are said to be shaped and molded by environmental conditions, not relations of production. The past is thus overly homogenized and romanticized, and pre-capitalist forms of exploitation ignored.

Relations of production only become pertinent with the introduction of market relations, which are viewed as an exterior imposition on the "closed corporate peasant community".

It is possible to discern this same pattern in the modernization paradigm of Clifford Geertz; a model Scott cites with approval (1976: 36). The polarization of ideal types of market and moral economies is implicit in the dual economy model made popular by Geertz in *Agricultural Involution* (1963). Importantly, this work is subtitled "the processes of ecological change in Indonesia", again pointing to the role of environmental determinants rather than relations of production in shaping "traditional" peasant economic systems. Drawing on the theories of Boeke and Furnival, Geertz hypothesized that Javanese peasant agriculture was molded by the "Culture System" of export crop exactions in the early to mid-nineteenth century: the Culture System "stabilized and accentuated the dual economy pattern of a capital-intensive Western sector and a labor-intensive Eastern one by rapidly developing the first and rigorously stereotyping the second" (1963: 53). This bald assertion that the "Eastern" economy was "rigorously stereotyped" is essentially correct, though not in the sense Geertz intended it. Geertz's interpretation of Javanese peasant economic rationality as static, coddled within a closed corporate community isolated from an externally introduced, capital-intensive colonial economy, has increasingly been challenged. Since Geertz first developed the model, his critics have questioned the persistence of this moral economy, and emphasized the capitalist transformation of the peasantry. They have pushed

> the temporal boundaries of the 'traditional' economy further and further back in history. Indeed the break with Geertz is less radical than his critics would have us believe, since almost all those attempting to demonstrate the demise of involution appear to posit an historical period in which Geertz's model, or something quite close to it, is said to be pertinent (Kahn 1985: 78).

What these critics have actually shown is that a pure moral economy matching its inductively reasoned characteristics has never existed independent of a capitalist market in Java.

It is thus possible to challenge dualistic, oppositional models on both historical and theoretical grounds. The critics of Geertz noted above have empirically questioned the persistence of economic duality without challenging the underlying assumption of an historically prior "natural" economy fitting the criteria of the moral economy model. Breman (1988), in contrast, has argued that the closed corporate communities of Java are a colonial creation. Combining both insights, a pattern in which state-sponsored "development" breeds an under-capitalized peasant sector dependent upon the "free" inputs provided by a "moral economy" becomes widely evident in Indonesia. This is clearly seen in the case of the To Pamona, who have been marginalized by development policies which pass the bulk of economic benefits of double cropping to an intermediary merchant class. The failure of the To Pamona to take advantage of the development "opportunities" offered them is viewed as evidence of their "communalism" and "traditionalism". This tradition has, of course, been selectively supported by the state, which alternately uses it to explain the continued underdevelopment of the ethnic group. Locals, in turn, are unwilling to criticize this official discursive traditionalism because of their own dependence on the practical kinship relations which flow from it.

Even the most successful peasant farmers of Tentena live near the margins, rarely having more than a hectare of wet-rice fields. Standing up to their necks in water, it takes only a ripple to drown them. In their situation, however, it is important to note that the provision of "subsistence insurance" takes place within the context of the differentiation of land holdings; faced with disaster, they are likely to lose both their land and their dependents. The provisioning of subsistence insurance in Tentena does not guarantee a sufficiency of productive resources for an unviable household. Rather, when in need, kin are transferred from unviable to wealthier households. These richer patrons respond in this way because they are living close to the margins themselves, and hence depend upon the exploitation of "free" labor inputs for marginal improvements in productivity. They generally cannot afford to pay market prices for agricultural inputs.

Keeping this labor "free" has involved the ideological elaboration of a kinship sphere within which it is said, rational economic calculation should not enter. Farmers actually make closely-calculated choices based on access to and availability of resources in a commodified market and their consequent investment in a non-market moral economy is often the result. As my discussion of posintuwu exchange has shown, rational calculation is purposely *not* applied to kinship because "it's not economical". That is, attempts to maximize benefits from this kinship sphere through rational economic calculation only begets further calculation, undermining the benefits they derive from "free" kin labor. At an ideological level then, market and moral economies seem in opposition, the one dominated by the ethnic Chinese merchants, the other by the To Pamona peasants. Yet at a practical level this dualism is undermined by the constant calculation of the costs and benefits of market or moral economy strategies of production and reproduction by the To Pamona themselves.

I would like to return then, to the larger theme with which I began. To the Dutch colonial government, and now the Indonesian state, this moral economy is a hindrance to "development". "The persistence of tradition" appears to be obstructing the creation of an economically rational peasantry, making the To Pamona themselves the source of their underdevelopment. In contrast, I have tried to emphasize the economic rationality of the To Pamona. Production has been commodified, and the choices of these farmers are consonant with their resources and subject to rigorous calculation. It is precisely this calculation which necessitates their investment in a non-market, moral economy. This moral economy is a response to differentiation and commodification, not some natural antecedent rooted in subsistence production. In fact, it is the moral economy which makes production for the market possible at all, by expanding the amount of available domestic labor that can be exploited for only marginal productive gains which can be sold, or which reduce household expenses. At a second level, the state has selectively supported elements of "tradition" as a means of defusing class antagonisms, leading to a relatively pacific accommodation to change. The creation of the "transactional order" of posintuwu exchange with state support makes the practical strategies of the moral economy possible, but also insulates state development policies from criticism. Since To Pamona farmers are dependent upon moral economy strategies for survival, they cannot easily refute state arguments that their own "traditions" are at the root of their own poverty.

ACKNOWLEDGEMENTS

The research on which this article is based was sponsored by the *Puslitbang Kemasyarakatan*, Indonesian Academy of Sciences, Jakarta, and funded by the Social Science and Research Council of Canada. I would like to thank Profs. Shuichi Nagata, Gavin Smith, Joel Kahn, Tania Li, and an unnamed reviewer for their helpful comments on early versions of this paper.

NOTES

[1] In Central Sulawesi Li (1997) notes that the highland Lauje "far from being 'penetrated' by capitalism [have] participated in the transformation process" (1997:141). She describes the relative absence of "community", collective action or communal ownership of resources, except under government tutelage and shows how "tradition" is being selectively reworked within state agendas, with the complicity of those local leaders with the symbolic capital to convert "traditional" land rights into individual property.

References

Adriani, N. and Kruyt, A.C., 1950, *De Bare'e Sprekende Toradjas van Midden Celebes (de Oost Toradjas)*, 3 vols., second edition. Verhandelingen der Koninklijke Nederlandse Akademie van Wetenschappen, Afdeling Letterkunde, new series, Vol. LIV. Amsterdam: Noord Hollandsche Uitgevers Maatschappij.

Attwood, D.W., 1997, "The Invisible Peasant" in *Economic Analysis Beyond the Local System*, Monographs in Economic Anthropology, No. 13, edited by Richard Blanton, *et al.*, pp. 147-170. Langham: University Press of America.

Bailey, C., 1991, "Class Differentiation and Erosion of a Moral Economy in Rural Malaysia" *Research in Economic Anthropology*, 13, 119-142.

Bourdieu, Pierre, 1977, *Outline of a Theory of Practice*, Richard Nice trans. Cambridge: Cambridge University Press.

Breman, J., 1988, *The Shattered Image: Construction and Deconstruction of the Village in Colonial Asia*. Dordrecht: Foris Publications.

Carsten, J., 1989, "Cooking Money: gender and the symbolic transformation of means of exchange in a Malay fishing community" in *Money and the Morality of Exchange*, edited by J. Parry and M. Bloch, pp. 117-141. Cambridge, Cambridge University Press.

Geertz, C., 1963, *Agricultural Involution: The Process of Ecological Change in Indonesia*. Berkeley, University of California Press.

Kahn, Joel, 1985, "Indonesia after the demise of Involution: critique of a debate". *Critique of Anthropology*, 5(1), 69-96.

1993, *Constituting the Minangkabau: Peasants, Culture and Modernity in Colonial Indonesia*. Providence: Berg.

Kruyt, A.C., 1924, "De beteekenis van den natten Rijstbouw voor de Possoers". *Koloniale Studien*, 8(2), 33-53.

Li, T.M., 1997, "Producing Agrarian Transformation at the Indonesian Periphery" in *Economic Analysis Beyond the Local System*, Monographs in Economic Anthropology, No. 13, edited by Richard Blanton, *et al.*, pp. 125-146. Langham: University Press of America.

Parry, J., 1986, "The *Gift*, the Indian Gift and the 'Indian Gift'". *Man* (N.S.), 21, 453-73.

Parry, J. and M. Bloch, 1989, "Introduction: Money and the morality of exchange" in *Money and the Morality of Exchange*, edited by J. Parry and M. Bloch, pp. 1-32. Cambridge: Cambridge University Press.

Peletz, M.G., 1983, "Moral and Political Economies in Rural Southeast Asia: A Review Article". *Comparative Studies in Society and History*, 25(4), 731-39.

Popkin, S.L., 1979, *The Rational Peasant: The Political Economy of Rural Society in Vietnam*. Berkeley: University of California Press.

Roseberry, W., 1989, *Anthropologies and Histories: Essays in Culture, History and Political Economy*. New Brunswick, NJ: Rutgers University Press.

Schrauwers, A., 1995a, *In Whose Image? Religious Rationalization and the Ethnic Identity of the To Pamona of Central Sulawesi*. Ph.D. thesis, Department of Anthropology, University of Toronto.

1995b, "The Household and Shared Poverty in the Highlands of Central Sulawesi". *Journal of the Royal Anthropological Institute, incorporating Man*, n.s. 1(2), 337-57.

Scott, J.C., 1976, *The Moral Economy of the Peasant: Rebellion and Subsistence in Southeast Asia*. New Haven: Yale University Press.

Wolf, E., 1966, *Peasants*. Englewood Cliffs, NJ: Prentice-Hall.

Wong, D., 1991, "Kinship and the Domestic Development Cycle in Kedah Village, Malaysia" in *Cognation and Social Organization in Southeast Asia*, edited by Husken, Frans and Jeremy Kemp, pp. 193-201. Leiden: KITLV Press.

Chapter 5

FOREST KNOWLEDGE, FOREST TRANSFORMATION: POLITICAL CONTINGENCY, HISTORICAL ECOLOGY AND THE RENEGOTIATION OF NATURE IN CENTRAL SERAM*

Roy Ellen

Since the mid nineteen-eighties Nuaulu living on the edge of lowland rainforest[1] in central Seram, Maluku, have become increasingly active in countering threats to their traditional resource base. This latter has been dramatically eroded, mainly through government-sponsored settlement and logging. Nuaulu have successfully defended land claims in the courts, there have been violent incidents at a nearby transmigration area leading to their imprisonment, and in their representations to outsiders they have become articulate about the damage done to their environment. However, Nuaulu have a long history of interaction with 'the outside world', of forest modification and participation in the market. They were politically engaged as early as the Dutch wars of the late seventeenth century and have been indirectly, and, more recently, directly subject to the oscillations and economic fall-out of the spice trade ever since. The seventies and eighties of the present century have seen the expansion of cash-cropping, together with accelerated rates of land sale and forest extraction.

I shall argue in this chapter that as different material and social changes take place, so Nuaulu have renegotiated their conceptual relationship with forest. In particular, I seek to ask why, given an apparent historic readiness to accept environmental change, they have now adopted a rhetoric which we would recognise as 'environmentalist'. I claim that part of the explanation is that older, local forms of knowledge which underpin subsistence strategies are qualitatively different from knowledge of macro-level processes — 'environmental consciousness' in the abstract — which only comes with a widening of political and ecological horizons.

THE NUAULU IN THE WORLD SYSTEM

The patterns of ecological change which have accompanied Nuaulu interaction with the rainforest cannot be understood properly except in relation to the history of contact (direct and indirect) between the forest peoples of Seram and various groups of outsiders: the rulers and subjects of various traditional coastal polities; the Dutch East India Company, its heirs and successors; various agencies of the colonial Dutch government, and thereafter of the government of an independent Indonesia (local district officers, police, military, and the personnel of assorted provincial level departments); and finally traders and settlers of diverse ethnic origins, but predominantly Chinese, Butonese and Ambonese.

The details of the early phase of the movement of biological species in and out of Seram (Ellen 1993b) is not relevant to the specific argument put forward in this chapter, but that it happened is a part of the general background picture. Thus, the circulation of valuables, upon which the reproduction of Nuaulu social structure became effectively dependent over several hundreds of years (Ellen 1988a) was based on articles traded in from the Asian mainland (porcelain from China and elsewhere, and cloth from India) and from other parts of the archipelago (including textiles from Timor and Java); and what we know of the dynamics of the regional Moluccan system suggests contact which goes back much further than this, and which must have involved the export of forest products.

The most important single factor affecting Moluccan forests during the early period was the rise in the international demand for spices, which by the early sixteenth century had led to the spread of production from the northern to the central Moluccan islands. Expansion and fluctuation in growing clove in particular from this time onwards (Ellen 1985, 1987: 39–41) played a crucial role — both directly and indirectly — in the lives of inland and coastal peoples alike. Although there is no evidence that the Nuaulu planted cloves or collected wild cloves for sale until the twentieth century, they did have an identifiable role in relations with politically significant trading polities and Europeans as early as the Dutch wars of the late seventeenth century, as we know from the VOC archives and from the *Landbeschrijving* of Rumphius (Ellen 1988b: 118, 132n2). We have a remarkably clear idea of the general location of their settlements in the mountains of central Seram from this time to the end of the nineteenth century through oral histories, corroborated by surface archaeology, botanical evidence and eighteenth century maps (Ellen 1978, Ellen 1993a: frontispiece). By the end of the nineteenth century, most Nuaulu clans had relocated around Sepa on the south coast [map 1], largely as a result of Dutch pressure, though they have continued an essentially highland, interior-oriented way of life down to the present, relying on historic zones of extraction. In the eyes of official agents of the present Indonesian government, other coastal peoples, and in terms of their own self-definition, they have never ceased being uplanders and people of the forests.

During the twentieth century there has been renewed clearance, on Seram as a whole, for clove, nutmeg and other tree crops, such as coconut,

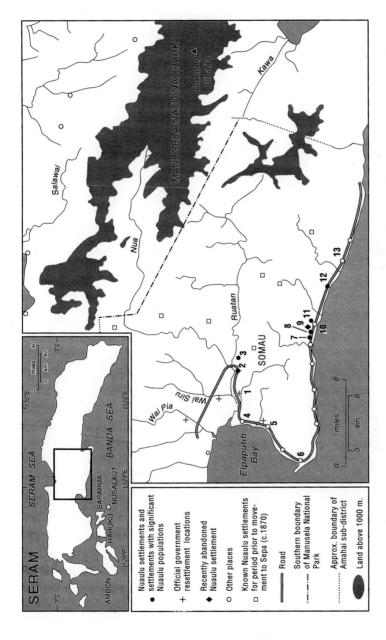

MAP 1 THE EASTERN PART OF THE AMAHAI SUB-DISTRICT, SERAM, SHOWING HISTORICAL, RECENT AND PRESENT NUAULU SETTLEMENTS, AND OTHER PLACES MENTIONED IN THE TEXT (AS OF 1994). THE NUMBERED LOCATIONS ARE AS FOLLOWS (NUAULU SETTLEMENTS IN ITALICS): 1. WAI RUATAN TRANSMIGRATION ZONE (KILO 5 — KILO 11), 2. *SIMALOUW* (KILO 9), 3. *TAHENA UKUNA* (KILO 12), 4. MAKARIKI, 5. MASOHI, 6. AMAHAI, 7. *BUNARA*, 8. *WATANE*, 9. AIHISURU, 10. SEPA, 11. *HAHUWALAN*, 12. *ROHUA*, 13. TAMILOUW.

cacao and coffee. The seventies and eighties have seen the expansion of market participation and cash-cropping (of clove, nutmeg and copra in particular), the planting of fast-growing pulp trees, together with accelerated rates of land sale and forest extraction. This has mainly taken place through logging and in-migration, first spontaneous and then official. Forest is being destroyed through unplanned slash and burn cultivation by non-indigenous pioneer settlers, and by the expansion of transmigration settlements into surrounding areas. There is no doubt that rapid forest clearance of this kind is damaging, and that long-standing swiddening practices which modify the forest, increase its genetic diversity and usefulness, and permit extraction on a sustainable basis, are being eroded by technological innovation, population pressure and market forces. Local populations are encouraged by government to deliberately cut mature forest for cash crops, and commercial estate plantations are spreading widely. Logging is a particularly serious threat in the area where the Manusela National Park meets the Samal transmigration zone. Here and elsewhere so-called 'selective' logging of *Shorea selanica* has led to water shortages, serious gully erosion and soil compaction. It has undermined existing forest ecology, resulting in more open canopy structures, *Macaranga* dominance, a greater proportion of dead wood, and herbaceous and *Imperata* invasions. In terms of fauna, there has been an obvious reduction in game animals. These effects have been systematically inventoried in the Wahai area by Ian Darwin Edwards (1993: 9, 11), but it is instructive to compare his description with that provided in the Nuaulu text discussed later, and which is appended to this chapter. However, it has been transmigration and its various knock-on effects which — more than anything else — have been responsible for forest transformation

NUAULU RESPONSES TO INTRUSION SINCE 1970

The phasing and character of indigenous responses to the kinds of change I have highlighted have depended very much on local perceptions of government policy and on the ways in which law and policy are interpreted by officials and translated into action. It is now widely acknowledged, for example, that the Basic Agrarian Law of 1960 and the Basic Forestry Law of 1967 are fundamentally contradictory and overlapping, and are viewed differently by different government departments and in different situations. Sometimes they are used to defend the rights of indigenous peoples, but more often they override *adat*, legitimating the confiscation of land, and criminalising those local inhabitants who insist on asserting long-established rights of use (Colchester 1993: 75; Hurst

1990; MacAndrews 1986; Moniaga 1991; SKEPHI 1992; SKEPHI and Kiddell-Monroe 1993; Zerner 1990). Where there are doubts, national interest is invariably placed above local interests (Hardjono 1991: 9). Up until recently, Nuaulu have been beneficiaries of an, on the whole, advantageous interpretation of the law (Ellen 1993c), though as I go on to explain, this may now be changing.

During the period covered by my own fieldwork, the Nuaulu population has continued to grow dramatically: from 496 in 1971 to an estimated 1256 in 1990. This has led to greater pressure on existing land, intensified by competition along the south Seram littoral with people from traditional non-Nuaulu villages, and due to unplanned immigration, mainly of Butonese. Growth along the south coast has been facilitated by the extension of a metalled road during the early eighties. At about the same time the government began to establish transmigration settlements along the Ruatan valley (map 1).

The overtures by provincial government authorities to the Nuaulu with respect to these developments, were, at least initially, benign and paternalistic. In part they have been guided by the special administrative status of the Nuaulu as *masyarakat terasing* (Koentjaraningrat 1993: 9–16; Persoon 1994: 65–7). Thus, local government officers (*camat, bupati*) have recognised uncut forest in the vicinity of transmigration settlements as 'belonging' to the Nuaulu, following the widely-held view of many non-Nuaulu inhabitants of south Seram. They then encouraged them to move into one of the new transmigration zone settlements along the Ruatan river, at Simalouw (map 1), an area which abutted sago swamps long claimed and utilised by Nuaulu. Although by 1990 only the villages of Watane and Aihisuru had moved permanently from their earlier locations on the south coast (about a quarter to one-third of all Nuaulu households), many Nuaulu established temporary dwellings, used the improved transport facilities to reach ancestral sago areas, and began to cut land for cash crop plantations. Moreover, two clans (Matoke-hanaie and Sonawe-ainakahata) moved even further inland, out of the original transmigration zone altogether to a place called Tahena Ukuna. Many Nuaulu saw these shifts as a return to traditional land, and for outsiders it confirmed Nuaulu status as upland forest peoples rather than lowland and coastal. Although Nuaulu had been located around the Muslim coastal domain of Sepa for the best part of one hundred years, and subject to the tutelage of its Raja, their self-image and the image of them held by non-Nuaulu, had never been otherwise. Moreover, implicit government recognition of Nuaulu preferential rights to over one-and-a-half thousand square kilometres enabled them to sell land in the Ruatan area to other incomers. This unusually positive approach was reflected in a successfully defended land claim in the courts at Masohi, the capital of Kabupaten Maluku Tengah.

The practical consequences of all this were alleviation of the growing pressure on Nuaulu land generally, and an opportunity to sell land along the more crowded south coast, most of which was sold to the inhabitants of Sepa itself and to incoming Butonese. This latter land, mainly old garden land and secondary forest, was a mixture of land gifted by the Raja of Sepa since the late nineteenth century, and land further inland which had always been regarded as Nuaulu. As I have argued elsewhere (Ellen 1993c), altogether, this created a rarely reported situation whereby an indigenous forest people appeared to be endorsing further forest destruction (both in the interior and along the south coast) by themselves and by others, for short-term gain.

Nuaulu cash incomes certainly increased through sale of land and trade with immigrants. Moreover, the practices which accompanied this were not dramatically contrary to any locally-asserted principles of indigenous ecological wisdom. However, there has recently been increased conflict with other autochthonous villages over rights to land, disenchantment with the effects of logging, and, since 1990, serious conflict with settlers resulting in convictions for the murders of two Saparuan migrants being brought against three residents of Rohua. This incident was widely reported in the local press, who made much of the manner of death (decapitation), and of removal of the heads back to the village and their burial near a *rumah adat*. The episode has understandably been viewed by some government officials and other observers as a reversion to head hunting, or confirmation that it had never ceased, though the protagonists themselves strenuously deny such interpretations. Whatever the case, this narrative amply highlights the fundamental ambiguity in the concept masyarakat terasing, seemingly indicating both the vulnerability of a people so labelled, their need of special protection and advancement by the state, as well as their primitive threatening character, which the state must subject and change. Either way, Nuaulu are frequently viewed as prime candidates for *pembangunan* (development) in its moral and ideological sense (Grzimek 1991: 263–83). Moreover, recent events reinforce a particularly pejorative local Ambonese stereotype of interior peoples as *Alifuru*, and have made it easier for the government to explicitly expropriate territory when the occasion arises.

HOW THE NUAULU HAVE CHANGED THEIR ENVIRONMENT

Conventional Western conceptions of nature are usually of some unaltered other, of wilderness; and conventional views of traditional peoples living on forest margins or biotopes, of tribes benignly extracting from an essentially pristine ecosystem. Such a view is, of course, now wholly unacceptable and there is mounting evidence of the ways in which humans dependent on forest actively change it. Much tropical lowland rainforest — in Indonesia as elsewhere — is the product of many generations of selective human interaction and modification (deliberate and inadvertent), optimising its usefulness and enhancing biodiversity. The outcome is a co-evolutionary process to which human populations are crucial. Indeed, particular patterns of forest extraction and modification are often seen as integral to its sustainable future. For some authorities, the evidence for intentional rather than serendipitous human influence is so compelling as to invite the description of 'managed' forest (Clay 1988; Schmink, Redford and Padoch 1992: 7–8).

The empirical work supporting these claims comes mainly from the Amazon (e.g. Balée 1993, 1994; Posey 1988; Prance, Balée, Boom and Carneiro 1987); but there is emerging evidence that it also applies to large parts of Malaysia and the Western Indonesian archipelago (Aumeeruddy and Bakels 1994; Dove 1983; Maloney 1993; Peluso and Padoch 1996; Rambo 1979). My own work, supported by recent botanical research, suggests that it is no less true for the forests of Seram, which have long been a focus of subsistence extraction, and where human agency has had decisive consequences for ecology. This has been largely through the long-term impact of small-scale forest-fallow swiddening and the extraction of palm sago over many hundreds of years (Ellen 1988a), but also through the introduction and hunting of deer, selective logging and collection for exchange in more recent centuries (Ellen 1985: 563). Since sago is a frequent reason for venturing into forest beyond the limits of the most distant gardens, and since it illustrates so well the kind of co-evolutionary relationship I have just been discussing, it is helpful to say a bit more about it here.

Sago (*Metroxylon sagu*) is currently extracted by Nuaulu both from extensive swamp forest reserves along major rivers and from planted groves much nearer to settlements. Certain swamp forest zones, such as at Somau, appear to have been continuously important for several hundred years, though smaller patches in the vicinity of the south coast villages may be the artifacts of more recent settlement histories. Smaller inland sago groves have been abandoned since coming to the coast, or are extracted from only occasionally.

Nuaulu manipulate vegetative reproduction of sago by replanting and protecting suckers from recently cut palms, selecting suckers from some palms rather than others, and transferring root stocks to village groves. The result is an interchange of genetic material between cultivated and 'non-cultivated' areas, even though there is no particular evidence of domestication through selective planting of seeds. Although most reproduction of sago palms in the lowland riverine forest areas of Seram occurs quite independently of human interference, in certain areas human involvement is highly significant, and the contemporary phenotypes of Southeast Asian sago palms are best seen as the outcome of a long-term process of human-plant interaction. Indeed, the historic spread of *Metroxylon* from its assumed centres of dispersal in New Guinea or Maluku suggest very strongly anthropogenic factors. Ecologically, the heavy reliance placed by Nuaulu and other indigenous peoples of Seram on sago has, over some hundreds of years, reduced the necessity to cut forest for swiddens. This has an important bearing on Nuaulu changing conceptions of their environment, as we shall see.

The distribution of many other useful trees throughout the lowland forests of Seram reflects patterns of human modification, and serves as a convenient botanical indicator of settlement histories. Many are certain or probable domesticates and semi-domesticates. One of the most culturally salient of these is the *kenari* (*Canarium indicum = commune*). This is found so widely in lowland areas, and in particular configurations, that its distribution must almost certainly be explained as a consequence of human interference, both motivated and inadvertent (Ian Edwards, personal communication). Kenari provides nuts rich in protein and essential oils, which are an important ingredient in local diet, but which for the Nuaulu also have a salient symbolic role, the precise character of which I shall return to later[2].

Nuaulu practices of swidden cultivation and movement have, over several centuries, altered the character of forest vegetation in measurable ways: increasing the proportion of useful species, increasing the numbers of stands of particular useful species, decreasing the proportion of easily-extracted timber trees against those which are resistant to extraction, creating patches of culturally productive forest in more accessible areas, and creating dense groves of fruit trees in old village sites. Many of the trees nowadays found in areas otherwise not obviously modified by humans represent species introduced historically, and even prehistorically, for their useful timber, fruits, and other properties (Ellen 1985). Indeed, approximately 78% of the 319 or more forest trees identified by the

Nuaulu have particular human uses which make them potentially subject to manipulation through forms of protection and selective extraction. No wonder, then, that the distinctions between mature forest, different kinds and degrees of secondary regrowth and grove land are often difficult to establish. Although the contribution of non-agricultural activities, narrowly-defined, to overall Nuaulu energy expenditure and production is not to be under-estimated, and by comparison with other Indonesian swiddening peoples is rather high, my earlier contrast (Ellen 1975) between 'domesticated' and 'non-domesticated' resources was, in retrospect, drawn too starkly.

RENEGOTIATING NATURE

My main argument in this chapter is that as different material and social changes have occurred — changes which have accelerated over the last 20 years — so Nuaulu have renegotiated their relationship with forest, and with 'nature' more generally. How people conceptualise nature depends on how they use it, how they transform it, and how, in so doing, they invest knowledge in different parts of it. I have argued in another paper that concepts of nature have underlying pan-human cognitive roots, all people appearing to derive them from imperatives to identify 'things' in their field of perception, situate these in terms of a calculus of self and other, and identify in discrete bits and aggregations essential inner properties (Ellen 1996). However, identifying these commonalities is not to deny that such concepts are everywhere ambiguous, intrinsically moral in character and a *condition* of knowledge (Strathern 1992: 194). Nature is not a *basic* category in the sense specified by Pascal Boyer (1993), and means different — often contradictory — things in different contexts. It is constantly being reworked as people respond to new social and environmental situations (Croll and Parkin: 1992: 16), and provides in the guise of something all-encompassing what I have elsewhere (Ellen 1986: 24) called a 'theory of selective representations'. Ambiguity itself, as Bloch (1974) has pointed out, can be socially useful. In the Nuaulu case there is an evident underlying tension between an oppositional calculus of forest and 'village' or 'house', and a non-oppositional calculus which draws much more on the lived experience of particular strategies of subsistence which unite what we loosely call nature and culture. Such an ambivalent conception of nature is wholly consistent with the difficulties faced in classifying the Nuaulu mode of subsistence according to conventional anthropological criteria (Ellen 1988a).

Before examining how these different concepts and their relative balance might be the outcome of a particular sequence of past events, and before highlighting contemporary patterns of change, it is necessary to sketch out in general terms the substance of the two apparently competing models or orientations. I do so on the basis of ethnographic data acquired by me at various times between 1970 and 1990. Since it is so obviously central, I start with the Nuaulu category of forest.

The Nuaulu use the term *wesie* to refer to forest of most kinds, but the term belies a complex categorical construction. Nuaulu relate to different parts of the forest — indeed to different species — in different ways. This mode of interaction is inimical to a concept of forest as some kind of void or homogeneous entity, and certain parts require different responses and evince different conceptualisations. Some bits of forest are protected, others destroyed without thought. Forest is never experienced as homogeneous, but is much more of a combination (rather than a *mixture*) of different biotopes and patches. As such it well reflects the complex historical ecology which I referred to at the beginning of this paper. With its emphasis on human acculturation, it fits comfortably into a non-oppositional model of the kind we more usually associate with hunting and gathering peoples (Ingold 1996).

On the other hand, the generic term *wesie* exists, and is linked into general symbolic schemes such that it stands for some kind of conceptual exterior, a natural other. In some significant respects it is rather like the received twentieth century English concept of nature. Although subject to degrees of effective control through practical and supernatural mastery, *wesie* is associated with essential qualities of danger and otherness, and opposed to an unmarked category of 'culture', most palpably evident in — but certainly not restricted to — the category *numa*, 'house'. As such it is intricately linked with gender imagery (c.f. Valeri 1990). This forest : house : : nature : culture logic is evident in a whole raft of rituals, and in the symbolic organisation of space. In some ways it is not what we might expect given Nuaulu lived subsistence, with its heavy reliance on extracting forest resources, where gardening is traditionally rudimentary, swiddening practised on a forest-fallow basis, where regenerated growth supplies many 'forest' resources over the longer term, and where — consequently — there is a definite blurring of anthropogenic and other forest.

The two somewhat contradictory models we find with respect to forest are repeated at the level of interactions with specific parts of nature. Thus, Nuaulu are (and have been continuously so for many centuries) primarily vegetative rather than seed propagators, and most of their starchy garden crops are tubers (taro, manioc, yam, *Xanthosoma*). Such agricultural regimes are widely associated in the ethnographic literature with notions of continuity between nature and culture, in contrast to seed propagators who tend to emphasise a sudden transition between nature and culture (Coursey 1978; Haudricourt 1962, 1964). In particular, Nuaulu place great practical and symbolic emphasis on sago palm starch extraction, and as we have seen, this species is ambiguously wild and domesticated. Such a view is reinforced by the highly reliable character of palm starch as a staple, with a stable output subject to little fluctuation, lack of economically significant pests (Flach 1976) and considerable potential as a food reserve. In these ways, not only does sago contrast with grain domesticates, but is superior to tubers such as yams and taro, and is, therefore, an even better symbol of the continuity between nature and culture.

Given that many 'forest' trees show evidence of human manipulation, occur simultaneously in cultivated and uncultivated areas, and provide long-term supplies of particular resources without continuous human attention and susceptibility to hazard, they too reinforce the applicability of the non-oppositional model. However, 'trees' are only homogeneous as a category if we ruthlessly simplify it to some common cognitive morphotype (woody, foliaceous, rigid). Different modes of extraction, use and characteristics involve different relationships with people, different social profiles and potential symbolic values. This often leads to classificatory patterns which appear to cut across conventional logics, and which are almost provocatively ambiguous. I have already indicated that two extremely important sources of food — the sago palm and the kenari tree — are ambiguous in terms of the forest : village (house) logic, and in terms of the unlabelled 'nature/wild' and 'culture/tame' categories of which forest and house are, respectively, the most dominant expression. Both species show evidence of proto-domestication, incipient cultivation, and their distribution is heavily affected by human use, despite the fact that they are for the most part culturally 'of the forest' and reproduce without much human interference. The problem is accentuated by the symbolic complementarity of the two: sago is the everyday starch staple and the product of — almost always — male labour, while kenari is collected for special festive occasions, when it is combined with sago by females to make *maiea* (Ellen and Goward 1984: 32). Thus, in certain contexts sago and kenari are linked together in opposition to products of the garden; in others they are contrasted in terms of an implicit gender distinction. Similarly, in the sphere of interaction with forest animals, I have

(Ellen 1996: 116–118) been able to demonstrate how a single ritual associated with killing (*asumate*) can simultaneously reflect a perspective which stresses the unity of all living things, and one which stresses human opposition through killing (c.f. Wazir-Jahan Karim 1981: 188). Nature, I repeat, is not a basic category in the sense that it has a rooted perceptual salience, but though it may be symbolically deployed in radically different ways, it is still able to convey notions of logical primacy.

In developing a model which will help us understand how social and ecological changes have influenced Nuaulu conceptions and representations of forest and nature, we also need to recognise that in almost every instance this will have been motivated by an alteration in the character and intensity of relationships with non-Nuaulu, and how the Nuaulu deal with this socially. As I have indicated, ecological change has almost always been a consequence of exogenous factors: whether this involves the introduction of new species, outside appropriation of endemic resources or clearance of forest for extraction, or agriculture. But whenever there is an environmental interface of this kind, there is also a cultural and social one. Transfer of new cultigens is not just about the movement of genetic material, but of cultural knowledge as well, knowledge which always carries a social burden. Contact with outsiders, in particular, seldom involves actors operating on equal terms, and the relationship is always mediated by considerations of power and control. For their part, the Nuaulu repeatedly represent changes of all kinds in terms of the interplay of principles of opposition and continuity, complementarity and hierarchy[3], symbolic schemes as opposed to practical experiences, outside influence versus persisting tradition. To show how this might work, we can, I think, provisionally identify three historical periods which are likely to have been associated with somewhat different conceptualisations of the natural world: pre-European contact, the VOC and early colonial period until about 1880, 1880 to 1980, and 1980 to the present.

From what we can reconstruct of pre-European Nuaulu social organisation, clans appear to have occupied separate dispersed settlements and had considerable autonomy, entering into loose alliances only for the purpose of intermittent political negotiation and to manage hostilities with outsiders. Thus, that subsistence placed less stress on gardening than became the case later on was wholly in keeping with what we know of political arrangements. We might, therefore, expect here a concept of nature which focuses much more on the symbolic logic of vegetative propagation and the systematic harvesting of forest trees, and which involves a less oppositional conception of *wesie*. Moving around in forest is not conducive, after all, to developing an enduring opposition with it. Historically, we know gardening on Seram to be very underdeveloped, and even at the present time gardens are relatively unimportant in many areas, while in describing Nuaulu subsistence the distinction between 'gathering' and 'cultivation' is very fuzzy (Ellen 1988a: 117, 119, 123, 126–7). There is no new evidence, as yet, ethnobotanical or archaeological (Stark and Latinus 1992), to suggest that horticulture amongst the native peoples of Seram was once more important than it is now (c.f. Balée 1992), except the general ethnological observation that pioneer migrant Austronesian speakers, their linguistic if not directly genetic precursors, depended on domesticates, including — in all probability — seed cultigens (Blust 1976, Bellwood 1978: 141).

The new embeddedness in the world system which developed from the sixteenth century onwards opened-up new pan-Pacific links, cut-out intermediary connections, and intensified exchange with Oriental, Asiatic and European centres. It also had immediate economic consequences in terms of spice production, and longer term implications for subsistence ecology. With the introduction of maize, manioc, *Xanthosoma* and *Ipomea*, reliable garden yields increased making these cultigens competitive with sago in their reliability and superior in the effort required to harvest them. This appears to have led to a greater dependence on gardens (Ellen 1988a: 123). Almost all the new garden crops were vegetatively propagated tubers, therefore sustaining a pre-existing conceptualisation of reproductive process and its metaphoric transformations; but they were also the harbingers of a longer term process of decentering sago from peoples conceptions of nature. Although sago is still culturally salient for the Nuaulu, amongst many present-day peoples of the central Moluccas sago (an indigenous crop) is nutritionally crucial but widely seen as inferior to (imported) rice. The same crops, because they decreased dependency on sago and other forest resources, encouraged greater emphasis on the symbolic opposition between gardens and forest. Increasing attention to cash-cropping, which both required high yield cultigens to offset the reduction of time and land available for subsistence extraction, and which provided opportunities to purchase — for example — rice, further accentuated this division.

The next major change came when the peripheral areas of Seram were formally drawn into the administrative system of the Dutch East Indies in the eighteen-eighties. From this time onwards environmental and social distinctions which had hitherto been implicit became underscored by administrative fiat. We have seen that from at least the late seventeenth century, the Nuaulu have had a distinct political identity in the eyes of outsiders. They had identifiable leaders, and were drawn into various alliances, always including Sepa. Indeed, this long history of interaction has made Nuaulu ultra-sensitive to questions of identity vis-a-vis other cultural groups, even though that identity has not always been reflected in any degree of permanent political centralisation. Formal incorporation into the Dutch administrative system, however, required that this identity and arrangement of traditional alliances of mutual advantage be regularised (Ellen 1988b: 118–9), both for administrative convenience and to provide the Nuaulu themselves with an effective channel of political communication. It is not therefore surprising that, at the time when the Nuaulu clans were relocating around Sepa, when Sepa was — in Dutch eyes — becoming administratively responsible for Nuaulu *rust, orde en belasting*, there emerges a line of Nuaulu *rajas*. This, in turn, changes the terms of the oppositional relationship between Nuaulu and Sepa into a more hierarchical one. Clans begin to lose some of their autonomy, even though the line of rajas effectively terminated after only a few generations. And ever since, the question of a Nuaulu raja and his possible reinstatement has been an issue which has periodically become the subject of heated debate, most recently at the time of the establishment of the Nuaulu presence at Simalouw. The same necessity for formal mechanisms to communicate with the holders of administrative power in Sepa, Amahai, Masohi, Amboina or Jakarta is reflected in Nuaulu involvement in rituals of the Indonesian state (Ellen 1988b).

Nuaulu movement to the coast meant a shift from a pattern of dispersed clan-hamlets and swiddens to concentrated multi-clan villages with large connected areas of garden land. This, in turn, led to a reconceptualization of the forest : village (house) boundary, contrasting owned land (*wasi*) with unowned forest (*wesie*), and gardens (*nisi*) with uncleared forest (*wesie*); the first distinction juridical, the second technical. The changes in Nuaulu social relations of land use which accompanied this (Ellen 1977, 1993c) — land sale, cash-cropping, individualisation, permanent occupancy — emphasised still further a view of the natural world in which dualistic and contrastive properties predominated, even though sago continued to dominate their lives as their most important source of carbohydrate and as a cultural symbol.

So, it is at least plausible that the apparent contradiction between oppositional and non-oppositional models, the one more concordant with external relations of exchange, the other with internal subsistence experience, is a dialectical function of a particular transitional history. It might also be connected with the historic emphasis on exchange of valuables for forest products (see above, and Ellen 1988a), and the influence and internalisation of Austronesian symbolic schemes otherwise more amenable to seed-cultivation. Whatever the case, the balance is tipping in favour of an emergent, more oppositional, reified concept of 'forest/nature'. Amongst the coastal peoples of Seram (such as the inhabitants of Sepa) the enduring perception of the Nuaulu has been of a forest people — the opposite of themselves. Forest is a much stronger exteriority for coastal Muslims than it has traditionally been for animist Nuaulu, but it is towards this view that the Nuaulu are now progressing. Similarly, the Dutch colonial government, and thereafter the Indonesian government, created forest as a strong official category, establishing bureaucracies to manage it, a component in a wider state administrative division of labour which encouraged implicit linkages between the geographical designation of forest and the social category *masyarakat terasing*.

Moreover, as forest has been reduced in extent, so its representation as some kind of ether in which humans are suspended has been transformed into a much more restricted environmental category, as just one ever-diminishing part of a wider non-afforested dwelling space. Not only does the small size of Moluccan islands make the forest more vulnerable physically, but also, as forest disappears, so it is reconceived as a fundamentally limited, rather than limitless, good. Thus, both material experience of environmental change and the necessity to participate in a state level of discourse are reifying Nuaulu concepts of forest, just as environmental degradation and the ecological movement have done in the West. In order to protect their own lives, Nuaulu find themselves adopting the discourse of officialdom and national politics, responding to agendas dictated by the state. From a history of commitment to environmental change, they have now adopted a rhetoric which we would recognise as broadly 'environmentalist'.

NEW RHETORICS AND RAPID SOCIAL CHANGE

What I have in mind by this new Nuaulu conception of nature and its relation to a more reflexive, globally-situated understanding of their own identity, is well exemplified by two empirical cases: the first is a video-recording (cassette 90-2, 8-3-90) which I was asked to make by the people of Rohua in 1990 and which was prompted by Nuaulu concerns of state non-recognition of their religion; the second is a text recorded and transcribed in 1994 by Rosemary Bolton addressed as a personal appeal to me.

The first — the video recording — consists of three parts, all of which refer to performances which occurred on 8 March. The first is a formal address given by Komisi Soumori (the kepala kampung and most senior secular clan head). It is an impassioned assertion of the legitimacy of Nuaulu core beliefs, showing how many Nuaulu believe their cultural identity to be, quite literally, 'rooted' in land, forest and sago. The spoken words and the visual imagery used (and this would be well understood by the local Nuaulu witnessing the event) evoke — though not explicitly — widely-shared mythologies of origin. All this is unashamedly broadcast to an outside, unseen, audience. What is significant about the event is in part its presentation: it is given in Nuaulu, because to speak of such things in any other language is to deny Nuauluness, but also because Komisi is most comfortable in Nuaulu. But the oratorical style and the physical props — rostrum etc. — indicate the acceptance that discourse should assume formats appropriate to engagement with the state (figure 1), and a notion that it is possible to communicate with an unseen audience, not indirectly through a human mediator, but directly employing an electronic medium to which they have only recently had access. The second part is a short dramatic performance by adolescents about discrimination against Nuaulu customs and religion at school and in the labour market. This is conducted entirely in terms of the kind of performance rhetoric which is, again, associated with government institutions, and which is, appropriately, spoken in Indonesian — the language of the state. Paradoxically, such conventions (and the education through which they are acquired) inevitably result in the further attrition of Nuaulu distinctiveness as perceived by non-Nuaulu, and perhaps the eventual disappearance of certain cultural markers which were once salient. This is, of course, not to rule out the likelihood that Nuaulu 'cultural identity' is anyway in transformation, subject to continual re-negotiation, and might emerge as strong as ever, but in a slightly different guise. The third part of the video-recording is a speech by the *ia onate (kepala) pemuda*, Sonohue Soumori, again in Indonesian, which pulls the various themes together. Such reflections can also be cast in a more traditional idiom, such as the *kepata arariranae* (a ritual verse form associated with male-female tug-of-war) and *kepata Sepa* (a ritual verse form associated with workplace routines and domestic relaxation), though on this occasion they were not.

The transcribed text, the English language version of which is provided here as an appendix, is a rather different kind of document. It was dictated by a long-standing acquaintance to Rosemary Bolton, and is separated in time from the 1990 performance by the harrowing events of 1993 in the

FIGURE 1 THREE FRAMES FROM A VIDEOTAPE RECORDED ON 8 MARCH 1990 AND DISCUSSED IN THIS CHAPTER. KOMISI SOUMORI (*IA ONATE SOUMORI* AND *KEPALA KAMPUNG*) ADDRESSES THE WORLD. HIS PROPS INCLUDE A YOUNG CLOVE SEEDLING AND SAGO PALM. NOTE THE MAKESHIFT ROSTRUM BEHIND WHICH HE STANDS.

Ruatan transmigration area, to which I have already referred. These events are structurally significant in Nuaulu representations of themselves because an attempt to defend legitimate interests resulted in defeat. The rugged independence and assertiveness so typical of the seventies and eighties, and so well exemplified in the 1990 videotaped events, has — it would seem — been replaced by a new quiescence and passivity: 'we are quiet and obeying them' (section 5). From a position in which Nuaulu saw themselves negotiating *with* the Indonesian state, they are now simply citizens *of* that same state. There is an acceptance that events are no longer under their own control, that they can no longer take them or leave them. As it happens, Nuaulu have a history of accommodating certain kinds of pragmatic change. This may explain their cultural survival, when most other groups of tribal animists on Seram have all but disappeared. But Nuaulu now claim not to want anything to do with the outside agents of change: government or logging companies. There is a realisation that the government does not keep its promises (7).

We can also see from this text how it is that the rapidity of environmental change has forced the Nuaulu to redefine their relationship with the natural world, to see connections between microclimatic change, deforestation and erosion, and game depletion; between land clearance, river flow, impacting caused by logging vehicles, and fish depletion. We can see in it how Nuaulu now identify their forest as a whole as a commodity, something which has exchange value, when previously it was inalienable. We find an equation between big trees and profit (5, 6), and governmental prohibition on sale. To begin with, Nuaulu accepted the advantages brought by the lumber companies: vehicles used the tracks and kept them clear, the tracks and trucks facilitated hunting (1,2). We also find recognition that replacement of large stands is in a time scale that is beyond the use of Nuaulu, that sustainable use has been superseded by something which Nuaulu would never seek to sustain (6), that old secondary forest, based on the cutting of patches and individual stands (Ellen 1985) has been replaced by wholesale clearance, which results in quite different patterns of regeneration, including more noxious vegetation (e.g. thorns). And the blame for these changes is placed quite squarely at the feet of logging companies and the state

So, recent Nuaulu reworking of their conceptions and responses to those things which we designate as 'nature', show that the patronage of various government departments, levels of organisation, and types of parastatal agency, as well as official categories, are no less central to an understanding of what is going on at the forest frontier than they are for lowland agrarian processes (c.f. Hart 1989: 31). 'Bringing the state into the analysis....entails understanding how power struggles at different levels of society are connected with one another and related to access to and control over resources' (Hart 1989: 48). As the forest frontier reproduces the inequalities of the wider state and its economically dominant groups, and as short-term production for use arises and is sustained by production for exchange (Gudeman 1988: 216), as Nuaulu move from being semi-independent 'tribesmen', relying on sago and non-domesticated forest resources, to being dependent peasant farmers, increasingly reliant on introduced cultigens and cash crops, so their conceptions of nature reflect this. There is, in an important sense, an ecological, economic and conceptual continuity between forest modification and farming, and redefining forest extraction as a kind of farming may help us appreciate its similarities with the agrarian process.

In the Nuaulu case, intensification of subsistence agriculture, cash-cropping, forest extraction, commercial logging and transmigration combine to threaten an existing relationship with the forest. But Nuaulu attitudes have always been tactical, depending on their perceived material interests, and it is therefore not surprising that their conceptualisations of nature should mirror this. Their initial response to forest destruction and consequent land settlement reflected perceived advantages in terms of a traditional model of forest interaction, based on *implicit* notions of sustainability of reproductive cycles of tree growth and animal populations. When this logic failed, complacency was replaced by uncertainty and bewilderment, eventually translating into hostility and decisive actions to defend their subsistence interests. Punitive actions taken by the state in response to this have engendered further uncertainty and bewilderment.

CONCLUSION

What I have tried to demonstrate in this chapter is that there is a connection between shifting Nuaulu constructions and representations of nature (particularly of environmental change), their social identity and the way they interact with the outside world.

There is nothing intrinsically problematic about environmental change for the Nuaulu. As we have seen, their cultural history is full of it. There is no overarching 'ecocosmology' or 'cosmovision' which rules it to be culturally illegitimate. Indeed, during the early phase of transmigration and logging in the eighties it was regarded wholly positively. What we need to recognise, however, is that there are different kinds of environmental change. The crucial distinctions here are between change which you can control, and change which is outside your control (and more specifically, is controlled by outsiders); and between change which is readily recognised as bearing unacceptable detrimental risks and that which is not so recognised. In terms of both distinctions it is the *scale* of change which provokes direct or delayed political responses and conceptual rejigging. The older, local, embedded forms of knowledge which underpin Nuaulu subsistence strategies are qualitatively different from knowledge of higher-order processes, 'environmental consciousness' in the abstract, which only comes with a widening of political and ecological horizons to a national and global level. In some ways this process is similar to how articulate Nuaulu have come to re-conceptualise their ritual practice as *agama* (religion), and their distinct way of life as *kebudayaan* (culture); agama, kebudayaan, *lingkungan hidup* (living space, milieu, environment) are — in Indonesian officialese — secondary abstractions of a comparable order. Forest, they now understand, is subject to pressures of in-migration, expropriation and economic exploitation in many places other than their own.

This quasi-global[4] consciousness is no better symbolised than by the arrival of electronic means of communication in Nuaulu villages, first radio and then television. Television has not only enabled Nuaulu to keep in touch with the world by watching English league soccer matches and Thomas Cup badminton, but — and this is the reflexive twist — to watch David Attenborough eulogise tropical rainforest in its death throes. Despite a long history of interaction with outsiders of various origins, changing patterns of environmental modification, patterns of subsistence and the conceptual modulation of these things, it is the major changes associated with cultural globalisation which have forced a really radical response from them. It could be said that the aggressive individualism of the eighties, the selling of land and market engagement represented both the end of an old small-scale conception of nature in which resources and forest are infinite, and the beginning of a new conception of participation in an open global ecology of limited goods. The changes, therefore, are a response to a different problematic, to a different social and political agenda, rather than a rejection of environmental change itself or an a priori endorsement of ecological holism. Nuaulu constructions of environment are changing to accommodate a new *level* of discourse, and it is no coincidence that those who currently complain that their schooled children are unable to obtain appropriate employment in the Indonesian state because they are told that the doctrine of *Pancasila* is an impediment, also — though paradoxically — adopt an environmentalist rhetoric which seeks to keep the state from their land.

APPENDIX: THE CONSEQUENCES OF DEFORESTATION — A NUAULU TEXT FROM ROHUA, SERAM, 1994

1. About we Nuaulu people. Our own government here in Indonesia allowed large lumber companies to come here looking for timber. Like *onia* [Malay *kayu meranti*, *Shorea* spp.]. So they leveled the tops of mountains, digging them all up. At the heads of rivers they cut down *punara* [*Octomeles sumatrana*] trees, they cut down onia along the edges of rivers, vehicles leveled and filled in the heads of rivers. While they lived here it was still good. We got around well because they were working.

2. Vehicles went up and down the roads so they were clear. Or if we went hunting we rode on their vehicles with them. But when they went home, our roads were covered up, trees started to grow on them and then we couldn't travel about well because when it rained landslides covered the roads. Game animals moved far away as did cuscus. Land slid into the rivers because they cut down the big trees along the edges of the rivers.

3. Therefore we are really suffering because we have to go around the roads. Before they came here we knew when it was rainy season and when it was dry season. But when they leveled our lands and rivers here in our forest it wasn't the same when it rained and when it was sunny. It was sunny all the time so land slid into all the rivers. Therefore we do not feed good because it is no longer like before.

4. Before, the rivers flowed well and the sun shone well so they looked good to us. But now that the vehicles leveled them so much the fish in the rivers and the game animals in the forest have moved far away. They electrocuted all the fish in the rivers so there are no more fish. So where can we look for our food? Even if we look for our food in rivers that are far away we do not find any fish. We do not find any game animals. The deer have moved far away.

5. Therefore we want to ask for money to cover the price of our forest but the government in Masohi and Amahai forbid us from doing so. So we are quiet and are obeying them. But because of our village and forest we are suffering. We suffer when it is so difficult to go to our forest and look for our food because they leveled all the rivers. They leveled all the mountains. The rivers do not flow well. It is difficult to find game animals. Therefore, we do not feel well about this.

6. They destroyed the lands and rivers. They took away all the big trees. They sold them and made a profit but they did not give any of it to us. Therefore the Nuaulu elders do not want anything to do with

them because they did not think of us. We let them take the wood because they said that they would plant new trees to replace those they cut down. But when will those trees grow? They will never grow like the trees before. How will they grow like those big trees? And when will they plant the trees to replace them? It will be a long time before those little trees are big.

7. Therefore the elders do not want anything to do with them because the lumber companies came here making things difficult for us with our forest. Our lands and rivers are no good at all. They have been gone a long time like the Filipinos. When we go to the river Lata Nuaulu or Lata Tamilou we have to cut the thorns that have grown with our machetes until we are almost dead because they block the path. If it rains just a little there are landslides cutting off the path and then we have to go far around them before we can find a straight path. Therefore, we are suffering a lot just because of this.

8. Therefore if there is any help or any word that can be given here in Indonesia that would help the officials here in Indonesia. Help quickly so that they will not agree that all our lands, rivers, and trees be taken. So the heads of rivers would not be leveled so we cannot eat well or find food well.

9. We people find our food in the forest. There are a lot of Nuaulu people who do not fish well so they look for food in the forest. This is just us Nuaulu people. Other people look for food and have a lot of people who fish but there are only a few of us who fish. Therefore these people look for food but do not find it. We are all dead from hunger [hyperbole]. Before the lumber companies came we got around well. We found food well because the deer slept nearby, pigs lived nearby, and cassowaries lived nearby. But when they leveled and destroyed these animals' places and caves they ran away. So it is very hard for the Nuaulu people to find food because they chased away all the pigs and deer so that they are now far away.

10. Therefore if Roy (Ellen) can find a little help and wants to talk to the officials here in Indonesia I ask that he help us a little so that they do not come here and work again. We do not want them to because we are already suffering a lot.
    That is all.

[Text recorded and translated by Rosemary Bolton, 1994.]

ACKNOWLEDGEMENTS

* The writing of this paper was supported by ESRC grants R000 23 3028 and R000 23 6082, and EC contract B7-5041-94.08-VIII, *Avenir des peuples des Forêts Tropicales*. Administrative permissions and financial backing for fieldwork on Seram between 1970 and 1990 are fully acknowledged at Ellen 1993a: xv-xvi. I am indebted to Rosemary Bolton of the Summer Institute of Linguistics in Ambon, and to an anonymised Nuaulu author, for permission to reproduce the text presented here as an appendix. I am also grateful to Tania Li for some helpful editorial suggestions on an earlier version of this chapter, though any shortcomings in its final form are very much my own.

NOTES

[1] Lowland is used here to refer to a forest type generally dominated by the dipterocarp *Shorea selanica*, in contrast to the montane vegetation of higher altitudes. In fact, the lowland forest of Seram covers, on the whole, hilly country and may extend to an altitude of some 1000 meters.

[2] Another striking case of human management of forest trees (though not one which I have observed in the Nuaulu area) is reported by Soedjito *et al.* (1986) for higher altitude forests in west Seram. Here, seedlings of the resin-producing *Agathis dammara*, important as a source of cash, are systematically planted to replace older, less productive, trees.

[3] Nuaulu symbolically represent their relations with outsiders, dialectically, in two ways: in terms of relations of complementarity, and in terms of hierarchy. The first is exemplified in the relationship between most local clans, in *pela* partnerships (that is between individuals linked through historical blood siblinghoods between villages) and through common membership of the *patalima* grouping (Valeri 1986). The second is reflected in their relations with Sepa and the Indonesian state. Here they manage to assert, simultaneously, a mythic superiority (usually expressed in the conventional older-younger sibling metaphor) and a pragmatic political submissiveness. The articulation of the two principles, however, is on their terms. They insist that they are prepared to accept the benefits of a good raja, but equally prepared to withdraw into their own autonomy when it suits them.

[4] I use the term 'quasi-global' to avoid any accusation that Nuaulu consciously conceive of themselves as global actors and consumers in the sense which has entered the consciousness of many in the West. It would be more accurate to say that they have become increasingly conscious of the degree of connectedness between their lives and those in remote places with whom they share common experiences (such as televised football matches) and material products (such as cassette players).

References

Aumeeruddy, Y. and J. Bakels, 1994, "Management of a sacred forest in the Kerinci valley, central Sumatra: an example of conservation of biological diversity and its cultural basis". *Journal d'Agriculture Tropicale et de Botanique Appliquée*, 36(2), 39–65.

Balée, W., 1992, "People of the fallow: a historical ecology of foraging in lowland south America" in *Conservation of neotropical forests: working from traditional resource use*, edited by K.H. Redford and C. Padoch, pp. 35–57. New York: Columbia University Press.

—— 1993, "Indigenous transformation of Amazonian forests: an example from Maranao, Brazil". *L'Homme*, 33(2-4), 231–254.

—— 1994, *Footprints of the forest: Ka'apor ethnobotany — the historical ecology of plant utilization by an Amazonian people.* New York: Columbia University Press.

Bellwood, P., 1978, *Man's conquest of the Pacific: the prehistory of southeast Asia and Oceania.* London: Collins.

Bloch, M., 1974, "Symbols, song, dance and features of articulation: is religion an extreme form of traditional authority?" *European Journal of Sociology*, 15(1), 55–81 .

Blust, R., 1976, "Austronesian culture history: some linguistic inferences and their relations to the archaeological record". *World Archaeology*, 8(1), 19–43.

Boyer, P., 1993, *Cognitive aspects of religious symbolism.* Cambridge: Cambridge University Press.

Clay, J.W., 1988, *Indigenous peoples and tropical forests.* New York: Cultural Survival Inc.

Colchester, Marcus, 1993, "Forest peoples and sustainability" in *The struggle for land and the fate of the forests*, edited by Marcus Colchester and Larry Lohmann, pp. 61–95. Penang, Malaysia: World Rainforest Movement, The Ecologist, Zed Books.

Coursey, D.G., 1978, "Some ideological considerations relating to tropical root crop production" in *The adaptation of traditional agriculture: socio-economic problems of urbanisation*, edited by E.K. Fisk, pp. 131–141. Development Studies Centre Monogr. 11. Canberra: Australian National University.

Croll, E. and D. Parkin, 1992, "Cultural understandings of the environment" in *Bush base: forest farm : culture, environment and development*, edited by E. Croll and D. Parkin, pp. 11–36. London: Routledge.

Dove, M.R., 1983, "Theories of swidden agriculture and the political economy of ignorance". *Agroforestry Systems*, 1(3), 85–99.

Edwards, I.D., 1993, "Introduction" in *Natural history of Seram, Maluku, Indonesia*, edited by I.D. Edwards, A.A. Macdonald and J. Proctor, pp. 1–12. Andover: Intercept.

Ellen, R.F., 1975, "Non-domesticated resources in Nuaulu ecological relations". *Social Science Information*, 14(5), 51–61.

—— 1977, "Resource and commodity: problems in the analysis of the social relations of Nuaulu land use". *Journal of Anthropological Research*, 33, 50–72.

—— 1985, "Patterns of indigenous timber extraction from Moluccan rain forest fringes". *Journal of Biogeography*, 12, 559–587.

—— 1986, "Microcosm, macrocosm and the Nuaulu house: concerning the reductionist fallacy as applied to metaphorical levels". *Bijdragen tot de Taal-, Land- en Volkenkunde*, 142(1), 1–30.

—— 1988a, "Foraging, starch extraction and the sedentary lifestyle in the lowland rainforest of central Seram" in *History, evolution and social change in hunting and gathering societies*, edited by J. Woodburn, T. Ingold and D. Riches, pp. 117–34. London: Berg.

—— 1988b, "Ritual, identity and the management of interethnic relations on Seram" in *Time past, time present, time future: essays in honour of P. E. de Josselin de Jong*, edited by D.S. Moyer and H.J.M. Claessen, Verhandelingen van het Koninklijk Instituut voor Taal-, Land- en Volkenkunde 131, pp. 117–35. Dordrecht-Holland, Providence-U.S.A.: Foris.

1993a, *Nuaulu ethnozoology: a systematic inventory of categories*. CSAC Monogr. 6. Centre for Social Anthropology and Computing and Centre for Southeast Asian Studies: University of Kent at Canterbury.

1993b, "Human impact on the environment of Seram" in *Natural history of Seram, Maluku, Indonesia*, edited by I.D. Edwards, A.A. Macdonald and J. Proctor, pp. 191–205. Andover: Intercept.

1993c, "Rhetoric, practice and incentive in the face of the changing times: a case study of Nuaulu attitudes to conservation and deforestation" in *Environmentalism: the view from anthropology*, edited by K. Milton, pp. 126–43. London: Routledge.

1996, "The cognitive geometry of nature: a contextual approach" in *Nature and society: anthropological perspectives*, edited by G. Pálsson and P. Descola, pp. 103–123. London: Routledge.

Ellen, R.F. and N.J. Goward, 1984, "Papeda dingin, papeda dingin ... Notes on the culinary uses of palm sago in the central Moluccas". *Petits Propos Culinaires*, 16, 28–34.

Flach, M., 1976, "Yield potential of the sagopalm and its realisation" in *Sago-76: First international sago symposium: the equatorial swamp as a natural resource*, edited by K. Tan, pp. 157–77. Kuala Lumpur: Kemajuan Kanji.

Grzimek, Benno R., 1991, *Social change on Seram: a study of ideologies of development in eastern Indonesia*. Thesis submitted for the Degree of Doctor of Philosophy: London School of Economics and Political Science, University of London.

Gudeman, Stephen, 1988, "Frontiers as marginal economies" in *Production and autonomy: anthropological studies and critiques of development*, edited by John W. Bennett and John R. Bowen, Monographs in Economic Anthropology 5, pp. 213–216. Lanham: University Press of America.

Hardjono, J., 1991, "The dimensions of Indonesia's environmental problems" in *Indonesia: resources, ecology and environment*, edited by J. Hardjono, pp. 1–16. Oxford: Oxford University Press.

Hart, G., 1989, "Agrarian change in the context of state patronage" in *Agrarian transformations: local processes and the state in Southeast Asia*, edited by G. Hart, A. Turton and B. White, pp. 31–49. Berkeley: University of California Press.

Haudricourt, André, 1962, "Domestication des animaux, culture des plantes et traitement d'autrai". *L'Homme*, 2(1), 40–50.

Haudricourt, André, 1964, "Nature et culture dans la civilsations de l'igname: l'origine des clones et des clans". *L'Homme*, 4(1), 93–104.

Hurst, P., 1990, *Rainforest politics: ecological destruction in southeast Asia*. London: Zed Books.

Ingold, T., 1996, "Hunting and gathering as ways of perceiving the environment" in *Redefining nature: ecology, culture and domestication*, edited by R. Ellen and K. Fukui, pp. 117–155. Oxford, New York: Berg.

Karim, W.A., 1981, "Mah Betisek concepts of humans, plants and animals". *Bijdragen tot de Taal-, Land- en Volkenkunde*, 137, 135–60.

Koentjaraningrat, 1993, "Pendahuluan" in *Masyarakat terasing di Indonesia*, edited by Koentjaraningrat, Seri Etnographi Indonesia 4, pp. 1–18. Jakarta: Gramedia Pustaka Utama.

MacAndrews, C., 1986, *Land policy in modern Indonesia*. Boston: Lincoln Institute of Land Policy.

Maloney, B.K., 1993, "Climate, man and thirty thousand years of vegetation change in north Sumatra". *Indonesian Environmental History Newsletter*, 2, 3–4.

Moniaga, S., 1991, "Toward community-based forestry and recognition of adat property rights in the outer islands of Indonesia" in *Voices from the field: Fourth Annual Social Forestry Writing Workshop*, edited by J. Fox, O. Lynch, M. Zimsky and E. Moore, pp. 113–33. Honolulu: East-West Center.

Peluso, N. and C. Padoch, 1996, "Changing resource rights in managed forests of West Kalimantan" in *Borneo in transition: people, forests, conservation and development*, edited by C. Padoch and N. Peluso. Kuala Lumpur: Oxford University Press.

Persoon, G.A., 1994, *Vluchten of Veranderen: processen van verandering en ontwikkeling bij tribale groepen in Indonesie*. Leiden: Riksuniversiteit te Leiden, Faculteit der Sociale Wetenschappen.

Posey, Darrell, 1988, "Kayapo Indian natural-resource management" in *People of the tropical rainforest*, pp. 89–90. Berkeley: University of California Press.

Prance, Ghillean T., William Balée, B.M. Boom and Robert L. Carneiro, 1987, "Quantitative ethnobotany and the case for conservation in Amazonia". *Conservation Biology*, 1, 296–310.

Rambo, A. Terry, 1979, "Primitive man's impact on genetic resources of the Malaysian tropical rainforest". *Malaysian Applied Biology*, 8(1), 59–65.

Schmink, M., K.H. Redford and C. Padoch, 1992, "Traditional peoples and the biosphere: framing the issues and defining the terms" in *Conservation of neotropical forests: working from traditional resource use*, edited by K.H. Redford and C. Padoch, pp. 3–13. New York: Columbia University Press.

SKEPHI and R. Kiddell-Monroe, 1993, "Indonesia: land rights and development" in *The struggle for land and the fate of the forests*, edited by M. Colchester and L. Lohmann, pp. 228–63. Penang, Malaysia: World Rainforest Movement.

SKEPHI, 1992, "Logging and the sinking island". *Inside Indonesia*, 33, 23–5.

Soedjito, H., A. Suyanto and E. Sulaeman, 1986, *Sumber daya alam di pulau Seram Barat, Propinsi Maluku*. Jakarta: Lembaga Biologi Nasional.

Stark, Ken and Kyle Latinus, 1992, "Research report: the archaeology of sago economies in central Maluku". *Cakalele: Maluku Research Journal*, 3, 69–86.

Strathern, Marilyn, 1992, *After nature: English kinship in the late twentieth century*. Cambridge: Cambridge University Press.

Valeri, V., 1989, "Reciprocal centers: the Siwa-Lima system in the central Moluccas" in *The attraction of opposites*, edited by D. Maybury-Lewis and U. Almagor, pp. 117–41. Ann Arbor: University of Michigan Press.

Valeri, V., 1990, "Both nature and culture: reflections on menstrual and parturitional taboos in Huaulu (Seram)" in *Power and difference: gender in island Southeast Asia*, edited by J.M. Atkinson and S. Errington, pp. 235–72. Stanford, California: Stanford University Press.

Zerner, C., 1990, *Community rights, customary law and the law of timber concessions in Indonesia's forests: legal options and alternatives in designing the commons*. Jakarta: FAO Forestry Studies TF/INS/065.

Chapter 6

BECOMING A TRIBAL ELDER, AND OTHER GREEN DEVELOPMENT FANTASIES

Anna Lowenhaupt Tsing

How does a globally circulating social category come to mean something to people in a particular political context? Categories are dream machines as well as practical tools for seeing; the fantastic view we are offered and the familiar job at hand are inextricably related. This essay is a backhanded defense of environmentally-inflected rural policy, including that sometimes called "green" or sustainable development. I argue that at least in one village in the Meratus Mountains of Kalimantan, collaboration between urban environmentalists and village leaders offers promising possibilities for environmental and social justice, that is, for building a world in which we might want to live. Yet my argument is a planner's nightmare. The collaborations I describe are made possible only by clever engagements with green development fantasies of the rural, the backward, and the exotic. "Tribal elders" are made in the mobile spaces found within coercive international dreams of conservation and development, and these men and women — granted agency within the fantasies of their sponsors — are enabled to forge alliances that yet somehow present the hope of transforming top-down coercion into local empowerment. Categories often come to life in this round-about way. Yet we can only appreciate their creative intervention and their political charge if we move beyond a sociology of stable interest groups and hierarchies to investigate the social effects of shifting rhetorics and narratives and the reformulations of identity and community that they engender.

My argument is composed at a moment when many scholars have become critical of social movements committed to combining the protection of endangered environments and the empowerment of indigenous peoples (Brosius, Tsing, and Zerner 1998). Fearing simplistic representations of wild nature and tribal culture, scholars dismiss what in my opinion are some of the most promising social movements of our times. In contrast, my approach offers an alternative to the choice between unselfconscious stereotypes of nature and culture on the one hand, and ironic dismissals of environmental and indigenous politics on the other. I argue that our discussions might better begin with the circulation and use of "green development fantasies." My focus on collaboration — as opposed to contestation or misunderstanding — offers a methodological framework for facilitating this discussion. In the late 1990s, both scholars and activists know a lot about how to talk about contests; we have less precedence for discussing the awkward but necessary collaborations central to both intellectual and political work.

Several layers of context are necessary for my argument to emerge. I begin by locating my essay within the concerns about upland transformations in Indonesia that form the subject of this volume. I then turn to the Meratus village of Mangkiling, which, already the subject of many green development representations, seems well suited for a meditation on the dynamics of representation. The fantastic aspects of international thinking about exotic and backward rural communities (for which I deploy the term "tribe" as a kind of shorthand) are my guide to the field of attraction in which Mangkiling representatives are able to become potential collaborators and political actors. Beginning conventionally enough in a rural sociology, I draw my argument into the unstable realm of pathos and love in which things that did not exist before can emerge. For it is in that realm that metropolitan fantasies *both* fulfil themselves *and* take the dreamers they construct by surprise.

UPLAND TRANSFORMATIONS

The residents of uplands Indonesia have come into a new visibility. For many decades, lowland peasants were the only rural peoples to figure in those great narratives of nationalism and development that plotted the country's past and future. In recent years, however, international concerns with the degradation of fragile environments have focused attention on rainforests and mountains — and their long-time residents. Policy makers have been pressed to rethink the uplands as key sites of environmental sustainability and to consider the role of uplands communities within environmental conservation as well as development programs. Non-governmental organizations focusing on issues of conservation and development have joined state officials in negotiating the role of uplands communities. Social scientists have been drawn into practical discussion of upland futures. Upland village farmers are aware of a new sense of focus and urgency in their dealings with state officials, NGOs, and social scientists alike.

The new attention to upland communities does not present itself in the form of a consensus. Discussion ignites fierce debates (Who owns the forest?) as well as unstated disagreements (What is a community?). Central to all this are much disputed issues of representation. On one end of a continuum, upland communities are represented as closed and static repositories of custom and tradition; on the other end, uplanders are portrayed as hyper-rational, individualistic entrepreneurs with no commitments to local social life or culture. Either side of the continuum can be presented as politically promising or socially worthless: uplanders as cultural communities may be backward savages or guardians of the forest; as individualistic entrepreneurs they may be model citizens or undisciplined mobs. Both ends of this continuum of representation draw upon hoary historical roots as well as contemporary legitimacy. Terms are revitalized. International environmental and minority rights movements work to transform the assumption that "tribes" are backward remnants of archaic humanity to argue instead that the world needs tribal wisdom and tribal rights to preserve our endangered biological and cultural diversity. Other environmentalists celebrate the new hegemony of free trade by portraying a post-communal world of independent innovators and entrepreneurs. In Indonesia, both ends of the "individuals-or-communities" continuum, as well as many compromises and middle zones, engage some social scientists, some community leaders and advocates, some village farmers.

Given the variety of ways these dichotomous strategies of representation have been and are being used and abused, this does not seem a moment to decide once-and-for-all which one is really right. Instead, it seems an important time to analyze the dynamics of representation itself, and particularly to look at how representational categories come to mean something to farmers, community leaders, scholars, advocates, or development bureaucrats in a particular political moment. In this spirit, this essay discusses representational strategies, and the social categories on which they rely, as dreams and fantasies that grab people under certain circumstances. Preexisting complexities are of course important, but by thinking of them as gates to, rather than walls against the imagination, it is possible to trace the emergence of unexpected ingenuities. As we attend to both creativity and constraints in upland self-fashioning, a number of elements come into view.

First, a new role has become possible for rural minority leaders who convincingly "represent" the kind of community that environmentalists and green developers might choose for co-operation, learning, and alliance. These representatives take on the mediations that make collaborations between village people and advocates or policy makers possible. Their collaborations sustain and give life to concepts such as village development, tribal rights, sustainability, community-based conservation, or local culture. At the same time, these same concepts make political agency possible on both sides: they are the medium in which village leaders and those who study, supervise, and change them can imagine each other as strategic actors and thus can mold their own actions strategically. We might call these representatives "tribal elders" because it is they who, to hold the attention of potential rural-minority advocates, take responsibility for the fantasy of the tribe.

For tribal elders to flourish, it is not enough to posit the existence of "tribes"; a field of attraction must be created to nurture and maintain the relationship between the rural community and its experts. Without this field of attraction, the community will be abandoned to its own fate; neither mediation nor collaboration is possible. Thus the single most important sign of a community representative's success is his or her ability to conjure, and be conjured by, that emotionally-fraught space that keeps the experts coming back. In this space creative action is possible, and collaborations are forged.

Collaborations are the hopeful edge of a political project. To condemn a project, it is not enough to say that it engages in simplifications; all social categories simplify even as they bring us to appreciate new complexities. Instead, it seems more useful to judge the political valence of a project by the promise for remaking the world of the collaborations it has engendered. Thus "tribal" fantasies in South Kalimantan, combined in an ambivalent and ambiguous manner with rural development dreams and hierarchies, lead to collaborations between urban activists and village leaders that offer possibilities for building environmental and social justice in the countryside as exciting as any I have heard of on the contemporary scene. At the end of this essay, I turn to two promising initiatives, collaborations between urban environmentalists and Mangkiling leaders that developed in the early 1990s. First, I show how "nature" is made into a utopian space of collaboration through the practice of naming trees. Second, I examine the mapping projects that, instead of clarifying land claims, amplify ambiguity in the system — and thus open the confusion in which village claims over forested land might hold their own.

It is not useful to be complacent about these collaborations. Tribal elders have no particularly striking powers; nor do they represent homogeneous or unified communities or grass-roots movements. Their "community" representations are vulnerable and contested; even close kin and neighbors are not necessarily supporters. A few minutes' hike away, no one may know a thing about their projects. Furthermore, environmentalist and tribal collaborations with outside patrons are hardly the most powerful rural collaborations around. In Indonesia, development visions in which rapid environmental destruction is appreciated as progress or regulated as government-endorsed "sustainability" continue to be much more powerful than emergent "tribal" environmentalisms. Song-and-dance tourism predominates over ecotourism. The role of environmentally-friendly tribal elder deserves special attention because it is new and promising, but it does speak for either long-standing culture or newly-made hegemony, whether locally, regionally, or nationally.

Then, too, there is nothing here to suggest the kinds of progressive politics we most easily imagine: coalitions of "interest groups"; workers and peasants and intellectuals in league. Instead, here are moments of creative intervention and the making of new identities. Ordinary villagers may or may not get involved; it is unclear how many will see their interests as being advanced. Yet the space is cleared for the tribal elder and for the field of attraction that makes his or her agency possible. The enactment of the tribe is, to use a term from the International Situationists (writing about the very different context of metropolitan spectacle), the making of a tribal "situation"; it is the recharging of political possibility through staging the fantastical realities of everyday life (Debord 1983).[1]

THE TRIBAL SITUATION

Let me turn to a particular tribal "situation." Consider a fragment from a document written by Musa, a Meratus Dayak elder of the village of Pantai Mangkiling.

> [W]e, as Indigenous Original Peoples of the Local Area, for the sake of guarding our Livelihood Rights and Environmental Conservation, as well as from our Culture, state as follows:
>
> 1. Our livelihood is to work the soil by DIBBLE-STICK PLANTING, and our care for our local natural world's plants from generation to generation has been as a productive garden, thus THERE IS NO WILD FOREST in our area.
> 2. We will not condone it if there is a destruction of our local natural environment, because this interferes with OUR BASIC HUMAN RIGHTS.
> 3. If someone destroys our local natural environment, this means they destroy our Basic Human Rights, and thus the destroyer will be confirmed as Violator of the Law of the Indigenous Original Local People.

In the Meratus Mountains of South Kalimantan, shifting cultivators have created socially-marked forest territories in which planted, encouraged, named, and closely watched trees signal the economic claims and social affiliations of particular individuals and groups. As Musa states, "There is no wild forest" in this area. Yet since the late 1970s, timber companies, transmigration projects, plantations, and migrant pioneer farmers from the Banjar plains have made increasing claims on Meratus forests. None of these claimants recognize Meratus Dayak customary rights to the forests; instead, the forests are seen as uninhabited, wild territories to be assigned to various users by the state.

Meratus Dayak responses have been various. Stories circulate about violence and the burning of timber company bridges. At the same time, people retreat farther into the hills, discouragement spreads, and young men sell trees to illegal loggers before the "legitimate" companies can take them without compensation. This has been a challenging time for community leaders, who maneuver within the government regulations and rhetorics that both disenfranchise their communities and provide the only legitimate channels for protest. Creative responses have been necessary to hold on to any community land and resources; the threat of involuntary resettlement in government camps for "isolated tribes" looms. It is in this context that Musa has composed this document.

The document is a land-rights claim of sorts. It makes its claim by overlapping three divergent streams of political culture that, outside of this text, have rather separate spheres of existence. First, regional administration: the typed document is an official statement (*surat keterangan*) signed by Musa "on behalf of the Committee of the Traditional Hall of the People of Pantai Mangkiling," as "acknowledged" by the village head and district military officer and "verified" by the district head. The stamps of various district officers occupy the bottom third of the page; the formality is recognizable and appropriate within the regional bureaucracy.

Second, international environmentalism: the document uses every globally circulating jargon word in the social ecologist's 1980s agenda. The author writes for indigenous people, original people, people who for generations have guarded and protected their natural environment. Their traditional conservation strategies are being threatened, and with them their human rights. To destroy the forest — as the unmentioned timber companies and plantations want to do — is against traditional law. Instead, as he explains later in the text, the forests must be used by village cooperatives. Where did Musa get this rhetoric? These are not terms that Meratus Dayaks ordinarily use; furthermore, neither district nor regency bureaucrats in South Kalimantan know much about this kind of talk. The Indonesian language of the text is official and elegant — much more so than either my translation or Musa's ordinary speech. Presumably there was collaboration here, and maybe collaboration with someone from outside South Kalimantan. However, this is not just a transplanted text, and there is a third stream evident: Meratus cultural ecology. For example, rather than engaging government problematics of shifting cultivation (*berladang berpindah-pindah*) or environmentalist endorsements of forest love and lore, the document goes straight to the cultural practice of dibble-stick planting (*menugal*), a much more locally relevant sign of social habitation.

Musa's tribal situation depends on his ability to evoke all three of these strands of political culture simultaneously. As a community representative, he can afford to show some agility with local knowledge. But he must articulate this knowledge within the discursive categories that make his community appear as an identifiable object to environmentalists, on the one hand, and government administrators and developers, on the other. His document is recognizable as a claim only to the extent that he evokes NGO and official ideas about rural minority communities. Thus my account detours momentarily from his text to introduce the community-like objects of environmentalism and development. I begin with the "tribe."

Until quite recently, tribes were supposed to represent our planet's past — the part of human evolution that city people were done with; tribal remnants were irrelevant to our times except as museum pieces. Suddenly, tribes have reentered stories of the future. The rainforests were shrinking; the ozone hole growing; the progress of *progress* looked terrifying. As the millennium drew to a close, the suggestion appeared that we had better pay attention to the wisdom of the tribes, since, after all, they are the ones who know how to maintain nature over the long haul. Attentive to the alternatives, a cosmopolitan audience looked up and listened. Tribes, it was argued, could be the guardians of the biological, pharmaceutical, cultural, and aesthetic-spiritual diversity that would make our future on earth possible. Even the most hard-headed of futurists, development planners, were forced to pay attention to this refigured planetary trajectory. The figure of the tribal elder became a small but insistent presence in the emergent rhetoric of sustainable — that is, environmentally sound — development.

Like any other political rhetoric, sustainable development plans can be idealistic and utopian or cynical and practical; they can be a tool in the hands of national military forces and transnational corporations or a rallying cry for community rights and social justice. Tribal rights is only one thin strand in an emerging "sustainability" rhetoric that more commonly takes for granted transnational capitalism and neocolonial management as it counts board feet, parts per million, growth rates, and the bottom line. Sustainability means different things to different groups. In the Indonesian context, sustainability has been debated in Jakarta by government bureaus and non-governmental organizations: conservation areas, laws, and goals have been proposed and sometimes adopted; the question of tribal rights has even garnered some interest.[2] However, at least in rural areas, attempts to deepen national commitments to environmental conservation have been impeded by the presence of an enormously bureaucratized, subsidized, and militarized machinery of non-sustainable development. This is a machinery not easily converted to new purposes. It is not just that administration and planning occur through this machinery; the ruling concepts and institutions of government, economy, culture, and citizenship in rural areas have been tailored within its workings. Attempts to ignore or evade this machinery are quickly labeled subversive, a label made serious by the pervasive presence of arms. Any suggestions about forest conservation or tribal rights in South Kalimantan must somehow make their way around or through the national and regional development apparatus.

In South Kalimantan, the goal of rural development is understood to be the management of rural peoples and places for the advancement of national priorities. Development is a top-down project for expanding administration; development brings villages and forests into line with national standards.[3] As with all administrative projects, there is negotiation of just what will count as locally appropriate. Yet I heard little disagreement about the importance of externally imposed directives in the administration of regional minorities, who are completely missing within the ranks of provincial administrators — and who are sitting on the province's most valuable forest resources. Development for them involves independent plans for forests and for people; the goal of development is to make the people orderly while simultaneously redirecting their forest resources to national priorities such as patronage, profit, and export production. Villages are to be units of administration; forests are national resource domains; there is no legitimate connection between the two. Thus, most regional development administrators have never given consideration to concepts of tribal rights or community-based forest management, each of which — whatever their constituency in Jakarta — contradicts the hegemonic logics of provincial development.[4]

In this context, Musa's endorsement of indigenous peoples' conservation is not a mimicry of ruling ideas; within provincial political culture, it is an innovative challenge. Musa's text argues that the traditional values of his village are not in need of development; they are the basis of the people's own equitable and sustainable development plans. Furthermore, even if Musa learned or copied the terms of his text from a Jakarta or Geneva visitor, to merely restate them in South Kalimantan could mean little, unless he could create a "situation" — that is, a dramatic enactment of phantasmic realities — in which these terms could come to mean something to the regional officials who control whether or not the village continues to exist.

How this situation was created is the subject of the rest of this essay. In the next sections, I examine a series of documents about the village of Mangkiling to look at how Mangkiling representatives became positioned as spokespersons for community conservation and development, or, in the shorthand I have been using here, as "tribal elders." On the one hand, Mangkiling can be said to be gifted with smart leaders who have been able to transform a regional development rhetoric of backward status and exotic culture into community entrepreneurship and self-representation. This requires that they engage the textual intricacies of the discourse of development administration to find what literary critic Ross Chambers (1991) might identify as its "room for maneuver." Their tricky transformations and revisions of regional development make local initiatives possible. But Mangkiling representatives cannot strategize as if they were generals on a battlefield in which opposing armies and objectives are clearly demarcated and unchanging. Instead, they are produced as representatives by outsiders' standards of representation. They enact a fantasy in which whether they play themselves or someone else's understanding of themselves is ambiguous; the community they can represent is produced in their development-directed performances of "community."[5]

To make sense of this double-sided agency, so much their own and so much not their own, I show the importance of what I have been calling "fields of attraction," for it is the longings, the broken promises, the erotic draw, and the magic of that Mangkiling enacted in the tribal situation that makes the tribal elder emerge as a politically active and creative figure. To the extent that conservation and development discourses can be engaged through these fields of attraction, local initiatives — whether for better or worse — become possible.

THE NATIVE IN THE DOCUMENT

If Musa's testament was an isolated object, it would be inspired but socially insignificant. However, Musa and his associates in Pantai Mangkiling have done more than write this text, and their ingenuity and persistence and sheer luck have paid off in making the village of Mangkiling a place that cannot be rolled over and erased easily. Whether or not Musa is properly considered an elder of a "Committee of the Traditional Hall of Mangkiling," as he signs himself, he has effectively constituted the village as an object of attention and respect for those interested in the conjunction of forest protection, community resource management, and ethnic pride. Government officials, ecotourists, naturalists, social science researchers, environmental activists, and journalists have been attracted there. In the process, a small mound of documents about Mangkiling has been generated.[6] Pantai Mangkiling may be the best documented village in the Meratus Mountains. Most of these documents are about Musa and his fellow villagers, not *by* them. The portraits of the village and the villagers found in these documents serve the purposes of others. Yet reading them with my questions in mind, it is possible to find traces of the encounters in which Mangkiling representatives, empowered to be more than passive objects of study and command, have renegotiated the very purposes that gave them agency; they have turned regional dogma to unexpected ends. These traces guide us to appreciate the formation and deployment of tribal sensibilities in Mangkiling.[7]

Through the documents, I can ask how Musa and his fellow village leaders managed to get so much respect as "community spokespeople" while operating within the discursive and institutional constraints of expected village status — that is, as those with nothing to say. I can trace the transformations through which these leaders made the village a formidable ethnic-environmental object with forests under noticeable, if perhaps unenforceable, traditional claim. The documents can tell us something about how Mangkiling leaders positioned themselves to make more documents about them happen, that is, to keep the village a possible subject of tribal rights.

The documents generalize about the villagers, but, sometimes, too, they name individuals. Three leaders stand out: Musa, his sister Sumiati, who is the village head, and their brother Yuni, the village secretary. These three are consulted, profiled, and quoted extensively. My interviews confirm that they are major architects of the Mangkiling project. Let me begin with a document that features Yuni.

In April 1989 a one-day seminar was held in the provincial capital of South Kalimantan on the dilemmas of Mangkiling as an upland, forest village in an era of national development and change (Yayasan Kompas Borneo 1989). Organized by a provincial environmental group and sponsored by the Ford Foundation, the seminar was attended by regional officials, scholars, and environmentalists. As the Assistant Governor who introduced the proceedings pointed out, the seminar's focus on one village could be generalized to propose concepts for the development of interior populations throughout the nation. The seminar featured a series of papers on the social, ecological, and economic features and challenges of Mangkiling. The papers, which were distributed afterwards in bound form, present a variety of research methodologies and perspectives. Some are based on field research; others contextualize the Mangkiling situation or offer theoretical viewpoints. Many authors are careful to point out the preliminary nature of their assertions.

Yet, to some, the results of the seminar were definitive. One of the province's two daily newspapers published a report on the seminar under the title, "The Economic System of the People of Mangkiling is Extremely Simple" (*Dinamika Berita* 1989a). The article focuses on a paper presented by the head of the regional office of the Department of Social Welfare. The paragraph from the original paper that inspired the headline reads as follows:

> The isolated population group in South Kalimantan still holds to a simple economic system, that is, it still employs a barter system with other families but still within the group. The products that they are able to gain from their efforts are only enough to fulfill their own needs, such that the fulfillment of life needs in a proper manner, as with other peoples, is still far from the reach of their thought. (Mooduto 1989: 3, my translation)

The most amazing thing about this paragraph is that it is utterly and entirely untrue. It is not even a plausible interpretation of the Mangkiling economy or that of any other Meratus Dayaks for the last four centuries, at least. While subsistence and inter-family networking is an important concern within Meratus Dayak communities, they have long been involved in production for distant markets. The conditions of marketing have shifted over time, and the key products have changed. However, the idea that Mangkiling people are unfamiliar with cash and markets is absurd. (Other seminar papers describe the importance in Mangkiling of banana and chili production for regional markets; in the early 1990s, Mangkiling also produced a variety of cash crops besides these, including peanuts, mung beans, coffee, bamboo and light wood construction poles.) The fact that an important regional office with jurisdiction over Meratus Dayaks would promote the idea of a barter-and-subsistence economy in Mangkiling, and that the provincial newspaper would choose this item to report, suggests the blinding relevance of stereotypes about the backward and the primitive in regional development affairs. Because the persistent conviction of Meratus Dayak traditionalism seems so necessary to the trajectory of regional development, planners and their publicity-makers let stereotypes about tradition overcome their other forms of knowledge about the area.

These stereotypes lead to discrimination and persecution. Yet they cannot completely close off Meratus Dayak agency. To the extent that they stimulate research and administrative contact between Meratus Dayaks and development planners, they can even present, ironically, new opportunities for creative community leadership. The seminar documents themselves demonstrate such an opening.

The last section of the volume distributed after the seminar is a photographic essay documenting the proceedings. The heads of speakers rise over the podium out of official uniform shirts; the microphone arches toward each serious face. The audience sits in straight parallel lines along long tables draped neatly with cloth; the exact line of tea glasses before them marks out the orderliness of the row. Some audience members lean forward, taking notes; others lean back, listening or bored. No one leans to the side. But one page of photographs is different; it offers the "profile of a Mangkiling village member who attended the one-day seminar, Mr. Yuni, the Secretary of Mangkiling village" (Yayasan Kompas Borneo 1989: Appendix I). Yuni is shown in three photographs. He is serious and neatly dressed but awkward, innocently out of place, standing as if on display between the audience rows. In one picture, the seated audience appears to be teasing him, laughing at him. He leans precariously, off balance or in a gesture of undisciplined motion.

Through his profile, Yuni "represents" the village in a number of senses. His photographs legitimate the seminar proceedings, and their images of primitive Mangkiling, both through the truth value of his attendance and his inability to pass as just another seminar member. At the same time, his pose reveals traces of the kind of leadership he is able to forge from this position. That artless, off balance stance presents him as the open, desiring subject of an imagined modernity yet with the untutored simplicity of tradition in his background and breeding. He is a tribesman longing for change. Nor need he have been "plotting" to devise this pose; what alternatives are there for the bureaucratically-undisciplined body in the midst of lines of authority and order? Yet, ironically, his lack of bureaucratic experience opens the possibilities of a community leadership role that even development planners can begin to imagine. If the "tribal situation" is to be enacted on the regional development scene, it is the cosmopolitan tribesman, the representative of unfulfilled desire, who can enact it. This representative is created within the opportunity spaces of the development apparatus itself, as villagers are brought in to join its activities. It is negotiated within encounters such as that recorded in Yuni's photographic profile; it finds its subtle traces in the documents that propose and debate the categories of development. My next sections offer more of the context in which I interpret Yuni's pose.

VILLAGES AS FANTASIES AND FRAMEWORKS

In the 1970s and early 1980s, the resettlement and resocialization of "isolated populations" (*masyarakat terasing*), including Meratus Dayaks, became an important component of the regional development plan in South Kalimantan. By working to assimilate these peoples into normative Indonesian standards and grouping them into discipline-oriented villages, the program provided a striking and inexpensive model of how development was expected to operate at a national scale. The process of development could be imagined within the diorama of village resettlement, in which tribes — that is those who did not have the know-how to live in proper villages — were to become modern citizens. Development was the elimination of tribes and the creation of villages. Furthermore, because official definitions of "isolated populations" stressed an imagined landless nomadism (i.e., as an interpretation of shifting cultivation), tribal groups targeted by the "isolated populations" program were defined out of any land rights recognized by the state.

In the 1980s and 1990s, a new regional administrative initiative overcame and indeed reversed some of the consequences of the "isolated populations" program by disciplining existing settlements rather than creating new model sites. The regional government redoubled its administrative efforts in all rural areas — "isolated" and otherwise — by dividing its administrative units into smaller and more closely regulated districts and villages. Where once there was one "village" unit, three or four were created. Villages were to be further naturalized and normalized in the process; while still development models, they were also somehow to correspond to on-the-ground communities. At the same time, district and regency officials refocused their attempts to find and train appropriate village leaders. Instead of allowing older men with existing community status to assume official village positions, they appointed younger men with formal education and the ability to articulate commitments to the goals of development and orderly state administration. These new leaders were offered travel opportunities, gifts, and ceremonies; village subsidies controlled by these new leaders increased rapidly. Furthermore, subsidies were offered differentially, depending on leadership performance. Village leaders were pressed into a competitive relation with each other, in which pleasing regional officials, rather than cooperating with each other, paid off in personal and village benefits.

In contrast to the "isolated populations" program (which continues to operate simultaneously, with reduced resources, but remaining a significant threat), this administrative initiative has promised a new stability for Meratus Dayak groups, in relation to their lands and resources. However, the terms of this stability have been community leadership that articulates and demonstrates compliance with the goals of regional development. There is a contradiction here. "Communities" in the Meratus Mountains are contentious, unstable social groupings, forged through day-to-day local initiatives. Yet to the extent that communities reaffirm themselves as communities, with independent initiatives and resources to manage, they refuse the demands of development, which require that they give up their autonomy and their resources to national planning. However, to the extent that leaders merely confirm national planning by forming villages without locally autonomous communal concerns, their communities slip away and they find themselves treated as pompous ideologues.

This contradiction is rendered more intense by the competition among factions and leaders. Since most contemporary Meratus "villages" only gained their current status sometime in the last fifteen years, the possibility of rearranging administrative affiliations — and thus capturing regional development resource flows — is obvious. As in Eastern Europe after the collapse of the Soviet Union, the struggle to create new polities before the polity-making time is over is zealous. Furthermore, current village leaders create the impression that they are in competition with all others for the survival of their communities; one group's advancement could mean the dissolution of another group. By the early 1990s, it was clear that the most successful village leaders were becoming rich and powerful from development subsidies in ways never before possible in the Meratus Mountains; the closest constituents of these successful leaders were also gaining disproportional benefits. Village offices had never been so important. Family ties were rearticulated, as young men cajoled their elders, hoping to coalesce some community to lead, while old men flattered the young, desperately needing a channel to regional power. Around these rearticulated family ties, factions fight and reform, each trying to channel the differential flow of resources from regional centers. Village leaders sense that if they are not sufficiently creative and aggressive in holding on to their positions, they can be quickly displaced.

It is important to understand that the unit of the "village" has not always been the most relevant to Meratus Dayak sociality. Most all Meratus Dayaks are shifting cultivators who clear new fields every one to three years, while turning old fields into more or less managed garden and forest areas. Small family-like groups (*umbun*) make their own farms; these groups affiliate in clusters of some five to twenty-five umbun to form work groups, share meat, fruit, and fish, and hold festivals and healing rites together. Living arrangements vary across the mountains. In the area that includes Mangkiling, clusters construct a large multi-roomed *balai* hall as their central settlement; every umbun makes its own room around the central floor. Single-umbun houses are also built near the umbun's fields. Until recent development subsidies offered the possibility of making balai with long-lasting construction materials, the halls were repaired or rebuilt every few years. On these occasions, the hall might be relocated, and new umbun might join or split off to join other clusters. Decisions to affiliate into another balai hall generally took into account the location of an umbun's familiar forests, gardens, and fields; it was rare to relocate far from one's most well-managed livelihood resources. However, because in any balai, the territories with which each umbun was most associated radiated out in different directions, toward different balai, each umbun had a number of options of groups with which to live, without ever straying from its familiar territories.

The village (*desa*) is a government administrative unit that operates over, within, and around these shifting clusters. Until the 1980s, villages were huge, unwieldy units, and it was mainly their constituent neighborhoods (*RT*) that had much meaning for local clusters. Some neighborhood and village leaders were more successful than others in gathering and holding communities (Tsing 1993). Since the administrative reapportioning and the subsequent increase in development subsidies for successful village leaders, villages have become more significant. Village leaders have more tools with which to convince their constituents to stay; at the same time, factions attempting to displace those leaders abound. In this context, village leaders and would-be leaders need to find aggressive ways to articulate regional development goals without losing all local support.

Pantai Mangkiling has been one of the most successful models of this new kind of village. Pantai Mangkiling is the name of a place — a flat spot (*pantai*) along the Kapiau River. There has not always been a balai there, although fields, houses, and planted, productive trees have marked the spot continuously; between the late 1960s and early 1980s, the central balai in the area was at a place called Apurung. Yet Musa, Sumiati, and Yuni — all of whom lived in Apurung — had their familiar territories around Pantai Mangkiling. Musa was already a political mover and shaker by the early 1980s when the reapportioning happened there (he had once been village head of the much larger territory), and so it is not completely surprising that, in the competition for new political focal places, his home grounds became a village center. Mangkiling became a village in 1982. Musa's family gained control of village politics when the district officer accepted his sister Sumiati as village head, and his youngest brother Yuni as secretary. By the mid 1980s, the location boasted a balai, a village office, and a cluster of houses. Several other current balai were included in the village territory; and while each grumbled about Mangkiling's new dominance, none was strong enough to change the situation.

In consolidating a central position in the village reapportionments, Musa, Sumiati, and Yuni acted similarly to many successful Meratus leadership factions. However, over the next few years, their leadership became exceptional in making Mangkiling a strong village, one that attracted the attention of environmentalists, scholars, officials, and tourists. Between 1985 and 1990, Mangkiling's leaders consolidated a set of national and regional connections and ties that brought them out of mainstream Meratus invisibility to become a focus of regional attention. It was in this period that their leadership creatively engaged with metropolitan fantasies and created what I am calling a "tribal situation." The events that led to these regional connections are complicated, and I analyze them in detail elsewhere (Tsing n.d.a.). Suffice it to say here that they involved a set of disputes with a timber company over rights to forest land and trees. By chance, a provincial environmental group got involved with Mangkiling's cause and took it as a training exercise to a national environmental forum that was scheduled in the provincial capital. After that publicity, the regent refused to renew the timber company's concession; Mangkiling had won a very major (if, perhaps, tentative) victory for village land rights.

In the process, Mangkiling leaders met a variety of advocates and adversaries including environmentalists, forestry officials, foreign visitors, and timber company workers, as well as regional administrators. The provincial environmental group, *Kompas Borneo*, decided to pursue their relationship with Mangkiling and wrote a successful grant proposal to the Ford Foundation for a research project there that would last several years and involve a large, shifting group of researchers. The 1989 one-day seminar was the first major event in this research relationship; field research, mapping, and various kinds of reports followed. The relationship also attracted funding and support from regional government. In 1992, Kompas Borneo applied to US AID for further support, although their grant was not successful. Meanwhile, Mangkiling became the subject of several series of articles in the provincial newspapers. Ecotourists from Indonesian cities as well as from foreign countries began to make their way there. South Kalimantan was already organizing for ecotourism by the mid 1980s, although the focus of organized tours was the most "developed" and therefore presentable Meratus village of Loksado. Adventurous tours and individuals, however, found Mangkiling. In the early 1990s, a Chinese Indonesian entrepreneur married a Mangkiling woman and set up a hostel for ecotourists. He electrified the balai, using generator power, and built a huge, rickety guest house out of bamboo. Meanwhile development agencies and groups began various small model projects, digging fish ponds and planting cacao, coffee, and other "development-positive" trees. District officials assigned special funds to allow the villagers to repair the balai, build a generator-operated rice mill, and improve their trails and bridges. Islamic groups and health agencies visited. Journalists from a national women's magazine made a trip. A conservation education tour, sponsored by environmentalists from several Kalimantan provinces, made a long stop there. Mangkiling became a bright spot on the regional map.

To tell Mangkiling's recent history in this fashion highlights the contributions of outsiders, which, indeed, have piggy-backed upon each other to make Mangkiling a place of note. However, from the perspective of Mangkiling villagers, these outside contributions have been sporadic, short-lived, and often more ceremonial than substantive. Even the infrastructural improvements cannot be counted on. Thus, for example, in 1994 when I visited, the ecotourism entrepreneur and his wife had moved down to town, taking their generator; their guest hostel had deteriorated beyond use. Mangkiling's status as a "good" village, that is, a village that has the privilege to hold on to current leadership and resources, cannot rest on its past achievements; it depends on keeping a stream of these visitors and benefits coming. This, in turn, depends on the village representatives' continued ability to present the village as needy, that is, backward and primitive enough to require special development attention. At the same time, they must present themselves as open to change, such that development attention will not be lost on them. Mangkiling has continued to be successful because its leaders have figured out how to present it as a community caught between tradition and modernity — needing help, and ready to change, at the same time as entangled in primordial cultural values.

This leadership stance is recognized in documents about Mangkiling that label it a "transitional" village. In the 1989 one-day seminar, for example, the head of the provincial directorate of village development concluded the presentations with an evaluation of the village as already on the move: due to the guidance of outsiders, the villagers were already more able to solve their problems and to escape the "influences of traditional custom that have a negative quality" (Soemarsono 1989: 1). Most importantly, he found "a change in attitudes and an open perspective along with the desire for progress" (1989: 2). Similarly, the 1990 research report of the provincial environmental organization found the village in a "transitional phase." "On the one hand, they want to carry out innovations; however, on the other hand, they are still tied to a traditional culture that does not support innovative efforts. This situation represents at the very least a potential for efforts at guidance and development" (Kompas Borneo 1990: ii). These evaluations assume that development, for villagers, is mainly a psychological process. They must rid themselves of adherence to static tradition and open themselves to change, that is, national directives; then outsiders will be freer to come in and tell them what to do. The challenge to maintain this transitional status — this openness in the midst of tradition — while courting a long string of advice and "guidance" from many visitors is formidable. Yet it must be maintained to keep the village's privileged status. This returns my analysis to the awkwardly off-balance pose of the village secretary, Yuni, surrounded by so many orderly-development experts. The always-unrealized yearning for change of this stance is perhaps even easier to see in the ways his older sister Sumiati, Mangkiling village head, negotiates her presentation in a series of newspaper articles about the village.

BROKEN PROMISES AND UNFULFILLED DESIRES

In October 1989, a series of six articles about the village of Mangkiling appeared in the provincial daily newspaper *Dinamika Berita* ("News Dynamics"). The articles, by woman reporter Irma Suryani, focus particularly on Sumiati and pay considerable attention to issues of concern to village women. Suryani is open-minded and sympathetic; her writing is warm, straight-forward, and sometimes poetic. She has clearly worked to build rapport with Mangkiling people. These are rare traits for any non-Dayak writer to bring to reports of Dayak communities. As a result, the self-presentations of village leaders come through with startling distinctiveness. It is not that villagers presented themselves to her with more authenticity or cultural autonomy than to other interlocutors; rather, because she listened to them, their distortions of regionally self-evident truths seem unusually clear in her portrayals.

Suryani's first two articles (1989a and b) revolve directly around her discussions with village head Sumiati. The reporter is sympathetic, and respectful of Sumiati's double burden as a woman village leader; she must carry responsibility for her family and overcome assumptions of women's political irrelevance at the same time as keeping up her leadership training and doing her job. (Indeed, Sumiati is one of two women village heads that anyone I spoke to could remember ever taking office in that entire regency.) Perhaps, the reporter seems to imply, Sumiati's unusual status as a woman village head makes her leadership dilemmas that much more striking: as a woman, no one would distinguish her from any ordinary traditional villager, but, as a leader, she has a dream of progress beyond tradition.

From the outset, Sumiati tells the reporter of her "hopes," "dreams," and "longings": she dreams that the village might have the conveniences of the cities; she longs for a road to be built to the village; she yearns for proper educational facilities. She hopes to be a "light" within her village. (The term the reporter uses for "light," *pelita*, is especially laden because it is the acronym for national five-year development plans.) Sumiati is especially clear about roads: "I wish so much that Mangkiling would have a road so that it would be easy for motor vehicles to come to the village," she tells the reporter "in her plain words." Her longing looks less plain-spoken if we look back a few years to 1986, when a road constructed by a timber company did come through the village territory. (By 1989, the road had eroded away, taking large pieces of hillside with it.) At that time, another newspaper article recorded the experiences of Mangkiling villagers, the reporter again taking his cue from village head Sumiati. "Other problems have been faced precisely because of the presence of a company that has made roads in the area of managed orchards. The fruit orchards of Mangkiling have been destroyed because they were hit by the road-building project of a company working there. Efforts to ask for help [compensation] have been made but have not received a response" (Ihsan 1986). As this quotation suggests, the issues that arise around road-building are complex. However, longing for roads is key to the "openness" that development thinkers demand. Mangkiling is "isolated," that is, primitive, as long as it is not on a motor road. Almost every report on the village begins with the difficult experience of the outside experts getting there; as long as they cannot travel easily to the village, there is no way that it can qualify as up to national standards. Sumiati is not faking her opinions: to speak within the lines of intelligibility, she glosses over her knowledge of the village's history with roads to show plain, innocent longing.[8]

Furthermore, Sumiati describes her longing for roads as just another example of an unfulfilled promise.

> She has often taken up this matter by approaching the qualified officials, but evidently of Sumiati's wish, only hope remains. "Several times already we have submitted proposals to the district to improve our settlement; our requests have even been approved. But in reality, it's not our village that receives the help, but another village, and we feel that we have been patient enough, even weary from waiting for the reality from these promises," she says, half moving me to pity.

"The People of Mangkiling Wait on a Promise," proclaims the headline of the second article (Suryani 1989b). By the time Sumiati has finished her explanation of the village's problems, it appears that the village has been offered nothing but empty promises. Even when they are offered "help," it comes in pointless, ritualistic forms that may satisfy regional administrators but is of little use for the village. The village has school buildings but no regular teachers to staff them. They have been given a television but no electric generator to run it. They have been formally converted to Islam but offered no religious instruction to learn it. If it wasn't so sad, one might say, it would be funny.

Empty promises have some local uses. The conversion story can illustrate: in 1985, the Mangkiling villagers decided to convert en masse to Islam. Regionally, Islam is equated with civilization, and thus this was a major step toward their acceptance of development. Hundreds of people hiked up to a wide spot in the timber road to meet the Ulamas who (arriving by motor vehicle) staged an official ceremony and duly noted and photographed the event. Then everyone went home. Afterwards, Mangkiling people continued to practice shamanic ceremonies and raise pigs and dogs. With a few exceptions, such as village leaders during their sojourns in town, no one practiced any Islamic religious rites. But they were then able to benefit from their ambivalence. On the one hand, one of the major attractions of Meratus Dayak villages for outside visitors is their colorful festival life. (One of Suryani's articles, entitled "Dancing Until Dawn" (1989e), describes a festival she attended.) On the other hand, no one can accuse them of being closed to the more cosmopolitan religion, Islam.

They are not stuck in tradition, but they do not lose their enticements for visitors — or their well-loved local events.

Sumiati builds her leadership stance on this ambivalence by placing the blame for failed development on the regional authorities. In another newspaper article, she explains the lack of Islamic practices in Mangkiling after their conversions as due to the sporadic attention of the provincial religious apparatus. She begins with her own conversion in 1982 when she was chosen as village head. "At that time, the proselytizers came to our place, but after that they have only come a few times, and as a result we don't know how to do the devotional activities," she explains. The prayer house built for them is falling down, she adds, because it is inconveniently located and no one came to care for it (*Dinamika Berita* 1989c). Surprisingly to me, this placement of the blame was readily endorsed by the authorities. Instead of blaming the villagers for their indolence or greed, provincial religious leaders, challenged by the newspaper articles, agreed that they had not properly instructed Mangkiling villagers, and that they must work harder in extending their missionary efforts (*Dinamika Berita* 1989d). Similarly, the Education and Culture Department took full responsibility for not sending teachers in a regular enough manner, when they, too, were challenged by the newspaper's reporting of Mangkiling complaints about the schools (*Dinamika Berita* 1989b and e). This occurred in a context in which regional authorities routinely blame villagers for their ignorance, bad habits, and lack of initiative. However, these latter traits are rooted in the "static thinking" of traditional culture, the bane of development. In contrast, no one can fault longing for change; this is what development is meant to instill. The trick for Sumiati, then, is to make visible a trail of broken promises that can be seen to generate ever more intense forms of longing.

The danger looms: because most development inputs are, indeed, gaudy handouts and cheerful rites with little long-term value, most will not have the kinds of transformative effects development planners fantasize. To the extent that regional administrators can interpret failed development in the village as a resurgence of tradition, that is, static thinking, Mangkiling will lose its privileged status as a "transitional" village, worthy of special development inputs. To renew these inputs, and with them village identity and leadership, Sumiati must continually produce an insatiable development longing. The traditional village woman must always have hope in her eyes for the lights of the city.

Tradition is that which developers most despise; yet it is also that to which they are most attracted. Ordinary poverty is uninteresting to those who imagine themselves civilizing the tribes. (Besides, tribal peoples are often well-endowed with land and resources until these are stolen from them; they don't necessarily need a better livelihood situation until after they are "developed.") Even as she honestly longs for change, Sumiati must know that no one would come to the village if it wasn't "backward." Backwardness is her commodity for negotiation. My next section explores the ways Mangkiling leaders are caught up in a discourse on tradition and exotic culture as they create, and are created by, the tribal situation.

LOVE MAGIC

Every village leader who wants access to development funds in South Kalimantan must cultivate a longing for development. Only Dayak minorities, however, must learn to work with the stigma of being considered not just technologically and economically backward but also primitive and exotic. The stigma is terrible, and it is created together with economic, political, and cultural discrimination. However, particularly in the last decade, there have been some ways to use it. The alliances Mangkiling leaders have built with environmental activists and their appeal to ecotourists are two clear examples of opportunities that would not have been available to South Kalimantan villagers not marked by the classification "primitive." With this support, based largely on their ability to identify as "indigenous people," Mangkiling villagers can at least *try* to create legitimate claims over their forests. Here lies the difference between those who can only work to create a "village situation" — a demand for rural citizenship — and those who can aim for a "tribal situation" — a staging of community identity and resource rights. To transform exotic stereotypes into community designs, however, is a work of magic — and a work of seduction.

One beginning move for outside advocates of the tribal situation has been to take the most positive stereotypes they know of the primitive to try to build an alliance with those whom they imagine as tribes. In this spirit, journalist Irma Suryani portrays Mangkiling villagers as experts in traditional herbal medicine, especially that used for contraception (1989c). International interest in indigenous knowledge of rainforest pharmaceuticals has come together with Indonesian population control priorities to make contraceptive herbal knowledge one of the few most positive "traditions" a minority ethnic group can have in Indonesia. Thus, Suryani portrays village head Sumiati expertly explaining the names and uses of herbs to regulate women's fertility. "Mangkiling people don't have to hassle with birth control pills because our natural world has already prepared birth control for us," Sumiati says "with pride." The journalist even permits a little criticism of development expertise: "We are afraid of the side effects," says Sumiati of birth control pills. With traditional contraception in hand, the ground is relatively safe.

Watching over the shoulders of the Kompas Borneo researchers, the journalist learns the names of a variety of traditional medicinal herbs explained by Mangkiling villagers: earth axis tree; King Kahayan vine; white medicine root; King Hanoman vine. Reading through the article, these names did not catch my eye; while none of them were herbs I remembered from the villages I know better in the Meratus Mountains, I expect variation in terminology, knowledge, and flora across the mountains. Then I encountered the list again in Kompas Borneo's report (Yayasan Kompas Borneo 1990: 24). After the list, the report continues, casually, "These medicines are also known to city people." Suddenly I remembered these herbs from urban and rural markets. They are not particularly Meratus herbs but rather commodified, cosmopolitan medicinal herbs used throughout the region. The self-positioning of Mangkiling informants became blindingly clear: to forge the best relationship, given the circumstances, tell the researchers the traditional medicines they already know.[9]

There is something here of flattery and of submission, but it is also an enormously complex skill to reproduce the dominant group's stereotypes so beautifully that they only see their imagined Other. Perhaps it is helpful to think of it in relation to the skill that women in so many places have used to make themselves attractive to men, that is, to make themselves "feminine" as men see it. This is one way to understand the erotic charge that this strategy of sympathetic acquiescence appears to have for outsiders and experts. Suryani is an honest enough reporter to let the reader see the male research group's compulsion to draw the village girls into a web of flirtation: "Wah... even without being dressed up you are so beautiful, let alone if you were dressed up, the city girls would lose," the men tease; "This one's name is Lili Marlen but she is lost in the Mangkiling forest" (1989d). But she also sexualizes the girls, describing their imagined ethnic innocence as seductive. The girls are natural objects of enticement, with their lively smiles and "golden skins" (the description often used regionally for Dayak women). Their naive efforts to adorn themselves are "cute" or "amusing" (*lucu*): they wear lipstick and curl their hair without knowing how. They wear their shirts open, revealing black brassieres, which sparks jokes with the researchers about the popular song, "Under the Dark Glasses."

In the hands of village head Sumiati, the seductiveness of asymmetrical ethnic acquiescence is both useful and hard to control. The primitive summons outside expertise into the community, but it also hints at illicitness and disorder. In this context, Sumiati appears in the newspaper as an ordinary Dayak woman: like other Meratus Dayaks, we learn, she has been married too many times (Suryani 1989b). The woman journalist tells us that this is unfortunate; even naturally seductive women, she seems to imply, can be victims. But she cannot completely suppress the sense that this is uncivilized sexuality. Indeed, those town people who had heard of Sumiati, who after all is a Meratus Dayak leader of some repute, warned me with rolled eyes that she was married to four men, not sequentially, but simultaneously. Whatever Sumiati says about her life, they do not believe her. For them, the seductiveness of Dayak exoticism turns quickly into savagery. Mangkiling leaders must handle this with care — for the closer they get to claiming the autonomy of tribal distinctiveness, the more erotically dangerous their claims.

Thus, according to reporter Suryani, when the Kompas Borneo researchers pin down the site where eroticism is thickest, they find it precisely in the formative place of exoticism and ethnic difference: magic. Magic is key to regional images of Meratus Dayaks. According to the regional majority, Meratus are sorcerers and concocters of magic oils, and it is this power that makes Dayaks both primitive and frightening. In my research in the region, I found that sorcery and magic oils were most important to Meratus Dayaks precisely as part of a regional trade with those who named Dayaks as sorcerers (Tsing 1993). In villages such as Mangkiling, outsiders make demands for mystical expertise, and, indeed, this expertise is produced. The importance of magic in regional images of Dayak "difference" is so great that I was not surprised that Suryani chose to devote half of her final Mangkiling article to magic oils (1989f). The oils she describes are used for seduction and for healing the wounds of fighting. In learning about them, the journalist and the researchers she accompanied place themselves in the middle of an ethnic exchange in which the seductions and healed-over hostilities of both exoticization and self-representation become difficult to disentangle. To follow this process, the article is worth quoting at some length:

> The issue that the writer will discuss here is the strength of belief of the Bukit [Meratus Dayak] people toward what one would call magic. They tie everything to the power of "dewa" spirits in which, until now, they believe.
>
> This is also the case with sorcery, which they always connect to mystical power. For example, this writer and the research group had the opportunity to meet with a resident of Pantai Mangkiling village whose condition was rather

alarming because other than suffering from deafness, he also had a deformed body. However, from him we obtained information as well as research materials that could be used for our analysis. Although to communicate with him, we had to use "Tarzan" language (signs).

From him this writer and colleagues from the Institute "Kompas Borneo" obtained an account of several kinds of oils with special qualities. For example, there is the oil that they call "Unchaste Adam" that they use to entice someone. Usually it is used by a woman to entice a man or, in reverse, for a man to entice a woman.

There are also oils that cause a person to be able to stand blows or gashes, and according to Pak Sani (a pseudonym), he has already proved it himself. Indeed, we could see his misshapen bones that looked like the result of a break but evidently had connected again (There is also an oil for this). Concerning the truth of these special characteristics, as presented by Pak Sani, this writer does not know but can only say that this is what they use up until now if they encounter the difficulties I have explained.

It is hard to fathom why the research group decided to use an interview with a deaf man as their decisive entrée into traditional knowledge. It is quite a scene to imagine: the deaf man and the researchers each pointing and gesturing and mimicking each other enough to develop some communication. The reference to Tarzan calls up the colonial situation, in which Europeans and "natives" faced each other across such gaps of communication, and in which at least the Europeans thought they were communicating with animals. Ganneth Obeyesekere has argued that European ideas of cannibalism in the Pacific were in part conjured up by scenes in which Europeans and Pacific Islanders, unable to speak with each other, each mimed a fantasy of cannibalistic consumption, biting arms and legs while the other party copied the mime (1992). In the Mangkiling exchange, too, language was omitted, and the researchers, through mime, learned exactly what they hoped and feared: Dayaks have the power to entice and to heal injury; their magic entraps expert attention and reconnects the shards of modern alienation. A fantasy of seduction and erased violence was woven around the deaf man's signs; the indeterminability of who exactly wove this fantasy is the underlying "magic" of the situation.

Through this love magic, Mangkiling villagers attract a stream of visitors, experts, and tourists. The motivations of visitors range from development assistance to nature appreciation to personal adventure; but all are drawn by the magic of exotic nature and culture. One record of these seductions is the visitors' log that is kept in the village office, where Yuni, the secretary, sometimes resides. Besides their names and the dates of their visits, visitors are asked to enter their trip's purpose and their impressions. Many of those who wrote in the log that I copied in 1994 explained themselves in the language of development; they came, they said, to examine, criticize, and help the villagers. But others wrote love notes — to nature, to the people of Mangkiling, and even in reference to their private affairs. Nature hikers expressed a platonic attraction: "Beautiful nature, friendly people"; "*Refreshing* while enjoying the ambiance of nature in the mountains of Pantai Mangkiling"; or, fully in English, "We are remember to Mangkiling. We can't stop loving you to Mangkiling." More ambivalent, perhaps, were the lovers who came to the village after it became a weekend destination for town toughs to bring their girlfriends; they drew on the hint of promiscuity that always accompanies love magic. Yet when one of these casual guests wrote that s/he had come to Mangkiling "carrying a heart wounded by my angry, jealous lover," s/he hinted at the dialogue in which Mangkiling had become an appropriate site for erotic recharging. Another guest drew an outline of a heart in the log.

THE UTOPIAN PROJECT: NATURE

In the ways I have been describing, Mangkiling leaders make themselves available to work with agencies interested in community development, ecotourism, rainforest conservation, and tribal rights. It is not enough to live in the forest. One must have a stable village that can be identified and funded. One must have a distinctive culture worth studying and saving. And one must have a strong, visible leadership to articulate community concerns in ways that these agencies can understand. To craft each of these is a work of imagination and artistry. Only with these prerequisites can Mangkiling be part of the global "sustainability" question: how can we meet the needs of the present without jeopardizing the resources of future generations? In that question, "tribal" forest communities have a special niche. Everyone wants to know: do these communities protect and manage the forest or destroy it? When agencies and experts flock to Mangkiling, it is in part because they are thinking about this question.

Yet, amazingly enough, this question is investigated directly nowhere in the documents I found about Mangkiling. Occasionally, an author makes a wild stab from his prejudices. Thus, although no research of which I am aware has examined Mangkiling forest use, an economist interrupts his otherwise modest survey of Mangkiling incomes to rant about the huge amount of money lost every time a Mangkiling farmer clears a swidden.

> 2400 cubic meters – 2800 cubic meters [of timber wood] x Rp. 50,000 = Rp. 120,000,000 – Rp. 144,000,000. If the problem of shifting cultivation is allowed to continue in the next ten years, one could estimate that forest

products, especially wood logs, worth 12 to 14.4 billion rupiah will be thrown away, not to count the environmental destruction that this causes. (Siddik 1989: 3)

This kind of thinking would be very easy to refute (e.g., by questioning the truth of the assumption here that Mangkiling farmers regularly cut down mature dipterocarps, by studying patterns of post-swidden forest regrowth and tree management, by examining forest destruction in commercial timbering, or by questioning who benefits from timber versus swidden incomes). Yet, for some reason, none of the many advocates who have conducted research in the village — and who clearly don't believe this economist — have bothered to address this question in their studies. Instead, they offer traditional beliefs in support of the spirit of forest conservation:

> The view of the people of the village of Pantai Mangkiling toward the world around them, such as the forest, mountains, rivers, and animals, is that it is a materialization from themselves (as human beings), and because of this they treat it carefully. (Yayasan Kompas Borneo 1990:35–6)

In explaining advocates' turn to traditional beliefs rather than local resource management practices, one might posit that advocates can't imagine officials taking local practices seriously; perhaps the idea that tribal people conserve forests is just too far from regional development dogma to imbue its technical features with any legitimacy. Alternately, perhaps conventions of separating social science and natural science research have made it difficult for researchers to ask questions about the human management of the environment. Yet a third possibility presents itself along with these: advocates' focus on abstract beliefs rather than a history of forest management practices creates a connection between environmentalists and villagers. Many environmentalists base their own hopes for forest conservation on the ability of their abstract beliefs in conservation to prevail, rather than on particular management practices. If village conservation is also based on an ecological vision, then villagers and environmentalists are ideal working partners.

Whatever the cause, there has been a noticeable silence on questions of the construction of the Mangkiling forest. Although researchers are clearly interested in the trees, no one has examined tree management; although they are interested in wild animals, no research has asked about hunting or the making of food-rich forest niches. The cycle of shifting cultivation is discussed, but researchers do not continue their studies after the harvesting of rice to ask about long-term vegetables, shrubs, and tree crops. And while one might assume that I bring up this silence as a criticism, in fact I want to point first to its positive effects. By ignoring the specificity of Mangkiling nature-making practices, and thus the differences in how nature is appreciated that divide urban environmentalists and rural shifting cultivators, environmentalists are able to imagine a utopian space of overlap and collaboration in which they join Mangkiling villagers in cherishing the forest. In this imagined space, loving the forest — the business of urban nature appreciation — is conflated with living in the forest — the business of Mangkiling village existence. The project of protecting this space of "nature" is utopian in both the best and worst senses. It is idealistic, offering the hope for making a liveable world. It is single-minded, glossing over its own improbabilities.

Furthermore, it has developed around its own distinctive and collaborative practice of naming the elements of nature. Most Meratus Dayaks know a great deal about their natural environment, including many plant and animal names, and, in my experience, people enjoy explaining these names to curious outsiders. Similarly, environmentalists love to learn the names of the flora and fauna. From these mutual pleasures, a characteristic event of environmentalist visits to Mangkiling has developed: the shared experience of hiking around identifying natural organisms. Of course, there are great differences in the significance of these names as a component of forest-management practices. Indonesian environmentalists draw on the European natural history tradition in which to name nature is to know it in all its universal abstraction; they also practice a more recent kind of nature loving in which to identify a plant is to identify with it, that is, to feel a sense of communion and mutual belonging on earth. In contrast, Meratus Dayaks tend to be most interested in the specificity of plants and animals as they occur in particular landscape locations. To know a tree it is not enough to know its species name; one must be able to understand the complex of other plants as well as human claims and histories that put that tree into a socially meaningful landscape. Despite the need to ignore these differences, however, plant and animal identification is a truly collaborative practice. Both environmentalists and Mangkiling villagers with whom I spoke felt a sense of having shared important information with the other.

My interpretation of naming nature in Mangkiling as collaborative diverges from recent scholarship that identifies "botanizing" as among the most insidious of imperialist practices. Both Mary Pratt (1992) and Paul Carter (1989) argue that European colonization was brought to a new standard of control through natural history, which, they argue, taught Europeans to imagine Third World lands as entirely without inhabitants.

By describing landscapes full of plants and animals, but without humans, eighteenth century natural historians created narratives that facilitated colonial control. Recent events in Mangkiling do not tell us anything about the eighteenth century texts these authors analyze; however, they do suggest that natural history investigations can be more politically open-ended and flexible than these scholars imply. Mangkiling "botanizing" texts also do not contain any writing about Mangkiling people. They tend to focus on lists of plants and animals with perhaps short descriptions or discussions of the landscape. However, a closer reading of these texts suggests the way utopian collaboration peers out even from a list of trees.

NO.	AREA NAME	LATIN NAME	PLOTS					TOTAL
			1	2	3	4	5	
01	LANDUR	*LOPHOPETALUM JAVVANICUM* (ZOLL) TURZ	3	–	–	2	4	9
02	HAMAK		1	–	–	1	1	3
03	MAHANG	*MACARANGA HYPOLEUCA* MUELL. ARG	4	–	–	–	–	4
04	HUMBUT	*XYLOPIA* SP.	1	–	–	–	–	1
05	MINJURUNG		3	1	–	–	–	4
06	TIWADAK	*ARTOCARPUS RIGIDUS* BL	1	–	–	–	–	1
07	LURUS	*PERONEMA CANESCENS* JACK	2	–	–	–	–	2
08	RAMBUTAN	*NEPHELIUM* SP.	1	–	–	–	–	1

EXCERPT FROM "INVENTORY LIST OF FLORA," YAYASAN KOMPAS BORNEO 1990.

I believe this excerpt can stand in not only for the rest of that tree list, which goes on for pages, but also for other tree lists I have encountered, published and unpublished. It follows the convention of supplying two items: "area name" and "Latin name." The latter is the scientific, Linnean term that unites genus and species; presumably the botanist supplies this information after s/he sees the tree. But the former term, the local term, suggests that the botanist does not find and identify the tree alone; s/he is brought to the tree by a villager who serves not only as guide but also as first botanical identifier.

The priority of the Mangkiling identification is suggested by the fact that in two cases (#02,#05), an "area name" is not followed by any scientific identification. The villager appears to have shown the botanist a tree s/he did not know. (This is consistent with the rest of the list, in which there are many blank spaces in the "Latin names" column, but no blank spaces under "area name.") Sometimes, perhaps, the botanist asks for a name for a tree about which the villager is unsure. (I have my doubts about #04, *Humbut*, "palm heart," as the best possible Mangkiling name for this plant, which I assume to be a palm; Meratus palm classifications can be very detailed.) But the local name is never omitted; it forms the first line of knowledge about the tree.

Other minor collaborations are suggested. For example, slightly later in the list, there are fourteen trees identified as *Damar* (area name)/*Shorea* sp. (Latin name), suggesting a joint decision not to be too picky about identifications. Dipterocarpaceae, the big emergents of the forest, are notoriously divergent as well as hard to sort out — from the perspective of botanists as well as Meratus Dayaks. Yet both do sort them out for appropriate occasions. These fourteen trees may not have sparked that sense of occasion for either party to the identification. For other dipterocarp entries on the list, smaller divisions *are* made.

The inventory offers the chance for another collaboration, however, that is not pursued. If read with the right questions, the list is a striking testament to the managed nature of the Mangkiling forest. *Landur* (#01), *Tiwadak* (#06), *Rambutan* (#08), and *Siwau* (#25) are highly valued fruit trees; they were probably planted, or, at the least, claimed and managed carefully. *Kahingai* (#20), *Kembayau* (#21), and *Tarap* (#22) are less valuable fruit trees; while they may not have been planted, Mangkiling residents would certainly have their eye out for them. *Damar* (#16, #17, #18) and *Bangkiray* (#09) can become sites for honey bee nests, in which case, they become expensive and carefully guarded claimed trees. Even without bees' nests, the *damar* trees may have been saved in swidden-making, encouraged, or claimed for their bark, resin, or other uses. *Lurus* (#07) has become highly commercialized in this region, since its price for construction poles rose sharply in the 1980s; it is a quick-growing and easy-to-foster secondary forest species, claimed by those on whose old swiddens it is encouraged. One could continue. However, this is not the framework to which this inventory has so-far been deployed. Off the track of the utopian project, forest management raises difficult questions about nature's purity and purposes. While one must praise the inventory project for allowing this unbidden text to be recorded, one could also criticize it for not, or not yet, making it possible to discuss these issues. As Musa stated in the document with which I began this essay, "there is no wild forest here." Yet environmentalists still need the image of the wild with which to build their most promising alliances.

MAPS AND DREAMS

Instead of listening to Mangkiling villagers' histories of forest management and use, environmentalists build their practical project of advocacy on a different front: the mapping of village territory. Perhaps this is their most important work for Mangkiling villagers; at least potentially, it offers the possibility of making a case for village control of land and forest resources. It draws together all the imaginative frameworks for collaboration that I have been discussing to create what appears to be a singular joint project: the map. The lines of the map offer a "common sense" obviousness. Either this is your territory, or it is ours; any administrator should appreciate that. However, mapping a politically charged landscape is never so simple. Environmentalists and Mangkiling leaders work together, I will argue, to use the technology of precision to increase fertile ambiguity, multiplicity, and confusion. From ambiguity, the possibility of tribal rights might emerge.

The potential of Mangkiling maps to build tribal rights is based on the viability of attempts around the world to reclaim resources through what Nancy Peluso has called "counter-mapping," that is, the use of maps to argue against state claims by spatially depicting the explicitness and historical priority of local resource control (1995). (Peluso's term acknowledges that mapping has generally been the tool of colonial or state expropriation of local lands; as she explains, however, mapping can also become a strategy of local resistance and struggle.) In places across the Americas, Australia, and Southeast Asia, including Indonesia, the issue of tribal rights has been argued through mapping. Thus, for example, the title I have given this section, "maps and dreams," invokes one of the these projects: the customary-use mapping project of Northwest Canadian Native Americans, as described by Hugh Brody (1981). In this project, a key challenge was the forgetfulness of the white-settler majority that living communities of Native Americans continued to exist; thus, when Native Americans mapped the spots they had gone hunting, fishing, or berry picking, they reminded the white majority of their presence. The maps Brody records show entangled lines of personal and community use of land and forest resources. In contrast, the mapping challenges or "dreams" in Indonesia are different. Since colonial times, the geography of local Indonesian peoples has been imagined in generally non-overlapping, bounded territories; local groups have been identified in relation to such imagined territories, and "indigenous" advocacy has often begun with the notion of territory. These are the territories recognized as *adat* lands, that is, the lands acknowledged under customary law. Counter-mapping projects make these adat territories explicit; they generally do not, however, break with historically legitimate conventions for imagining space — for example, to show overlapping patches and entangled lines marking histories of individual and collective use, as in Brody's maps. To be effective, mapping for tribal rights must be convincing within regional and national histories of policy and politics.

This need to convince opens opportunities even as it imposes constraints. In their maps, environmentalists and Mangkiling leaders have adapted the colonial and national advocacy-through-adat tradition to make a joint statement about village lands and forests. Since adat is nationally understood as an indigenous conceptual system, to map adat lands is to articulate the inner logic of indigenous minds. Maps are not seen as analyses or even descriptions of tribal life; like folklore or cosmologies, they are supposed to be direct expressions of the native point of view. Collaboration between environmentalists and village leaders does not, then, produce "the native in the document" for which my earlier questions searched; instead, it aims for "the document in the native." Unlike lists of trees in which collaboration is made evident, the goal in making maps of adat lands is to create a single, seamless product in which the technological expertise of the map-maker seems only to enhance the traditional knowledge of village elders. To make this joint product, both environmentalists and village elders must imagine they are mapping the same thing: here the common space created by the utopian project of nature becomes crucial. The maps then superimpose and join the tactics of village leaders and environmentalists, as each aims to convince the authorities of the legitimacy of adat lands.

Making adat claims legitimate is no easy task, despite the long history of administrative discussion of adat lands in Indonesia. It is never enough merely to establish the status of a given plot as adat land in order to hold it; one must then argue against all the other classifications to which that same plot is assigned. First, adat land is an insecure classification. Since the colonial era, arguments for the recognition of adat lands have always been "counter-arguments" in a debate in which state domain over land and resources has been the opposing opinion.[10] In Indonesian national law, adat lands are sometimes recognized and sometimes not. In the Basic Agrarian Law, for example, adat is said to be the underlying law of the land. In the Basic Forestry Law, in contrast, all forests are said to be the domain of the state. The partial recognition of adat creates the possibilities for local arguments over the status of particular territories.

Second, official mapping offers contradictory views about the status of any given plot — whether or not adat status is at issue. Territories officially classified as "forests," i.e., government-controlled land, may include entire districts and multiple towns and villages with their agricultural terrains. Government departments often map areas differently, such that potential transmigration sites, production forests, and nature reserves may be found, in different maps, on precisely the same site. The forest in

Mangkiling is simultaneously classified as protection forest, production forest, a proposed nature reserve, and village territory.

How can village rights be established in this mess? The counter-mapping projects in Mangkiling do not clarify the situation; instead, advocates and village leaders add to the layers of ambiguity. Rather than making a single, clear-cut map, environmentalists and village leaders in Mangkiling have confused and layered conventions and land claims. First, they have conflated varied map-making standards to create complex products in which different kinds of land claims appear to garner the same legitimacy. Second, they have stacked overlapping, contradictory, and redundant maps. All the possible claims on the forest are shown, sometimes on top of each other, sometimes on separate pages. In the context of village powerlessness, clear and simple village claims would probably be officially dismissed, while adding layers to already recognized claims creates the potential for tentative local successes. By adding to the pool of overlaid possibilities, they make openings for local claims that cannot hold their own as singular logics.

The chain of village maps I have seen begins with two maps attributed to Musa and drawn sometime in the 1980s. I am unsure who else besides Musa worked on these maps; I assume they are the collaborative product of Musa and village advocates. I reproduce the first, the easier to read, as Map 1. At first glance, this is a nicely drawn but ordinary enough sketch map of village territory, as marked by the locations of the various constituent balai halls, as well as the village center and school buildings. Yet closer attention to the stylization of the map suggests that it offers more than the location of village settlement clusters; it creates the implication that these settlement clusters control territorial segments, which together constitute village land. In this sense, the map, like the written document to which it is attached, is a land rights claim. In order to achieve this effect, the map brings official mapping conventions to portray local conventions of land use and occupancy. However, neither mapping conventions nor local land-rights conventions go untransformed in the process. In order to make a hopeful village land claim, Map 1 overlaps, combines, and deforms both local and official understandings of landscape.

The map presents the local river system as if it were a set of boundary lines both drawing together and dividing up the land; tree-like, there is a straight, upright trunk stream — which defines the unity of village settlement — with branching arms that mark off village subsections. (The stylization becomes evident when Map 1 is compared to the more-standard geographic representation made by environmentalists in Map 2.) The

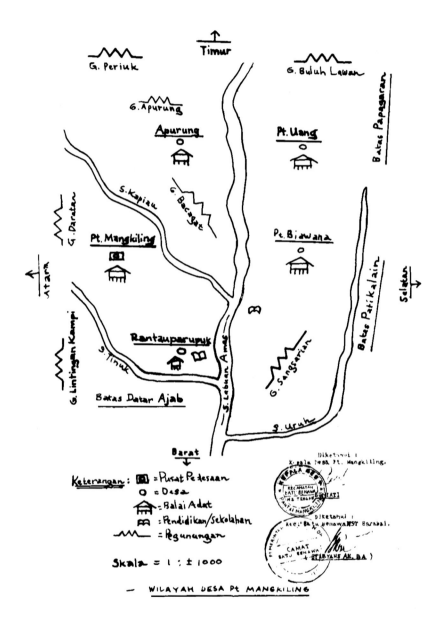

MAP 1 TERRITORY OF PT. MANGKILING VILLAGE.
ORIGINAL MAP BY MUSA. MAP REDRAWN BY BRIAN ROUNDS.

LEGEND: ▣ VILLAGE ADMINISTRATIVE CENTER   ○ VILLAGE SETTLEMENT   ⌂ TRADITIONAL HALL
🕮 EDUCATION/SCHOOL   ⋀⋀⋀ MOUNTAIN RANGE

UTARA "NORTH"; SELATAN "SOUTH"; TIMUR "EAST"; BARAT "WEST"; PT. [PANTAI] "FLAT"; BATAS "BOUNDARY"; S. [SUNGAI] "RIVER"; G. [GUNUNG] "MOUNTAIN"

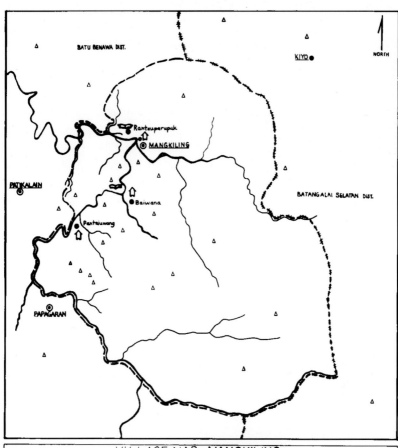
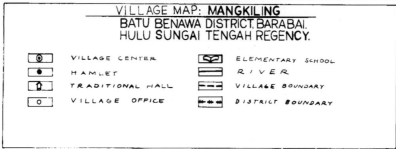

MAP 2 VILLAGE MAP: MANGKILING.
ORIGINAL MAP BY KOMPAS BORNEO INSTITUTE. MAP REDRAWN BY BRIAN ROUNDS.
ORIGINAL OFFERS ADDITIONAL GEOGRAPHIC DETAIL AND LEGEND IN INDONESIAN AT A SCALE OF 1:25,000.

river system appears to divide village land into discrete and somewhat equivalent chunks. Each chunk has a traditional hall, the heart of a community, at its center. Mountains bound the territories where they are not marked by streams. In this representation, then, community centers appear to preside over segmented territories, whose unity makes up the village.

The map's success in drawing the village in this way draws on two key features of the Meratus Dayak social landscape: the association of particular kin and neighborhood groups with particular areas of the forest, on the one hand, and the focus of social ties around particular leaders, groups, and central sites, on the other. Areas of the forest are associated with groups of people who once created swiddens there, and who continue to plant, encourage, harvest, and manage the forest there. Old living sites as well as farm sites become orchards and foraging grounds for those who know them best. The managed and well-used forest territories of different individuals overlap. However, group clustering around focal individuals, families, or sites creates the effect of center-controlled territories. When people live together in a balai hall, their familiar forest and swidden territories spread and radiate in each direction around the balai. It is these center-focused territories that are given the authority and permanence of graphic representation in this map. The mapped territories are not illusory; however, they stabilize and specify shifting aggregations.[11]

The map uses and confuses Mangkiling landscape conventions, but it does the same with official mapping conventions. Territorial domains claimed by settlements are never drawn in official maps in this region. Official maps offer a strict separation of settlement, on the one hand, and territorial divisions, on the other. They show settlement as a dot rather than a territory. Even huge villages with dozens of small, scattered settlements are depicted as a single dot. This dot represents the stability, and thus the administrative appropriateness, of settlement; no village can claim legitimacy without its dot. But a dot takes up no space. In contrast, territorial divisions are marked in official maps of land use, forest classification, concession areas, and the like. Settlements may be sketched in on these maps, but they are for place identification not territorial claims; these are maps of state and private domain. They offer villagers no rights. The Mangkiling map Musa sponsored conflates and combines these two bureaucratic conventions, to create an intelligibility that draws on and exceeds each. His map offers administerable centers yet implies territorial jurisdiction. It is a usefully confusing hybrid.

This kind of creative confusion was not the choice of the environmentalist mappers who followed Musa's lead to draw more maps of the village in the 1990s. These mappers show much more allegiance to official conventions; after all, they want their maps taken seriously in official circles. Thus, they reseparate out administrative and territorial maps. Their administrative map (Map 2) shows the familiar dots, as these guarantee that Mangkiling will be administratively recognized. Like Map 1, however, Map 2 shows all the constituent balai and settlement groupings of Mangkiling rather than just a single village center. It is a joint project of representation that employs local categories. It also includes village boundaries, but because of the irrelevance of their spatial relation to the settlements, it is hard to use this map to imagine that village people control all this territory.

The territorial maps produced in this project neatly depict village adat lands, including current and past swiddens and protected adat forest. Territorial maps insert Mangkiling claims into the realm of forestry department and land use planning representations; they argue for equal billing for village territories. The messiness and shifting status of forest territories thus must be eliminated; secondary forest and protected forest must be separated by neat lines. Here, too, village leaders and environmentalists must have worked together to form a joint product of hopefully-legitimate simplification.

The environmentalist maps, however, do not stop here; they proliferate in piles of overlapping territorial classifications. In showing all kinds of claims and classifications, these maps extend the concerns of village leaders into the agendas of environmentalists; they make it possible to imagine a democratic space of debate, that is, to make forest territories into a "public sphere" of pluralistic and open discussion. Kompas Borneo's Mangkiling project has produced not just maps of adat-protected forest and maps of village swidden areas, but also maps of production forest timber concessions and maps of nature reserve areas. And, most pointedly, there are maps in which many of these things are shown on top of each other. I reproduce one of the most intricate as Map 3. Map 3 shows the timber concession of the company that was logging Mangkiling trees in the 1980s. The concession neatly overlaps the zone of village territory, including both mature (*hutan*) and secondary (*belukar*) forest. The map, to me, is a *tour de force*. Village claims are given the same status as timber company claims — thus offering a sensitive official the chance to pick the villagers rather than the company as the appropriate local claimants, and all without having to uphold a general principle of adat rights. By showing overlap and contest in forest classifications, environmentalists add — rather than subtract — layers of possibility in policy discussions. The precise technologies of mapping do not narrow down the truth but instead open territorial classifications as a matter of democratic public debate. Indeed, this proliferation of options makes the alternative conventions of the map attributed to Musa also come alive as the map that could be made by the tribal elder, the indigenous map. Its collaborative layers disappear as it too becomes one perspective in this debate, the village text in the technical dossier.

REPRISE

What does it mean to speak of or for a "tribe" in the late twentieth century? The term has emerged in international movements for environmental conservation and minority rights to draw attention to the political and ecological importance of marginalized rural communities. At the same time, scholars have criticized the traditions of representation in which these communities have been understood to have backward customs and exotic cultures, that is, to be identified as tribes. The concept of the "tribe," recent scholars argue, calls up a history of metropolitan fantasies about the bizarre, the natural, or the originary lines of human evolution (e.g., Clifford 1988; Torgovnick 1990; Kuper 1988). It is never a simple descriptive term.

The political rehabilitation of the tribe and its scholarly rejection too often speak past each other. Instead, I have argued that we must begin both our political rapproachments and scholarly investigations with the question of how the concept of the tribe, with all its simplifications and codifications of metropolitan fantasy, comes to mean something to people caught in particular political dilemmas. The fantastic aspect of tribal identity does not make it irrelevant to marginalized people who pass as tribals; to the contrary, it is the fantasy of the tribe that becomes the source of engagement for both tribals and their metropolitan others. Both scholarship and advocacy deserve a closer look at such histories of engagement.

Recent cultural theorists have shown how cosmopolitan dreams and fantasies forge the categories and narratives through which central and peripheral social settings are segregated and aligned with each other. Emergent notions of polity and history — such as modernity (Foucault 1970), nationalism (Anderson 1983), colonial rule (Stoler 1991), or archaic folk traditions (Ivy 1995) — have rebuilt the framing architecture through which we organize and recognize the local, in city and countryside, lowlands and uplands. "Local" self-conceptions and notions of place, personhood, desire, marginalization, and resistance have changed

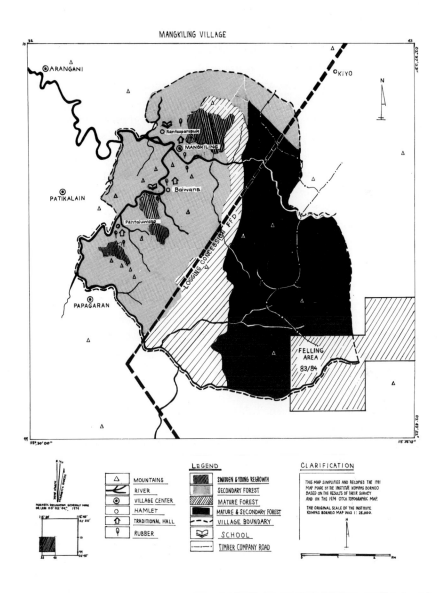

MAP 3 MANGKILING VILLAGE.
ORIGINAL MAP BY KOMPAS BORNEO INSTITUTE. MAP REDRAWN BY BRIAN ROUNDS.
ORIGINAL OFFERS ADDITIONAL GEOGRAPHIC DETAIL AND LEGEND IN INDONESIAN AT A SCALE OF
1:25,000.

to live within these emergent architectures. We assert ourselves as "rational men" as "citizens," as "natives," as "women," or as "community representatives" within the cosmopolitan dreams and schemes that make these self-imaginings possible. Yet these dreams and schemes never work out in the ways they are supposed to. Their formulations of difference get away from them, slipping into unexpected transformations and collaborations. No theory of resistance along the lines of already assumed, immutable material interests (workers on strike; peasants in community) can capture the nuances with which metropolitan desires fulfill themselves. What is needed is a theory of localization, in which attention can be focused on the ways categories become stretched beyond themselves in particular events and confrontations. Such a framework points us toward the situational deformation of globally circulating categories. In my examination of Mangkiling documents, I have focused on the staging of "situations" in which the categories of green development are creatively transformed to make Mangkiling a village, a tribal location, and a place on the map that cannot be erased.

The "tribal elder" is a position empowered by international concerns for environmental sustainability and community-based environmental justice. This is an agenda with powerful backers but also substantial enemies. Its local deployments, however, do not depend entirely on the international play of this agenda; instead, they involve attempts by would-be tribal leaders and their advocates to pick up on important local concerns, that is, to contextualize international agendas and shape them in new ways. Notions of community, territory, and culture are reconstructed around the new tribal discourse as it is interpolated with tribal deployments in government administration, commercial enterprise, regional religious doctrine, research, and tourism.

Local articulations of tribal autonomy and rights make use of "room for maneuver" within administrative categories for local people and activities. Even so, some creative transformations are needed to make the difference between resource loss and bureaucratic encompassment, on the one hand, and community initiative, on the other. In development programs that require local communities to function as docile administrative units, room for maneuver is particularly prominent in the community research components that readjust and align development initiatives at the regional level. Environmentalist concerns, which entered Indonesian regional development in the 1980s, increased this community research load and shifted some of it to non-governmental organizations, some of which thought of themselves as community and environmental advocates. Through this trajectory, tribalism entered within the program of development.

In Mangkiling, then, tribal elders long for development at the same time as they hold on to markers of tradition. The appearance of tradition draws the guests who hope to change them and offers them legitimacy among these guests as authentic community spokespersons. With the right leadership stance, it becomes possible to enter into collaborative projects in which Mangkiling concerns assume the aura of urban professional environmentalism, and vice versa. The more layers of alternative interpretations collaborators are able to add, the better the chances, one might argue, of successful Mangkiling advocacy. These collaborative layers then form the space of local articulation for so-called global environmentalism. They also transform it, as it becomes a tool within local negotiations of related, but not synonymous, makings of Mangkiling.

ACKNOWLEDGEMENTS

This paper draws from interviews and documents gathered in 1994. I would like to thank Mangkiling village leaders, past and present members of Yayasan Kompas Borneo, and the Jakarta Program Officers of the Ford Foundation for their generous willingness to discuss the issues and events described in this paper and their openness with documents about Mangkiling village. I am also extremely grateful to Tania Li and Nancy Peluso for careful readings of an earlier draft of this paper. I benefitted, too, from discussions with Mercedes Chavez, Paulla Ebron, Karen Gaul, Cori Hayden, and Bettina Stotzer. Brian Rounds produced the maps, working from South Kalimantan originals. Map 1 has been reproduced as exactly as possible; Maps 2 and 3 were redrawn to increase clarity by eliminating some detail, such as latitude and longitude grids.

NOTES

1. Nancy Peluso (personal communication) offers an important political contextualization: "These are clearly not inter-village movements because this would be politically impossible. Organizing across villages could raise various spectres; if not communist or "tribal" insurgency, any anti-government organization would be suspect. So the focussing of "development" on making documents happen, creating situations, etc., is of necessity focussed *on the village* and best served in the person of the village leaders."
2. Preliminary histories and analyses of the Indonesian environmental movement can be found in Belcher and Gennino 1993 and throughout the journal *Environesia*. For an account of environmentalists' attempts to use the concept of *adat* to build a national appreciation of tribal land and resource rights, see Tsing n.d.c.
3. Much of Escobar's (1995) analysis of development expertise is relevant to Indonesia.
4. There is a social forestry program with pilot projects in South Kalimantan, but the focus of this and other "participatory" efforts is to design model communities rather than to empower already existing community-based forest management.
5. If this seems odd, it may be useful to think of a woman's enactment of womanhood or an

Asian country's enactment of the Orient; where are the lines between player and role? Self-making here brings to life the powerful desires that define one's Otherness; and only by inciting those powerful desires can one act "as a woman" or "as the Orient." Other kinds of agency are, of course, possible for these actors, but these do not lead to collaborations on these lines of difference, here manhood and womanhood, East and West.

6   To analyze documents written by and about relatively uninfluential people raises important questions about confidentiality and exposure. Once the analysis refers to a public document, it becomes impossible to change the names and places referred to; yet, it seems proper to protect the strategies and reputations of both writers and their objects from undue prying. In this essay, I have tried to keep my analysis to documents that have been distributed, registered, or published in public places. Furthermore, I have tried to avoid attention to idiosyncratic foibles and mistakes to focus instead on systematic meanings and asymmetries as well as acts of courage and imagination.

7   Mangkiling was never the center of my ethnographic research. I have stayed in the village and talked with Musa and other key figures, and my understandings of our conversations are guided by research in other Meratus areas (see Tsing 1993). Even without extensive participant-observation, the documents are revealing; they offer the kinds of historical materials so often unavailable to an ethnographer of rural areas.

8   Tsing n.d.b. offers a complementary but rather different analysis of Meratus road-longing.

9   It is unclear from the texts whether the researchers asked about medicinal herbs by name or solicited these names from Mangkiling informants. In either case, it appears that the villagers did not challenge the researchers' ideas of what might constitute "traditional medicine." I imagine that the researchers were already so sure of the forms exoticism should take that a heroic effort to introduce new pharmaceutical models probably would still have been unsuccessful. Rather than intentional deceit, going along with researchers' preconceptions involved only villagers' willingness to avoid being annoying.

10 Potter (1988: 138–41) describes the debate among colonial officials over forest control in Borneo earlier this century; aspects of this debate are replicated in current controversies over Kalimantan forests.

11 Stabilization and specification began long before this map, and the map cites and rewrites other efforts. In particular, Musa's imaginative framework for the map appears to invoke an earlier document he helped design in 1967, when he was village head. This was a written text which put land rights on paper by assigning sectors of the village to particular neighborhood groups.

References

Anderson, Benedict, 1983, *Imagined Communities: Reflections on the Origin and Spread of Nationalism*. London: Verso.

Belcher, Martha and Angela Gennino, 1993, *Southeast Asian Rainforests: A Resource Guide and Directory*. San Francisco: Rainforest Action Network.

Brody, Hugh, 1981, *Maps and Dreams*. New York: Pantheon Books.

Brosius, J. Peter, Anna Tsing and Charles Zerner, 1998, "Representing Communities: History and Politics of Community-Based Resource Management". *Society and Natural Resources*, 11, 157–168.

Carter, Paul, 1989, *The Road to Botany Bay*. Chicago: University of Chicago Press.

Chambers, Ross, 1991, *Room for Maneuver: Reading the Oppositional in Narrative*. Chicago: University of Chicago Press.

Clifford, James, 1988, *The Predicament of Culture*. Cambridge: Harvard University Press.

Debord, Guy, 1983, *Society of the Spectacle*. Toronto: Black and Red Books.

Dinamika Berita, 1989a, "Sistem Perekonomian Masyarakat Mangkiling Sangat Sederhana", April 13.

Dinamika Berita, 1989b, "Hanyar Mangajar 2 Hari dalam Sebulan". October 16, 1+.

Dinamika Berita, 1989c, "Warga Mangkiling Kembali Anut Kaharingan". October 17, 1+.

Dinamika Berita, 1989d, "Perlu Keterlibatan Semua Pihak". October 18, 1+.

Dinamika Berita, 1989e, "Dinas P & K akan Ambil Tindakan Kepegawaian". October 19, 1+.

Escobar, Arturo, 1995, *Encountering Development: The Making and Unmaking of the Third World*. Princeton: Princeton University Press.

Foucault, Michel, 1970, *The Order of Things*. New York: Vintage.

Ihsan, A. Muhaimin, 1986, "Pantai Mangkiling, Di Antara Benturan Tradisional Dan Modernisasi". *Banjarmasin Post*. December 21, 4+.

Ivy, Marilyn, 1995, *Discourses of the Vanishing: Modernity, Phantasm, Japan*. Chicago: University of Chicago Press.

Kuper, Adam, 1988, *The Invention of Primitive Society: Transformations of an Illusion*. London: Routledge.

Mooduto, H.L.D, 1989, "Pokok-Pokok Pembinaan Masyarakat Terasing Daerah Kalimantan Selatan" in Yayasan Kompas Borneo 1989.

Obeyesekere, Ganneth, 1992, "'British Cannibals': Contemplation of an Event in the Death and Resurrection of James Cook, Explorer". *Critical Inquiry*, 18(Summer), 630–54.

Peluso, Nancy, 1995, "Whose Woods Are These? Counter-Mapping Forest Territories in Kalimantan, Indonesia". *Antipode*, 274(4), 383–406.

Potter, Leslie, 1988, "Indigenes and Colonisers: Dutch Forest Policy in South and East Borneo (Kalimantan) 1900 to 1950" in *Changing Tropical Forests*, edited by John Dargavel, Kay Dixon and Noel Semple, pp. 127–53. Canberra: Centre for Resource and Environmental Studies, Australian National University.

Pratt, Mary, 1992, *Imperial Eyes*. New York: Routledge.

Siddik, Abdullah, 1989, "Makalah Mengenal Kehidupan Orang Bukit Di Pedalaman Mangkiling Ditinjau Dari Aspek Sosial Ekonomi" in Yayasan Kompas Borneo 1989.

Soemarsono, 1989, "Tanggapan Berupa Catatan Dan Masukan" in Yayasan Kompas Borneo 1989.

Stoler, Ann, 1991, "Carnal Knowledge and Imperial Power: Gender, Race and Mobility in Colonial Asia" in *Gender at the Crossroads of Knowledge*, edited by Micaela di Leonardo, pp. 51–101. Berkeley: University of California Press.

Suryani, 1989a, "Sumiati, Profil Wanita Desa Mangkiling". *Dinamika Berita*. October 7, 1+.

1989b, "Masyarakat Mangkiling Menunggu Janji". *Dinamika Berita*. October 9, 1+.

1989c, "'Akar Rapat' Populer Sebagai Alat KB". *Dinamika Berita*. October 15, 1+.

1989d, "Keriting & Bibir Merah Jadi Model". *Dinamika Berita*. October 16, 1+.
1989e, "'Batandik Sampai Pagi'". *Dinamika Berita*. October 17, 1+.
1989f, "Minyak 'Sumbang Adam' Untuk Pemikat". *Dinamika Berita*. October 18, 1+.
Torgovnick, Marianna, 1990, *Gone Primitive: Savage Intellects, Modern Lives*. Chicago: University of Chicago Press.
Tsing, Anna, 1993, *In the Realm of the Diamond Queen: Marginality in an Out-of-the-Way Place*. Princeton: Princeton University Press.
n.d.a. "What Happened in Mangkiling: Localizing Global Environmentalism".
n.d.b. "The News in the Provinces" in *Cultural Citizenship in Southeast Asia*, edited by Renato Rosaldo.
n.d.c. "Land as Law: Negotiating the Meaning of Property in Indonesia".
Yayasan Kompas Borneo, 1989, *Laporan Seminar Sehari: Kamis, 6 April 1989*. Banjarmasin, Indonesia: bound photocopy.
Yayasan Kompas Borneo, 1990, *Studi Diagnosa Tentang Komunitas Orang Bukit di Desa Pantai Mangkiling Kalimantan Selatan*. Banjarmasin, Indonesia: bound photocopy.

Chapter 7

REPRESENTATIONS OF THE "OTHER" BY OTHERS: THE ETHNOGRAPHIC CHALLENGE POSED BY PLANTERS' VIEWS OF PEASANTS IN INDONESIA

Michael R. Dove

INTRODUCTION

Indonesian plantations were thrust into the global limelight in 1997 by dramatic episodes of social and environmental violence in Kalimantan. The former involved an outbreak of tribal warfare in the vicinity of huge oil and rubber plantations in West Kalimantan in January and February, and the latter involved the engulfment of plantation lands throughout Kalimantan by wildfires so great as to imperil human health in several neighboring countries. These developments, unique in both scope and character, can be interpreted as signs that something is fundamentally wrong with the social and environmental relations in the state plantation sector (Dove 1997). Yet public representation of these relations, which has been dominated by the official views of plantation managers, gave no prior hint of this. The social and environmental conflagrations of 1997 suggest that these official views have been distorted and self-serving. The purpose of this study is to examine how and why this distortion occurs.

Plantations and Peasant-Planter Rhetoric

State and para-statal agricultural enterprises constituted an important part of Indonesia's upland (and to a lesser extent, lowland) landscape in historical times and this is ever-more true in the contemporary era. The lives of significant numbers of the peasants and tribesmen who live in these uplands have been affected by these enterprises, either because they work for them or because they compete with them for local lands and other resources. Peasant-planter conflict stemming from these labor relations and resource competitions has become a salient component of the upland "ethnoscape." The national bureaucratic elite that manages the plantations attributes these conflicts to the cultural, economic, and political backwardness of the peasantry. Indeed, the rhetoric of the managerial elite characterizes this local population in "polar" terms: in industry, intelligence, and attitudes toward development, they are portrayed as being the opposite of the elite in every way.

One of the most distinctive features of this official characterization (and the feature that initially prompted this study) is its uniformity across different provinces, ethnic groups, and plantation types and crops (see Rosaldo 1978). These rhetorical continuities extend to Indonesia's timber concessions (*Hak Pengusahaan Hutan* [HPH]) and its now-proliferating timber plantations (*Hak Tanaman Industri* [HTI]). They also extend across time: the way that planters represent peasants today is much like the way that they represented them in the colonial era, even down to the use of the same pejorative terms and phrases. Finally, these continuities extend across uplands and lowlands: planters' characterizations of workers in lowland sugar plantations and mills are much the same as planters' characterizations of workers on upland plantations. Perhaps most "remarkable" of all, these characterizations even hold in lowland Java, where the Javanese (the nation's dominant ethnic group) are both the characterizing and the characterized. This spatial, temporal, sectoral, and cultural continuity merits our attention, because it suggests an important and abiding aspect of the logic of self and other in Indonesian governance.

This pattern of characterization or representation privileges, through symbolic inversion (Rosaldo 1978:254), planter interests and decisions at the expense of peasant interests and decisions. The issues at stake in this process of privileging are not merely rhetorical: as Berry (1988:66 cited in Li, this volume) writes: "Struggles over meaning are as much a part of the process of resource allocation as are struggles over surplus or the labour process." As my analysis of Indonesia's contemporary plantation sector will show, power in this case does not work solely or perhaps even largely through the obviously self-serving scheming of elites. Rather, it works more subtly through the conceptual structures by means of which the planters perceive and represent the plantation world. Relations of power between planter and peasant are embedded in these perceptions and representations.

There has been increasing interest over the past decade in the study of localized, informal expressions of state authority. Dillon (1995:333), following Foucault, calls this the study of "governmentality," defined as (the state's) "ways of doing things; ways of ordering things ... the peculiar ways both of speaking and doing ...." The study of governmentality is based on the premise that "the state as an officially coherent entity" is not experienced at the local level, only some discrete fragment of it; and yet it is through such local institutions that the state is created

(Gupta 1995:384). Hence the interest in studying "local centers of power-knowledge" (Escobar 1984:380–381). I suggest that Indonesia's state plantations are an influential, and largely overlooked, local center of this sort.

Generalization/Typification

In my approach to this study of Indonesia's state plantations, I have taken one "slice" of the subject and necessarily not taken others. I have focused on recurring patterns in the relations between planters and peasants, as opposed to nuances either within these relations or within one or the other group. I have been interested to ask, in particular, what is *accomplished* by the characteristic ways that plantation managers talk about plantation workers? In attempting to answer this question, I have had to *generalize* about plantation managers and workers, a move which runs somewhat counter to current sensitivities in anthropology.

Such sensitivities are exemplified by Abu-Lughod's (1991:149–151) statement that generalization in the social sciences is not "neutral description" but "is inevitably a language of power," the remedy for which is "ethnographies of the particular." A suspicion of generalization and typification is part of the wider, post-modern critique of representation in general. Post-modern anthropologists maintain that traditional ethnographic representation is a form of domination (Roth 1989:560), and they argue that today's ethnographic challenge is not to make better representations but rather to avoid representation altogether. Thus we should not exercise ethnographic authority but critique it (Tyler 1986:128). The post-modern critique is not entirely misplaced insofar as the traditional ethnographic subject is concerned. Ironically, it is less appropriate for the new ethnographic domain (expanded under some of the same theoretical influences), which may encompass not just a community but, for example, the state.

The need for generalization in new ethnographies of the state is suggested, in part, by the fact that generalization is both a characteristic of the state and a matter of concern to its officials. As noted earlier, it was evidence of over-generalization in the plantation managers' views of peasants that drew my attention to this subject in the first place. I was also alerted by evidence that managers were resistant to becoming the subject of generalizing: indeed they evinced at least as much distress at the specter of social science generalizing as Abu-Lughod. They insisted that their interpretations of peasant behavior were case-by-case responses to completely idiosyncratic circumstances, and they reacted with indignation and alarm to my attempt to generalize about these interpretations. For these reasons, among others, I believe that generalization is an analytic tool that should not be lightly abandoned.

Critiques of generalization which focus on excesses by scholars alone are an egocentric inversion of the real world and academe, and make an untenable equation of the real world with its study. Focusing overly much on the scholarly text, they imply that everything can be resolved in the text — as if when ethnography stops typifying, typifying stops. This of course is false: the threat posed by generalization or typification typically does not come from Western ethnographers determined to "ignore or reduce cultural diversity" (Marcus and Fischer 1986:32–33); rather it stems from national elites who misread local diversity in ways which support their own political-economic agendas. Some scholars have correctly noted that the need to "reenter the world" obliges us to go beyond an "excessive concern with how to 'represent' the 'other'" (Escobar 1993:378). I concur that we need to stop worrying about representation of the other by ethnographers, but I suggest that this does not relieve us of worry about representation of the other *by others*. This raises a final, fundamental question: how can we study representation or typification if we, as ethnographers, reject typification as a tool of analysis? This is a question about how to relate ethnography to the structures of power beyond the academe — a question about how to study power.

Outline of Study

I begin with an outline of plantation agriculture in Indonesia and my methodology for studying the views of plantation managers. Following this, I discuss plantation managers' views of the "strange native," the "lazy native," and the "obtuse native." I then present my interpretation of the respective views of plantation managers and workers. I conclude by discussing the implications of this analysis for the study of power in contemporary anthropology.

THE SUBJECT AND METHOD OF THE STUDY

Plantation Agriculture in Indonesia

Indonesia's history is in part a history of plantation agriculture. Many contemporary plantations originated as Dutch colonial enterprises that were nationalized in the late 1950s and early 1960s. Today, most plantations are para-statal organizations, including both *Perusahaan Negara Perkebunun* and *Perusahaan Terbatas Perkebunan*, respectively government and semi-private plantation corporations, as well as *Pabrik Gula*, semi-private sugar mills. (Hereafter, my use of the term "plantation" may be taken to include these sugar mills as well.) The development of these para-statal organizations was part of a wider effort by regional governments to exercise more direct and personal control over valuable commodity production (Koppel and Hawkins 1994:36).[1] The political importance of these estates has always exceeded their economic importance: in 1992, for example, whereas these estates contributed just 4.9% of agriculture's share of Indonesia's GDP, smallholdings — peasant enterprises run with little capital or technology — contributed two and one-half times as much, at 12.3% (Barlow 1996:8; cited in Li, this volume).

The drive by the Indonesian government (and international creditors like the World Bank) to develop these para-statal enterprises was stimulated in the late 1970s and early 1980s by projections of steadily decreasing oil-gas exports and concern to develop other sources of foreign exchange. When questions of socioeconomic equity were raised about the commitment of national and international capital to the plantation sector, some programs were developed with a purported orientation toward peasant smallholders. One of these, which figured prominently in this study, is called the *Perkebunan Inti Rakyat* "People's Nucleus Estate," which consists of smallholdings clustered around, and selling produce to, a government nucleus estate.[2] In the 1980s, some of these estates were developed in conjunction with outer island transmigration schemes, based on the theory that such schemes could not only produce exportable commodities but also promote social equity, regional development, and the redistribution of Java's population.

The nucleus estate projects were beset by difficulties (see Barlow and Tomich 1991; White, this volume), many of which involved relations between the estates and local communities, especially regarding land compensation. For example, if a local farmer's land was appropriated by a plantation that he or she subsequently joined as a smallholder, the farmer received no compensation for it. However, the amount of debt that the farmer had to assume in exchange for his or her smallholding was the same as that of a transmigrant who gave no land to the plantation. The inequity of this predisposed many local farmers against participation. If they did not participate, they were theoretically entitled to compensation for any land taken by a plantation, but such compensation was routinely inadequate and long in coming. Even in the absence of problems with local communities, many nucleus estate schemes foundered on their own internal economics or ecology. In one case, actual earnings averaged as little as 4% of official projections, forcing participants into prostitution to stave off hunger (Down-to-Earth 1990b:10). In another case in West Kalimantan, mortality rates for the high-yielding rubber clones provided by the scheme averaged 70%, in spite of which the government still insisted upon full repayment of the credit that was extended to the participants to purchase these clones in the first place.³

A second smallholder-oriented program that was much discussed by plantation managers during the period under review was called *Tebu Rakyat Intensifikasi* (People's Sugarcane Intensification). Beginning in 1975 the national government made a concerted effort to regulate and intensify sugar production in Indonesia (especially in the historic cane districts of Java).⁴ This involved, in part, an attempt to fix in advance the source of sugarcane to be delivered to each sugar mill each season. This was accomplished by dividing the privately owned wet rice lands surrounding each mill into blocks, and specifying which blocks would be planted in sugarcane during which seasons (which is called the *glebagan* "rotation" system).⁵ Many farmers resisted obligatory participation in the rotations, however, on the grounds that the opportunity costs of taking land out of rice to plant sugarcane far exceeded the returns.⁶ For example, a delegation of Central Javanese farmers, who were seeking legal redress in this matter, reported to the press that they each had to forego 200,000 rupiahs (equal to approximately $179) in earnings from wet rice, in exchange for just 34,000 rupiahs ($30) from planting sugarcane (Kedaulatan Rakyat 1985a). In fact, sugarcane promises higher earnings in theory — and this was cited as the official justification for the rotation program — but this was not the case in practice, because the government rotation system provided greater opportunities to outsiders to extract a portion of the farmer's income. This aspect of the rotation system, although resolutely ignored by government officials, was a prime source of conflict between sugar mill officials in charge of the program and local farmers.

Studying Planters' Views of Peasants

This study draws on diverse sources of data. My data on plantation workers are drawn from both short- and long-term studies of peasant and tribal communities carried out mostly between 1974 and 1985. My data on plantation managers are drawn primarily from my participation in a unique year-long study carried out for Indonesia's Institute for Plantation Training (*Lembaga Pendidikan Perkebunan*, LPP) in Yogyakarta, Central Java, in 1984–85. The study brought together a three-man team of social scientists (including myself) from Gadjah Mada University (where I worked as a visiting professor and research advisor for six years), senior officials in the Institute, and visiting plantation managers. The visiting managers came from state-owned plantations (and some sugar mills) all over Indonesia, but principally from Java, Sumatra, Sulawesi, and Kalimantan. The managers were mostly Javanese or Batak and in most cases they had to work with local communities belonging to other ethnic groups.[7] Three distinct categories of workers or peasants were involved in the study: (1) permanent plantation employees (*karyawan*), (2) day laborers (*buruh harian*) hired from nearby villages, and (3) other inhabitants of proximate communities who came into conflict with the plantation over land appropriation or other matters.[8]

The LPP study was structured around a visit to the Institute each month by a team of managers from a plantation somewhere in the archipelago which was experiencing a marked level of conflict with local workers or communities. The visitors would present written and oral accounts of these conflicts, upon which the university team and officials from the Institute then commented. In addition to participating in this commentary, I took notes during the formal presentations, analyzed the written texts, and gathered data as an institutional participant-observer. My premise in gathering and analyzing these data is that since these monthly seminars were not public, and since the visiting managers were coming to the Institute in the capacity of students coming to a school (and many of them had in fact studied formally at the Institute), the reports made by the visiting plantation teams accurately reflected the actual views and, thus, policies and practices of plantation managers. This premise was supported by subsequent field visits by the university team to specific plantations — in my case, to plantations in West Kalimantan and Southeast Sumatra — where this process of reporting and commentary was continued with local plantation officials and workers. At the end of one year, the university team presented summaries of our findings to a large assembly of plantation officials from all over Indonesia.

The expectation of the plantation officials regarding this presentation and the LPP study as a whole was that we (the university-based social science study team) would help them to better understand the behavior of workers and local peoples that they deemed inimical to plantation policies. Instead of problematizing the peasants' reality, however, we wound up problematizing the reality of the planters. In my own presentation, I attempted to transcend the particularistic constraints of individual peasant-planter conflicts by isolating what seemed to be common and distinctive about them all, namely devices that privilege the reality of the planters at the expense of the reality of the peasants.[9] The foremost such device was a premise of peasant cultural, economic, and mental backwardness — a premise which is as widespread in Indonesia's upper and middle classes as the ideology of *pembangunan* (development) that encompasses it. It was this premise which enabled managers to misinterpret conflicts over the economic and political rights of peasants as conflicts over their culture. I suggest that this premise is part of the institutional culture of para-statal plantations in Indonesia.

The belief systems of contemporary planters have received relatively little attention from Indonesianists, which is surprising because the subject of plantations (in particular the historical sugar industry) has otherwise drawn so much attention (e.g., Breman 1983, 1989; Elson 1979, 1984; Fasseur 1992; Knight 1980). There is a sizable literature on colonial planters and peasants, and also on contemporary peasants, but there is next-to-nothing on contemporary planters. This lacuna is not peculiar to the Indonesian literature but reflects a general absence of anthropological studies of the belief systems of government officials: systematic, theoretically informed studies of officialdom (e.g., like Ferguson 1990) are still more the exception than the rule.

PLANTATION MANAGERS' VIEWS OF PLANTATION WORKERS

It quickly became apparent during the LPP study that whenever plantation managers experienced a conflict with plantation workers or proximate farmers, there was a strong inclination to attribute it to some failing of the workers or farmers (as opposed to some failing in plantation policy or administration). Three purported failings of workers and farmers were consistently invoked: their strangeness, their laziness, and their obtuseness.

The Tea Party and Native Strangeness

The plantation managers who were interviewed in the LPP study reported numerous cases of peasant behavior that was deemed to be "disturbing." One of the most disturbing cases involved a tea party that was held in early 1984 along the banks of the Kapuas River in West Kalimantan. The party's hostesses were the Javanese and Sumatran wives of the managers of a plantation being developed there by the government, and the guests were the wives of local Dayak tribesmen. When the guests arrived, they reportedly gathered up the food that had been prepared and then left, taking the food with them and leaving their shocked hostesses behind. The managers of the plantation attributed this "strange behavior" (*perilaku aneh*) of the Dayak women to the problematic culture of the tribesmen; and they cited this tea party as incontrovertible evidence of how — even with the best of intentions — it was impossible to get along with them. Similar criticisms of local culture punctuated the oral and written reports of other plantation officials, especially those working in the outer islands. Thus, managers from Kalimantan referred to Dayak tribesmen as "backwards people" (*masyarakat terbelakang*) and criticized them for being "undisciplined because of a way of life that is free and nomadic."[10]

The plantation managers who reported the tea party incident said that it was not the loss of the food that bothered them but the fact that this behavior had "frightened" their wives. They regarded the incident as but one more example of the strangeness of Dayak culture and of the consequent — and, for them, *understandable* and *predictable* — difficulty in dealing with the Dayak. In actuality, their interpretation ignored the two central facts of the incident. First, the Dayak clearly *intended* their behavior at the tea party to be if not strange and frightening, at least offensive. When the Dayak of this region attend feasts in other households or villages, all food and drink are customarily consumed on the spot: the guests take nothing home for later consumption (Dove 1988:165n.24). (Ironically, it is the Javanese custom to take food home from rituals (Geertz 1960:13); but the reaction of the plantation hostesses — some of whom were Javanese — indicates that they did not expect the tea to follow Javanese norms in this respect.) The second salient fact of the incident is that the Dayak *forcibly* took food from the planters. Other reports from this plantation indicated that the Dayak did not believe that they received proper compensation for hereditary lands that had been appropriated by this plantation. The Dayak "appropriation" of plantation food must be viewed as a counterpoint to the plantation's unsanctioned appropriation of the source of their own subsistence, their land.[11] By saying that the loss of the food was unimportant and that the real problem was the bizarre culture of the Dayak, the plantation officials were (however unknowingly) diverting attention away from the economic dimensions of the conflict in favor of spurious cultural dimensions.

Sugarcane Tourism and Native Laziness

Plantation managers claimed that the tribesmen and peasants involved in their projects were not only culturally deficient, but economically deficient as well. Thus, the officials of a plantation in South Kalimantan described the local Banjarese and Dayak to the study team as "not diligent or persevering," (*kurang tekun dan kurang ulet*) while the staff of a plantation in West Kalimantan reported that the Dayak involved in their project "have low [work] output ... like to loaf around, and have no desire to work" (in both cases they were contrasted to Javanese transmigrants[12]). The plantation managers attributed this perceived low intensity of labor either to the tribesmen's lack of needs or to their lack of sufficient foresight to anticipate their needs. As an official of a plantation in West Kalimantan put it, "Once [the Dayak] have money, they do not want to work."[13] Speaking of Bugis plantation workers in South Sulawesi, another official said "They work in the sugar mill only to obtain cash to pay their taxes." This same manager further told the study team that the Bugis continue to cultivate their own fields while working on the plantation. As a result, "There is a problem: for the workers, working in the sugar mill is not a necessity."

The Dayak, Bugis, and many other indigenous groups in Indonesia work off-farm as part of a *portfolio* of activities designed to achieve an optimal balance between needs and resources, and risks and rewards. This portfolio may include wage labor on plantations, swidden cultivation for insert food, cash-cropping, and the gathering of non-timber forest products. The place of wage labor in the overall portfolio, as just one activity of many, is typically not as central as planters would like it to be. For example, when the goal of a given period of wage labor is met, the workers may cease their wage labor and turn their attention to other parts of their portfolio — thus leaving the planters without their labor force. It is possible for the fundamental conditions that underlie the portfolio to change, however, elevating the role of wage labor to a more determinant position.

This was illustrated by the case of one plantation on the middle reaches of the Kapuas River in West Kalimantan. The conversion of forests to state plantations and rising population/land ratios, among other factors, convinced many of the tribesmen living around this plantation that their best prospects for the future lay less in swidden-farming and more in plantation work (see Dove 1986:7–8; cf. Stoler 1985a:185–186). As a result, these tribesmen said that they felt "aggrieved" (*susah*) when the planters placed a 35–year age cap on candidates for full-time employment, leaving many of them with no option but to work as day laborers. The uncertainty of this work made it impossible, the tribesmen said, to "think about the future." In this case, because a deteriorating local subsistence base provided a better guarantee of a dependent supply of labor, the tribesmen's desire to make a long-term commitment to the plantation was not welcome to the planters. It is welcome only in circumstances of labor scarcity. The underlying issues, therefore, were control of the mobilization of labor and responsibility for the reproduction of labor. The issue was not, as Li points out in her contribution to this volume, one of tradition versus change — although this was how the plantation managers tried to construe it — but one of what kind of change there would be and how it would be achieved.

The issue of control dominated the plantation managers' discussions of plantation-peasant relations. For example, even the limited autonomy given to smallholders in the nucleus estate system was a matter of concern to plantation officials. A manager from a West Kalimantan oil palm plantation argued that the central mills that serve such smallholders should have some land of their own under palms, "So that [the mill] will not be dependent upon the people." In the thinking of this manager, if the government-owned mill had to depend entirely upon the production of smallholders — who could in theory withhold their crops if they were dissatisfied with the price offered by the mill — then mill profits could be jeopardized.

The aversion of plantation management to any dependence upon peasant decision making was clearest in the case of the sugar mills of Java. Mill officials had little sympathy for farmers who were unhappy about participating in the government's compulsory cane cultivation system. Thus, the officials of one mill in East Java suggested that the government should levy sanctions against any farmer refusing to plant sugarcane when his or her turn came within the rotation. These same officials further suggested that sanctions should be applied against any farmer who planted sugar *out* of rotation, and then shopped around (from one mill to the next) for the best price — a practice which the mill officials derisively referred to as "sugarcane tourism" (*pariwisata tebu*).[14] Any planting or selling outside official rotations and channels decreases the mill's control over its supply of cane, whereas the entire purpose of the compulsory cane-planting program was to increase this control. The planters' underlying and motivating concerns do not emerge in their rhetoric, however; rather they are obfuscated by it. Thus, the planters' rhetoric portrays the farmers' behavior not as a quest for economic independence but as a quest for something foolish. The phrase employed by the planters is not "competitive price shopping," which is hard to criticize, but "sugarcane tourism," which invites ridicule.

The Misunderstood Official and the Misunderstanding Native

A third common thread in the plantation managers' criticisms of their workers was the idea that the workers' "mentality" was at fault. Thus, plantation managers attributed any reluctance on the part of laborers or locals to embrace plantation priorities to thinking that was "negative" or in which there was "asymmetry" (*kesengajaan*). Thus, one plantation manager suggested that the Dayak objected to his appropriation of their land because "they do not understand" the agrarian laws that purportedly permit this (but that in fact do not (Dove 1987)). Other managers resorted to a similar explanation of complaints about inadequate housing and the difference in lifestyle between laborers and managers on a nucleus estate project in Sumatra: "The fuss about housing for local participants in the nucleus estate ... in fact did not need to occur, if the people had understood the spirit of the project .... Probably that [envy] arose because the people misunderstood the role of the national plantation" (Sinar Harapan 1984, my translation). In short, in the eyes of the plantation officials, one of their main problems is that they are misunderstood by their workers.[15]

The solution to this misunderstanding, all plantation officials agreed, was mental change on the part of the peasants.[16] As an official of a plantation in South Kalimantan stated: "The main problem is how to change the [peasants'] mental attitude and pattern of thinking, so that they have the attitude and thinking of entrepreneurial farmers in their farming activities." A sugar mill official in East Java made a near-identical observation concerning the participation of Javanese farmers in the government's compulsory scheme for intensifying cane cultivation: "Continued effort is needed in order to change the attitudes and pattern of thinking of the farmers, so that the farmers are prepared to increase their participation."

In a particularly egregious case, after officials of a nucleus estate project in Sumatra's Riau province reneged on promises to include local farmers in the project despite having appropriated their land, destroyed 1,000 hectares of their rubber, and shot one farmer during an ensuing protest, the head of a government investigating commission concluded that if the local people were still "stupid" (*bodoh*) then they should be taught the "right skills" (*Merdeka* 12 January 1990, cited in *Down to Earth* 1990a:3).

A typical proposal for changing the peasants' incorrect attitudes and pattern of thinking, by a provincial official in West Kalimantan, was for "Mental-spiritual indoctrination, training and education, and courses and indoctrination in the state ideology" (*Pembinaan mental-spiritual, latihan dan pendidikan, kursus-kursus serta P4*).[17] The sweep of this proposal reflects not only official bewilderment over how to deal with the problem, but the fact that the problem itself — the crippled mentality of the peasants — is a conceit of the plantation imagination.

ANALYSIS: PEASANT-PLANTER RELATIONS

Analysis of the plantation managers' views of plantation workers and peasants suggests the presence of a shared and embedded view of the native "other."

Planters' Views of Peasants

The data just presented suggest that Indonesian plantation managers view plantation workers and local peasants as quintessential "others." Leach (1982) (cited in Borsboom 1988:429) notes that one social group distinguishes itself from another according to food, sex, and attire. To these distinctions, Borsboom adds reason versus emotion and communal versus private ownership. The examples taken from the planters' anti-peasant rhetoric — involving criticism of peasant culture, emotions, intellect, labor, settlement pattern, temporal horizon, clothing, and eating customs — cover most of these points of discrimination.

There was a remarkable consistency in the planters' views of the worker or peasant "other." In every case of conflict examined in this study, problems that I attributed to differences between planter and peasant in economic self-interest were attributed by plantation managers to differences in culture and personality. There was no geographic variation in this regard: plantation managers all over the Indonesian archipelago, despite great local variation in ethnicity, economics, ecology, history, and politics, made the same attribution, often using the same words and phrases.[18] There was historical consistency as well: colonial managers in Indonesia explained their problems with workers in near-identical terms.[19] As Alatas (1977:62) writes, "the theme of the lazy Javanese ... functioned as a major constituent of the colonial ideology" in support of coercive agricultural policies. Similarly, Stoler shows that colonial planters in Sumatra blamed peasant resistance on either social causes — chiefly the "child-like," irrational and rapacious character of the coolie (1985a:48,51) — or political ones, depending upon whether their priority lay in discouraging or encouraging interference by the colonial government in planter-peasant relations; but in neither case did they blame resistance on the economic self-interests of the embattled peasants. In the same vein, Elson (1979) notes reluctance on the part of colonial officials to link intentional burning of sugarcane fields to the deleterious impact of sugarcane cultivation on the economy of the Javanese peasantry.

The persistence of these views of the peasantry, in spite of changes in time, place, and culture (of the planters as well as the peasants), suggests that they are not sociological but ideological in origin. It suggests that they are based less on social reality, the local variation in which they would otherwise reflect, than on an ideological reality, consisting of a pan-plantation political and economic agenda. The planters' views of peasants support this agenda by misconstruing (what I interpreted as) economic conflicts as cultural conflicts. This is a misconstrual of workers' intentions, in particular. In the case of the tea party, for example, the plantation managers correctly perceived that the behavior of the Dayak women was extraordinary, but they failed to perceive that it was extraordinary to the Dayak as well as themselves and, hence, it was intended to convey a message (in this case an economic and political one). By virtue of their misconstrual, the plantation managers not only do not have to respond to this message, they do not even have to receive it; thus they do not have to acknowledge the conditions that prompted its sending.

Peasants' Views of Planters

Workers' and peasants' views of plantation officials did not mirror the officials' views of them: they were structurally dissimilar. In none of the cases examined did peasants impugn the culture, intelligence, or emotions of the officials. The peasants said things like "We are afraid that the Batak [managers] will become lords" (*Kami takut Batak jadi Rajah*) or, as some disgruntled plantation workers said to me with regard to Batak officials on a plantation in West Kalimantan, "We are afraid of deception [on the part of the managers]" (*Kami takut penipuan*). But the peasants did not say that they were afraid of the managers doing something stupid, irrational, or otherwise reflective of an unfathomably different cultural tradition. The peasants believed that the managers shared the same general interests and values as they did themselves. The peasants worried that the managers would maximize their own interests to the detriment of the peasants', but they did not worry that the managers would maximize some alien and incomprehensible interest. The peasants viewed the managers as potential adversaries, therefore, but *not* as alien "others."

Peasants' views of planters also differed from planters' views of peasants in being less consistent. Peasants did not characterize plantation officials in the same way, much less with identical terms or figures of speech. Rather, the peasants' characterizations varied according to their own ethnic group, the character of individual plantations and, especially, the personalities of particular plantation managers. Stoler (1985a:197) similarly notes that in her contemporary field site in North Sumatra, "Specific estate managers are singled out as individuals for being especially demanding, aloof, or compassionate, without any generic attributes commonly and consistently applied to them." While planters clearly thought of peasants as a class, therefore (in the sense that they saw most peasants as sharing certain problematic characteristics), the reverse was not true.[20] This does not mean that the peasants mystified the conditions of their oppression, only that they used an idiom that was culturally and politically more appropriate. Scott (1984:209) argues that it is both convenient and strategic for oppressed groups to focus on the "local and personal" causes of distress: it is convenient "to blame those who are most immediately and directly responsible for ... reverses," and it is strategic "because it focuses on precisely those human agents which are plausibly within their sphere of social action."

The idiom of resistance was drawn from the value systems of the local peasant and tribal communities. For example, a Dayak whose land was appropriated by a nucleus estate project in West Kalimantan said "As long as [everyone] eats [profits] the same and works the same, there is no one who will not want [to cooperate with the plantation program]" (*Asal sama makan, sama kerja, tidak ada yang tidak mau*). This comment reflected the fiercely egalitarian values of the Dayak groups in this part of Kalimantan (see Dove 1986:14,17). It is deviation from such local values that causes concern, as reflected in the fear, noted earlier, that the Batak managers were trying to become their "lords" (*rajah*).[21] The tribesman who made this comment did not say he was afraid that the Batak managers would become "rich" (*kaya*). Acquiring wealth is an acceptable (if not always socially esteemed) goal in this tribal culture, but acquiring lord-like power over the lives of others is not.

Source of Peasant-Planter Conflict

The immediate focus of the peasant-planter conflict was rhetorical in nature: it was a battle for the moral high ground. The plantation managers' rhetoric suggested that peasant resistance to plantation plans and policies was due to problems originating with the peasants, in particular their irrationality. The peasants' rhetoric, in contrast, suggested that their resistance was provoked by problems originating with the managers, such as their desire for too much power. The peasants' rhetoric implied that there was a basic difference in self-interest between themselves and the managers, whereas the managers' rhetoric rejected the concept of such difference. Any public construal of the conflict as one based on competing interests was inimical to the managers' position. The plantation managers, as functionaries in para-statal enterprises, needed to act in a manner seen as consistent with state ideology regarding commitment to the welfare of the common citizen. The managers' rhetoric preserved the illusion of this commitment by focusing attention on alleged peasant shortcomings and deflecting attention from plantation policies that were not in the peasants' best interests.[22]

The source of the conflict between peasant and planter views of reality was not rhetorical in nature, however, but political and economic. The plantation managers wanted the peasants to be completely dependent on the plantation, insofar as this would facilitate the mobilization of labor on the plantation's own terms; but they did not want the plantation to be completely responsible for reproducing the conditions of the peasants' existence (e.g., housing and social security, in addition to reasonable hours and wages). Stoler (1985a) also found this to be the case in colonial and post-colonial plantations in Sumatra. The peasants, on the other hand, wanted the reverse: they wanted the plantations to make more of a commitment to them and assume more responsibility for their welfare; but they still did not want to become completely dependent upon the plantations. This difference in views was reflected in the rhetoric of the two sides.

The plantation managers complained that the peasants were not always ready and willing to work in the plantations, and the peasants complained that the work was not always available when they wanted it. The managers worried that the alternate sources of income available to the peasants would give them *too much* independence, and the peasants worried that if they lost these sources (viz., dry-rice fields and rubber smallholdings (cf. Dove 1986:8–9,11–13)) they would have *too little* independence.[23]

Maintenance of Official Beliefs

Although the plantation managers' rhetoric supported plantation policy and honored the niceties of state ideology at the expense of peasant welfare, it was not ironic or cynical in tone, no matter what the audience or location.[24] The managers' belief, that peasant resistance to their policies was based on mental and cultural backwardness, appeared to be entirely sincere. This sincerity was made possible by a system of plantation administration that implicitly minimized and misconstrued peasant feedback.[25] Plantation officials demonstrated great aversion to critical feedback, especially when delivered directly to the person responsible. Among all of the examples of "untoward" peasant behavior reported by plantation officials during this study, the one that disturbed them most was a "demonstration" by disgruntled Dayak tribesmen in front of a plantation manager in West Kalimantan. The prevention of further demonstrations — which were termed "mental threats" (*ancaman mental*) — became a primary goal not only of the plantation management but also of the local government, which proposed the following "solution": "Prevent the holding of demonstrations by people who want to make known their desires directly (*langsung*) to the plantation and the district officer, or directly to (*langsung*) plantation officials, in a harsh way".

Direct communication to higher officials, thus, was intolerable. The implication was that any plantation worker with a complaint should, at most, make it to his immediate superior; the latter might then pass it on to his immediate superior, and so on, until — and if — the upper levels of plantation management and government were reached.[26] This hierarchical structure shields upper level officials from contact with the peasants who bear the consequences of their management decisions and it thereby shields their views of peasants from confrontation with a contrary peasant reality.

The aversion to feedback was implicit. There was explicit support in plantation administration for a variety of information-gathering mechanisms, but in practice they were all subverted. For example, when meetings were held between local peasants and visiting officials or researchers (such as myself), the local plantation managers invariably attended and their presence — especially given the Indonesian reticence for direct confrontation — effectively suppressed any possibility for candid feedback from the peasants (cf. Dove 1986:23–26).[27] Whatever research was carried out on peasants by the plantation establishment was structured to ensure that nothing untoward was discovered. For example, the aforementioned Dayak demonstration prompted the management of the plantation in question to carry out a survey of local farmers' attitudes toward plantation policies. The farmers were asked to rank their attitudes on the following three-part scale:

worst	(1)	*Belum pasti senang* "Not yet satisfied for sure"
	(2)	*Lumayan* "Relatively satisfied"
best	(3)	*Senang* "Satisfied"

Thus, even on a plantation on which demonstrations and "terrorism" on the part of the local tribesmen were being reported, the structure of the study did not allow for the possibility that dissatisfaction with plantation policies ran deeper than "Not yet satisfied for sure." My point is *not*, of course, that the planters simply needed to add one or two further steps to their measurement scale; it is that the way they structure this scale reflects the institutional limitations on their ability to accept candid feedback. A steadfast aversion to recognizing dissatisfaction with plantation projects is ubiquitous: a local government official told critics that a much-troubled nucleus estate project in West Java could not be canceled and disagreed that the project had "failed," preferring instead to say it had "not yet achieved ideal results" (*Kompas* 25 August 1990 cited in *Down to Earth* 1990b:11).

The plantation managers were able to maintain their aversion to feedback from dissatisfied peasants by a willingness to seek recourse to force if necessary.[28] The use of force as a substitute for dialogue was explicit in the response of an official of the national sugar board to the refusal of some Javanese rice farmers to plant sugarcane in accordance with the block rotation pattern of their local sugar mill:[29] "Regarding persuasion, it is necessary to consider what level the farmers are at. This can be compared with the development of children, where at a certain level the use of force (*paksaan*) is still necessary, to be followed by the use of persuasion (*bujukan*) when adulthood is reached."

DISCUSSION AND CONCLUSION

A comparison of responses to this study by plantation workers, plantation managers, and fellow anthropologists has important implications for the study of power.[30]

Responses to the Study by Peasants, Planters, and Anthropologists

When I described this study to a gathering of tribesmen in West Kalimantan who were losing their lands to a state plantation and asked them how they wanted the plantation managers to address their concerns, their response was an explosion of laughter.[31] The idea that the way to resolve conflict between government and tribesmen was to ask the tribesmen what they wanted was perceived as hilariously ingenuous. Clearly, no one in any official capacity had ever asked the tribesmen this before. This question, and the dialogic relationship with the state that it implied, left not just the tribesmen but the government officials accompanying me completely nonplussed. Both sides were used to beginning with what the government wants and then fashioning an acceptable local adaptation to these wants; they were not used to beginning with local wants. This was reflected in the subsequent *non-sequitur* of a conclusion by the manager of the government plantation in question, who said that what these tribesmen needed was "extension that is really extension" (*penyuluhan yang betul-betul*). *Penyuluhan* (extension) is the standard Indonesian gloss for imparting government information and instructions to the rural population.[32] Thus, the manager's concern was not that the government was not getting the tribesmen's message but that the tribesmen were not getting the government's message — which needed, therefore, to be repeated more loudly and more often.

When I presented the major points from this analysis to a meeting of plantation managers in Central Java, the public response was polite; but the real response took place the previous day, when the meeting organizers saw the text of my talk for the first time. The response involved a hasty re-shuffling of the meeting program, moving my presentation from a prominent position at the beginning to a less prominent position in the middle, and closure of all but the opening ceremony to the press. The meeting organizers had expected a synthesis of, and response to, the plantation managers' views of the problematic peasantry, not a critique of these views. They had expected a study of peasants, not a study of managers. By presenting the latter, I challenged the managers' privileged position in their dialogue with the peasants.[33] More generally, I challenged the fundamental principles that (1) plantation managers are the sponsors not the subjects of objective study, and (2) managers solve problems, they are not the source of problems.

In short, the response to this study from the peasants reflected the fact that they are not normally invited to participate in their representation; and the response from the plantation managers reflected the fact that they are not normally studied or represented. When I presented this study to fellow anthropologists, however, the dominant response was to question the nature and purpose of representation itself. A common response was to ask me for a more nuanced, in-depth study of a single plantation, village, or incident. Those making this response appeared to be oblivious of the possibility (raised in this study) that the most problematic aspect of real-world representation is not its theoretical possibility but rather its everyday contestation. I suggest that the focus of many anthropologists on ontology rather than politics is itself *part* of the politics.

The Importance of Studying Power

In her famous twenty-six year-old challenge to "study up," Nader (1972:289) said that one of the benefits from studying power — from studying "the colonizers rather than the colonized, the culture of power rather than the culture of the powerless" — is that this "would lead us to ask many 'common sense' questions in reverse" (compare Rosaldo 1978:241). In this spirit, my study did not focus upon peasants perceptions of the state but rather on state perceptions of peasants. This led to further questions about the role of representation in state projects, and the potential complicity of ethnographers. On the latter count, the findings of this study suggest that the critique of ethnographic authority for its past support of oppressive political-economic formations actually undermines ethnography's current ability to counter such formations.

Anthropology's current concern about generalization and typification may be another example of what Watson (1991:89) (initially referring to anthropology's preoccupation with the exotic) has called "institutional incompetence." There seems to be an almost reflexive tendency within anthropology to shift theoretical paradigms in a manner destined to limit the discipline's relevance to the world, thereby preserving the boundary between theory and praxis. Thus, just at the historical moment when it becomes possible to extend representation beyond the local community to the political-economic structures that exercise control over that community, many anthropologists are choosing — instead of doing counter-representation — to problematize the issue of representation *per se*.[34] At this point, it is necessary to return to the question I raised in the introduction to this study: What is the significance of abjuring generalization, typification, and simplification when these are the *tools* of power?

The Study of Power

There is confusion within anthropology regarding the significance of being the subject versus object of typification. Whereas power *works* by typifying others, its work is *obstructed* when it is itself typified. If so, then what is the *purpose* of doing the "nuanced" study of the hearts and minds of plantation managers (or even plantation workers), that was suggested in some of the collegial responses to this study? What are the implications of this scholarly movement toward nuance in a world in which power is expressed through eliding nuance? In the recent rush to critique ethnographic authority the role of ethnographic representation in countering representation by repressive institutions has perhaps been forgotten (see Rappaport 1993:302; Sangren 1988:406).[35]

Astute observers have correctly noted that those who wish to oppose repressive states are obliged, to some extent, to employ in their opposition the same simplistic images as the state itself employs (Tsing 1993; Li this volume). Increasingly, to avoid these simplistic images is to *not* write against the state and even perhaps to write *for* it. Said (1983:7) heightened our critical self-awareness by reminding us that we do not write for ourselves and enjoining us to always ask, Who do we write for? I would add that it is equally important to ask about the negative case, namely, Who are we *not* writing for, and who are we not writing *against*? According to Said (1983:9) "There is always an Other; and this Other willy-nilly turns interpretation into a social activity, albeit with unforeseen consequences, audiences, constituencies, and so on." I suggest that in the study under discussion, this "Other" is the state, and that its presence makes a "social activity" of any of our interpretations that reject typification, especially of the state itself. And I ask, in any ethnographic context dominated by state typification, is not an interpretation that abjures typification socially complicit with that domination?

The Significance of Not Studying Power

What would it mean for ethnographers to give away the power of typification, a power that we used to exercise ourselves? To answer this question, we must first ask, What are the implications of rejecting typification when power is expressed by typifying? More specifically, What are the implications of rejecting typification just when the ethnographic "Other" is being projected beyond peasants to the state? The ethnographic "Other" long consisted of marginal, powerless groups; increasingly today, the "Other" consists of non-marginal groups. Yet this is the moment when many scholars have decided to abandon typification: precisely at the time when typification would have real consequences for their own power relations.

Some observers have suggested that, far from being coincidental, the current rejection of typification (or representation more generally) makes perfect sense as part of the "logic of late capitalism" (Jameson 1984).[36] It is suggested, that is, that the postmodern critique of representation indirectly naturalizes the status quo, with the *de facto* effect of supporting the global economic order of late capitalism.[37] This is itself merely one part of the more general epistemological challenge faced by any critical project, namely, to construct a critique that is not complicit with, but somehow transcendent of, the implicit logic of the overarching authority. There are many formulations of this challenge (e.g., Bateson 1958:298–302), but Jameson's (1984:86) is as good as any:

> In a well-known passage, Marx powerfully urges us to do the impossible, namely to think this development [of capitalism through history] positively and negatively all at once; to achieve, in other words, a type of thinking that would be capable of grasping the demonstrably baleful features of capitalism along with its extraordinary and liberating dynamism simultaneously, within a single thought, and without attenuating any of the force of either judgement.

It is to the progression toward (if not attainment) this impossible end that my study has been dedicated.

ACKNOWLEDGEMENTS

Research in Indonesia was initially carried out between 1974 and 1976 in Kalimantan, with support from the National Science Foundation (Grant #GS-42605) and with sponsorship from the Indonesian Academy of Science (LIPI). Additional data were gathered during six years of subsequent work in Java between 1979 and 1985 (with periodic field trips to Kalimantan), with support from the Rockefeller and Ford foundations and the East-West Center and with sponsorship from Gadjah Mada University. A recent (and ongoing) series of field trips to Java and Kalimantan, beginning in 1992, have been supported by the Ford Foundation, the United Nations Development Programme, and the John D. and Catherine T. MacArthur Foundation, with sponsorship from Padjadjaran University and BAPPENAS. Special thanks are due to the Institute for Plantation Education (LPP), which invited me to participate in its 1984–85 seminar. I am grateful to Carol Carpenter and Tania Li for constructive suggestions on an earlier version, and to Helen Takeuchi for editorial assistance. None of the aforementioned people or organizations necessarily agrees with the conclusions in this paper, for which the author is alone responsible.

NOTES

1. These para-statal enterprises have played a significant role in the "crony capitalism" of Southeast Asia's economies; see Koppel (1995:21n. 19).
2. See Barlow (1991:100) and Syarifuddin and Soetatwo (1982).
3. Institute of Dayakology Research (Pontianak, Indonesia), personal communication. See also Gouyon, De Foresta, and Levang (1993:198–200).
4. See *Instruksi Presiden no. 9/1975* (Presidential Instruction Number Nine).
5. *Glebagan* historically referred to the rotation of communal land among villagers to ensure equality of access over time to the best-irrigated parcels (Moertono 1981:125). Now it refers, in an irony characteristic of development policy, to equality of participation in an unpopular government program.
6. Compulsory sugar cultivation was similarly disadvantageous to peasants in the colonial period (Gordon 1979).
7. To protect the anonymity of my subjects, neither the exact names and locations of the plantations involved nor the names of their managers will be given. Where I have quoted from their verbal and written statements, translations are my own.
8. The fact that conflict was expressed at all, and that managers were concerned enough to attend our seminar indicates, perhaps, a bias in our sample towards plantations with less repressive conditions. These were more likely to be found in non-traditional plantation areas (e.g. Kalimantan) than in long-established plantations areas (e.g. Sumatra) with long histories of worker oppression. I am grateful to George McTurnan Kahin for bringing this point to my attention.
9. I treat the oral and written accounts of the plantation managers as "rhetorical," in the same sense used by Stoler (1985b:643) with respect to accounts of nineteenth-century Dutch colonial planters.
10. Similarly, managers from sugar mills in Sulawesi referred to Bugis workers as having "low IQS" and being "jealous," *(serik)*, "quick-to-anger," *(emosionil)*, and "hard-headed," *(keras watak)*.

11 Colonial planters also fed their workers before negotiations in an attempt to improve their negotiating position (Stoler 1985a:147). The behavior of the Dayak women can be seen as a counter to this ploy. It rejected the social playing field selected by the plantation elite.
12 There *may* be some empirical basis for this contrast, since Dayak are accustomed to shifting cultivation which produces higher returns to labour than Javanese wet-rice (Dove 1985). Thus, whereas plantation labor often does not represent an intensification of labor for the Javanese, it typically does for the Dayak.
13 Plantation agriculture's requirement for an unvarying labor supply has been widely noted (e.g. Murray 1992:52–53); as has the importance of the continuity of work in a capitalist system (Wolf 1982:275–276). Concern with limited native needs extends back to colonial times (Boeke 1953:40), although what it typically reflected was concern with limited native desires for what colonial regimes had to offer.
14 Efforts by government and plantation management to avoid the operation of free market forces on plantation inputs and outputs is an old theme in plantation studies.
15 There were only two possible explanations in the eyes of plantation managers for opposition to plantation policy: political subversion or lack of understanding (cf. Lucas 1992:87). Peluso and Poffenberger (1989:338) note the same phenomenon in forestry.
16 Compare McTaggart (1982:55).
17 Alatas (1977:8) sees the image of the developmentally needy native as the contemporary equivalent of the colonial image of the lazy native.
18 Shared planter world views arise from their centralized training and subsequent mobility as they rotate between centers of plantation administration on Java and Sumatra and field postings elsewhere in the country. See Anderson (1983:55–57) on the effect of these "secular pilgrimages".
19 The similarity is not surprising since, to use Anderson's words (1983:145), the government plantations inherited most of the "wiring" of the plantation system directly from the Dutch.
20 Stoler (1985a:196–197) also found this to be the case on colonial Sumatran plantations. In the years since my study, peasant views of planters have become more ideological and generic as a result of the development of more cohesive resistance to plantation policies and the growing involvement of local, national and even international activist organizations.
21 Compare the stress on equality in this Indonesian tribal idiom with the Malaysian peasant idiom analyzed by Scott (1984), in which it is deviation from the traditional obligations of the rich to the poor that is feared.
22 Compare Stoler (1985b:652).
23 Stoler describes how North Sumatran plantations minimized their responsibility and maximized their control by relocating their workforce to pseudo-agricultural communities on the plantations' peripheries. She regards this "part-proletarian, part-peasant positioning of workers as an ingenious cost-cutting device on the part of capital" (Stoler 1985a:6), but she acknowledges that in other cases it may represent a bid for self-sufficiency on the part of the peasants. I suggest that both forces were at work in my study.
24 In contrast, Scott (1985:233) notes that the rhetoric of Malaysian landlords and village officials varies with the audience.
25 McTaggart (1982), in one of the few references to this topic, acknowledges the problem of information flow through the Indonesian government, but he fails to see that it is structurally embedded, blaming it instead on a simple overabundance of data. See Ascher (n.d.) for a more insightful analysis into the institutional forces behind information flows and non-flows in Indonesia.
26 This hierarchical principle is highly important to the Javanese and they are extremely sensitive to deviations from it. Thus, the aboriginal Badui of West Java gained island-wide notoriety in the 1970s when a delegation of them left their mountain forests, walked to Jakarta, and presented a petition *directly* to President Suharto.
27 Stoler (1985a:58,79) documented the same practice in the colonial plantations of North Sumatra. In my study the suppression of feedback was blamed not on plantation norms or personnel but, predictably, on the peasants and their culture. Thus, a government minister,

discussing farmer resistance to the sugar intensification program in Java, suggested that one of the problems was the lack of a "culture of openness" (*kebudayaan terbuka*) among the farmers (*Kedaulatan Rakyat* 6 May 1985b).

28 A plantation manager in West Kalimantan told the study team that whenever his Dayak laborers cause him any trouble he reports them to the local military garrison, which sends a patrol to intimidate them. As he ingenuously put it: "I know the weakness of the Dayak tribes: they are afraid of the green shirts [viz., the military]."

29 Compare Nandy (1987) on analogies between the human growth cycle and social developmental cycles.

30 See Pierce's (1995:96) analysis of the "outlaw" ethnographer, "whose movement between positions proves to be a critical advantage in uncovering "regimes of power."

31 This study is reported in Dove (1986).

32 *Penyuluhan* also can have connotations of coercion (Hansen 1973:5–6), as the response of the afore-mentioned plantation manager suggests.

33 My challenge to the plantation managers' representations of workers and peasants has extended beyond this meeting to numerous other presentations to planters, civil servants, and university students as well as publications in Indonesian-language journals. Today I continue to publish on this topic for international academic and policy-oriented audiences and I consult on this and related topics for both governmental and non-governmental agencies working in Indonesia.

34 An element of self-preservation may also be at work here: some feminist scholars (e.g. Parpart 1993:442) suggest that self-preservation is what motivates male anthropologists to devalue the importance of textual authority just at the moment that they are losing this authority.

35 Compare Rappaport (1993:302):

> We should not forget that we are citizens as well as anthropologists. We should not, any more than anyone else, stay out of public arenas or check our professional modes of understanding when we enter them, nor should we forget that public approaches to public problems are now informed by views of the world, its ills, and ways to cure its ills provided by other, narrower disciplines no better founded than our own, and considerably less humane.

36 Jameson (1984:57) writes:

> What happens is that the more powerful the vision of some increasingly total system or logic — the Foucault of the prisons book is the obvious example — the more powerless the reader comes to feel. Insofar as the theorist wins, therefore, by constructing an increasingly closed and terrifying machine, to that very degree he loses, since the critical capacity of his work is hereby paralyzed, and the impulses of negation and revolt, not to speak of those of social transformation, are increasingly perceived as vain and trivial in the face of the model itself.

37 Cf. Polier and Roseberry (1989:259) "Extreme versions of post-modern thought *had the effect* of denying a world of politics and economics as both became more threatening;" see also Best (1991:223) and Dillon (1995:340).

References

Abu-Lughod, Lila, 1991, "Writing Against Culture" in *Recapturing Anthropology: Working in the Present*, edited by Richard G. Fox, pp. 137–162. Santa Fe: School of American Research.

Alatas, Syed Hussain, 1977, *The Myth of the Lazy Native: A Study of the Malays, Filipinos and Javanese from the 16th to the 20th Century and Its Function in the Ideology of Capitalism*. London: Frank Cass.

Anderson, Benedict, 1983, *Imagined Communities: Reflections on the Origin and Spread of Nationalism*. London: Verso.

Ascher, William, n.d., Political Economy and Problematic Forestry Policies in Indonesia: Obstacles to Incorporating Sound Economies and Science (unpublished).

Barlow, Colin, 1991, "Developments in Plantation Agriculture and Smallholder Cash-Crop Production" in *Indonesia: Resources, Ecology and Environment*, edited by Joan Hardjono, pp. 85–103. Singapore: Oxford University Press.

Barlow, Colin and Thomas Tomich, 1991, "Indonesian Agricultural Development: The Awkward Case of Smallholder Tree Crops". *Bulletin of Indonesian Economic Studies*, 27(3), 29–54.

Bateson, Gregory, 1958, *Naven*. Stanford: Stanford University Press.

Best, Steven, 1991, "Chaos and Entropy: Metaphors in Postmodern Science and Social Theory". *Science as Culture*, 2(11), 188–226.

Boeke, J.H., 1953, *Economics and Economic Policy of Dual Societies: As Exemplified by Indonesia*. New York: Institute of Pacific Relations.

Borsboom, Ad, 1988, "The Savage in European Thought: A Prelude to the Conceptualization of the Divergent Peoples and Cultures of Australia and Oceania". *Bijdragen*, 144(4), 420–432.

Breman, Jan, 1983, *Control of Land and Labour in Colonial Java: A Case Study of Agrarian Crisis and Reform in the Region of Cirebon During the First Decades of the 20th Century*, Verhandelingen No. 101. Dordrecht: Foris Publications.

1989, *Taming the Coolie Beast: Plantation Society and the Colonial Order in Southeast Asia*. New Delhi: Oxford University Press.

Dillon, Michael, 1995, "Sovereignty and Govermentality: From the Problematics of the "New World Order" to the Ethical Problematic of the World Order". *Alternatives*, 20, 323–368.

Dove, Michael R., 1985, "The Agroecological Mythology of the Javanese and the Political Economy of Indonesia". *Indonesia*, 39, 1–36.

1986, "Plantation Development in West Kalimantan: Perceptions of the Indigenous Population". *Borneo Research Bulletin*, 18(1), 3–27.

1987, "The Perception of Peasant Land Rights in Indonesian Development" in *Land, Trees and Tenure*, edited by John Raintree, pp. 265–271. Nairobi/Madison: ICRAF/Land Tenure Center.

1988, "The Ecology of Intoxication among the Kantu of West Kalimantan" in *The Real and Imagined Role of Culture in Development: Case Studies from Indonesia*, edited by Michael R. Dove, pp. 139–182. Honolulu: University of Hawaii Press.

1997, "Dayak Anger Ignored: Michael Dove Traces Dayak Unhappiness to Inequities in State Development". *Inside Indonesia*, (July-September) 51, 13–14.

Down to Earth, 1990a, "Farmer Shot in Plantation Dispute", 7 (March), 3.

1990b, "Farmers Impoverished by World Bank-Funded Plantation project", 11 (November), 10–11.

Elson, Robert E., 1979, "Cane-Burning in the Pasuruan Area: An Expression of Social Discontent" in *Between People and Statistics: Essays on Modern Indonesian History*, edited by Francien van Anrooij *et al.*, pp. 219–233. The Hague: Martinus Nijhoff.

1984, *Javanese Peasants and the Colonial Sugar Industry: Impact and Changes in an East Java Residency, 1830–1940*. Asian Studies Association of Australia. Singapore: Oxford University Press.

Escobar, Arturo, 1984, "Discourse and Power in Development: Michel Foucault and the Relevance of His Work tothe Third World". *Alternatives*, 10, 377–400.

1993, "The Limits of Reflexivity: Politics in Anthropology's Post-Writing Culture Era". *Journal of Anthropological Research*, 49, 377–391.

Fasseur, Cornelis, 1992, *The Politics of Colonial Exploitation: Java, the Dutch and the Cultivation System*, Studies on Southeast Asia, Trans R.E. Elson and Ary Kraal, edited by R.E. Elson. Ithaca: Cornell University Southeast Asia Program.

Ferguson, James, 1990, *The Anti-Politics Machine: "Development," Depoliticization and Bureaucratic Power in Lesotho*. Cambridge: Cambridge University Press.

Geertz, Clifford, 1960, *The Religion of Java*. Chicago: University of Chicago Press.

Gordon, Alec, 1979, "The Collapse of Java's Colonial Sugar System and the Breakdown of Independent Indonesia's Economy" in *Between People and Statistics: Essays on Modern Indonesian History*, edited by Francien van Anrooij *et al.*, pp. 251–265. The Hague: Martinus Nijhoff.

Gouyon, A., H. De Foresta and P. Levang, 1993, "Does 'Jungle Rubber' Deserve Its name? An Analysis of Rubber Agroforestry Systems in Southeast Sumatra". *Agroforestry Systems*, 22, 181–206.

Gupta, Akhil, 1995, "Blurred Boundaries: The Discourse of Corruption, the Culture of Politics and the Imagined State". *American Ethnologist*, 22(2), 375–402.

Hansen, Gary E., 1973, "The Politics and Administration of Rural Development in Indonesia: The Case of Agriculture, Research Monographs". No. 9, Center for South and Southeast Asian Studies, University of California, Berkeley.

Jameson, Frederick, 1984, "Postmodernism, or the Cultural Logic of Late Capitalism". *New Left Review*, 146, 53–92.

Kedaulatan Rakyat ("The People's Sovereignty") (Newspaper), 1985a, "*20 Petani TRI Banyuraden Sleman adukan Nasibnya ke LBH Yogya*" (Twenty Sugar Cane Intensification Farmers from Banyuraden, Sleman, Contest Their Lot via the Institute of Legal Assistance, Yogyakarta). (16 April.)

1985b, "*Petani tebu berhasil jika memakai sistem penelolaan koperatif*" (The sugar cane farmers would succeed if they used a cooperative system of exploitation). (6 May.)

Knight, G.R., 1980, "From Plantation to Padi-Field: The Origins of the Nineteenth Century Transformation of Java's Sugar Industry". *Modern Asian Studies*, 14(2), 177–204.

Koppel, Bruce, 1995, "Why a Reassessment?" in *Induced Innovation Theory and International Agricultural Development: A Reassessment*, edited by Bruce M. Koppel, pp. 3–21. Baltimore: The Johns Hopkins University Press.

Koppel, Bruce and John Hawkins, 1994, "Rural Transformation and the Future of Work in Rural Asia" in *Development or Deterioration? Work in Rural Asia*, edited by Bruce Koppel, John Hawkins and William James, pp. 1–46. Boulder: Lynne Rienner Publishers.

Leach, E.R., 1982, *Social Anthropology*. Glasgow: Fontana Paperbacks.

Lucas, Anton, 1992, "Land Disputes in Indonesia: Some Current Perspectives". *Indonesia*, 53, 79–92.

Marcus, George E. and Michael M.J. Fischer, 1986, *Anthropology as Cultural Critique: An Experimental Moment in the Human Sciences*. Chicago: The University of Chicago Press.

McTaggart, W. Donald, 1982, "Some Characteristics of Government and Quasi-Government Writings Dealing with South Sulawesi". *Indonesian Quarterly*, 10(2), 44–62.

Moertono, Soemarsaid, 1981, *State and Statecraft in Old Java: A Study of the Later Mataram Period, 16th to 19th Century*. Monograph Series No. 43, Rev. ed. N.Y.: Cornell University Southeast Asia Program, Ithaca.

Murray, Martin, 1992, "'White Gold' or 'White Blood'? The Rubber Plantations of Colonial Indochina, 1910–40". *Journal of Peasant Studies*, 19(3–4), 41–67.

Nader, Laura, 1972, "Up the Anthropologist — Perspectives Gained from Studying Up" in *Reinventing Anthropology*, edited by Dell Hymes, pp. 284–311. New York: Pantheon.

Nandy, Ashis, 1987, *Traditions, Tyranny and Utopias: Essays on the Politics of Awareness*. New Delhi: Oxford University Press

Parpart, Jane L., 1993, "Who is the 'Other'? A Postmodern Feminist Critique of Women and Development Theory". *Development and Change*, 24, 439–464.

Peluso, Nancy and Mark Poffenberger, 1989, "Social Forestry in Java: Reorienting Management Systems". *Human Organization*, 48(4), 333–344.

Pierce, Jennifer, 1995, "Reflections on Fieldwork in a Complex Organization: Lawyers, Ethnographic Authority and Lethal Weapons" in *Studying Elites Using Qualitative Methods*, edited by Rosanna Herz and Jonathan B. Imber, pp. 94–110. Thousand Oaks (Ca.): Sage Publications.

Polier, Nicole and William Roseberry, 1989, "Tristes Tropes: Post-modern Anthropologists Encounter the Other and Discover Themselves". *Economy and Society*, 18(2), 245–264.

Rappaport, Roy A., 1993, "Distinguished Lecture in General Anthropology: The Anthropology of Trouble". *American Anthropologist*, 95(2), 295–303.

Rosaldo, Renato, 1978, "The Rhetoric of Control: Ilongots Viewed as Natural Bandits and Wild Indians" in *The Reversible World: Symbolic Inversion in Nature and Society*, edited by Barbara A. Babcock, pp. 240–257. Ithaca: Cornell University Press.

Roth, Paul A., 1989, "Ethnography without Tears". *Current Anthropology*, 30(5), 555–569.

Said, Edward, 1983, "Opponents, Audiences, Constituencies and Community" in *The Politics of Interpretation*, edited by W.J.T. Mitchell, pp. 7–32. Chicago: University of Chicago Press.

Sangren, P. Steven, 1988, "Rhetoric and the Authority of Ethnography". *Current Anthropology*, 29(3), 405–430.

Scott, James C., 1984, "History According to Winners and Losers" in *History and Peasant Consciousness in South East Asia*, edited by Andrew Turton and Shigeharu Tanabe, pp. 161–210. Senri Ethnological Studies 13. Osaka: National Museum of Ethnology.

1985, *Weapons of the Weak: Everyday Forms of Peasant Resistance*. New Haven: Yale University Press.

Sinar Harapan ("Ray of Hope") (Newspaper), 1984, *A.M. Nasution Lebih Suka Memilih Kelapa Sawit* (A.M. Nasution Prefers to Choose Oil Palm). (11 July.)

Stoler, Ann, 1985a, *Capitalism and Confrontation in Sumatra's Plantation Belt, 1870–1979*. New Haven: Yale University Press.

1985b, "Perceptions of Protest: Defining the Dangerous in Colonial Sumatra". *American Ethnologist*, 12(4), 642–658.

Syarifuddin, Baharsyah and Soetatwo Hadiwigeno, 1982, "The Development of Commercial Crop Farming" in *Growth and Equity in Indonesian Agricultural Development*, edited by Mubyarto, pp. 144–178. Jakarta: Yayasan Agro-Ekonomika.

Tsing, Anna L, 1993, *In the Realm of the Diamond Queen: Marginality in an Out-of-the-Way Place*. Princeton: Princeton University Press.

Tyler, Stephen, 1986, "Post-Modern Ethnography: From Document of the Occult to Occult Document" in *Writing Culture: The Poetics and Politics of Ethnography*, edited by James Clifford and George E. Marcus, pp. 122–140. Berkeley: University of California Press.

Watson, Graham, 1991, "Rewriting Culture" in *Recapturing Anthropology: Working in the Present*, edited by Richard G. Fox, pp. 73–92. Santa Fe: School of American Research.

Wolf, Eric R., 1982, *Europe and the People without History*. Berkeley: University of California Press.

Chapter 8

NUCLEUS AND PLASMA: CONTRACT FARMING AND THE EXERCISE OF POWER IN UPLAND WEST JAVA*

Ben White

Plasma. 1712. 1. Form, mould, shape (rare) ... 3. *Phys.* The colourless coagulable liquid part of blood, lymph or milk, in which the corpuscles (or in milk, the oil-globules) float (Shorter Oxford English Dictionary).

Historical studies of agrarian change in Java have concentrated mainly on the impact of colonial systems for the production of export crops, and their interaction with a peasant "subsistence" sector; sugar cane seems to have captured the main attention of the agrarian historians, with upland crops being relatively neglected. From the early 1970s onwards, as noted in the introduction to this volume, research has tended to concentrate more on "green revolution" studies in the many regions of irrigated rice production in which there is no significant involvement of peasants in export or commercial production of other crops.

Changing policies and production relations in the rice sector are of course important, paddy being still the largest single branch of agricultural production, and the one which involves the most people as farmers and labourers, as well as being Indonesia's most politically sensitive crop (Hüsken and White 1989). Besides this, however, a balanced picture of contemporary agrarian change in Java requires a broader range of research focuses, with more attention to changing agrarian relations in forest and upland regions (as in the pioneering studies of Peluso (1992) and Hefner (1990)), and — in both lowland and upland regions — the old (rehabilitated) and new branches of commercial agro-production for export and/or urban markets.

The development of these various new forms of agro-commodity production (and the intensification of some old ones) has been accompanied by the official promotion (and in some cases, the autonomous emergence) of new organizational forms in production, processing and distribution. Much of this production is no longer organized along classical plantation lines but through schemes which link smallholders by contract to a larger agribusiness core which exercises varying forms of control (and varying degrees of coercion) over the labour process. These forms of production are of course not new; contracted export-crop cultivation by peasants, whether forced or "free", was a historical forerunner of plantations in many cases, and was the basis of coffee, indigo, sugarcane and tobacco production under the Cultivation System (Elson 1994). However, they are growing in importance in Indonesia (WIM 1994; Bachriadi 1995) and Southeast Asia (Glover & Teck Ghee 1992) as in many other world regions (Glover & Kusterer 1990; Little & Watts eds. 1994). These developments have profound implications for questions of economic opportunity and welfare, and equally for the structure of rural society itself: " the dispersion of contract farming marks ... a watershed in the transformation of rural life and agrarian systems in the Third World in general" (Watts 1994: 24). This aspect of agro-industrial development has been relatively neglected in research in Asia: while some recent studies have focused on the economic aspects of contract farming systems (for example, Glover & Teck Ghee 1992) very few have attempted to integrate analysis of the social, cultural, political and economic aspects of contract farming and the specific forms of social change which it implies, in contrast to work on this topic in Sub-Saharan Africa (Little & Watts eds. 1994).

This chapter examines the experience of state-sponsored contract farming in Indonesia, using as illustration a hybrid coconut Nucleus Estate Smallholder scheme in the hilly southern part of West Java. First however I will consider some general issues on the role of contract farming in agrarian transitions, and also provide a brief introduction to the national and regional context of the case study.

CONTRACT FARMING AND AGRARIAN TRANSITIONS

Contract farming is a particular way of linking commercial agro-production and agro-industry in which primary production (of annual or tree crops, livestock, dairy, poultry, eggs, fish, shrimps etc.) is not concentrated in large capitalist (or socialist!) production units but remains in the hands of "smallholders", linked institutionally through contracts to a larger "nucleus" enterprise which handles one or more of the upstream and downstream activities such as input supply, output processing and marketing. As has been noted by several authors, if plantation agriculture is the agrarian analogue of a large factory, contract farming is the agrarian equivalent of family-based industrial sub-contracting. These parallels, and some of their implications, are drawn out further by Watts:

> The deepening of contract production in agriculture bears striking resemblances to so-called neofordism or 'flexible accumulation' in sectors of industrial capitalism with a growing reliance on multiple outsourcing through industrial

subcontracts [...] Under contract in centralized systems, peasants work as *de facto* piece workers often labouring more intensively (i.e. longer hours) and extensively (i.e. using children and other non-paid household labour) to increase output or quality (Watts 1992:95)

There are several schools of thought on the pros and cons of such schemes and their ability to provide a path away from rural poverty through rural employment and income generation. Not surprisingly, it is a field of agricultural policy in which opinions tend to be strongly held and firmly expressed. Williams and Karen's *Agribusiness and the Small-Scale Farmer* (1985), a study commissioned by the Bureau of Private Enterprise of the United States Agency for International Development, concludes for example that

> agribusiness has found a fit in practically every social structure, every stage of human development, and where the industry has prospered, the people involved have begun to prosper, [...] agro-industrial enterprises with a satellite farming procurement system, sometimes called 'nucleus estates', have a unique capability to transfer technology very rapidly and to generate widespread participation by local people (Williams and Karen 1985:1,8)[1]

On the other side, some authors argue that the interest of large-scale capital and international agencies in "small peasant" (family-farm) forms of production represents an attempt to subjugate them to capital in a form which allows the surplus profit from agricultural modernization to be captured not as profits for direct producers but as profits for the "core", and transforms peasants into a class of virtual "development peons" (Payer 1980). Watts, in a series of publications since the late 1980s has explored the "disguised proletarian" character of contract farming: "peasants produce under contract in varying positions of unfreedom and accordingly constitute a distinct class that may be seen as a fraction of an emerging global proletariat" (1994:71; cf. also Watts 1990, 1992). This view echoes Chayanov's observation seventy years earlier that "new ways in which capitalism penetrates agriculture ... convert the farmers into a labour force working with other people's means of production" (Chayanov 1966:262). It needs however to be combined with awareness of the tendencies to differentiation and wage labour within contract-farming communities, as will be argued further below.

For smallholders such arrangements do offer some potential advantages compared to production for an "open" market. To some extent they can more reliably forecast their incomes (if the "core" provides price guarantees, and keeps to its side of the agreement); through the link to the "core" they can indirectly capture economies of scale in access to material inputs and support services (but again, only if these economies are passed on to them in the form of input prices lower, and output prices higher than those offered on the open market), and they can gain access to larger markets for their output (Goldsmith 1985:1127). Contracting *in itself* does not necessarily spell hardship or doom for smallholders, and in fact, all over the world, contracting of some kind is a necessity for many or most forms of modern commercial agriculture. Certainly growers of non-staple crops destined for distant or export markets will nearly always have entered some kind of advance contract with a buyer, and the buyer will in most cases have included one or more specifications as to production conditions.

The crucial (potential) problem for contracting smallholders in their insulation from "open" markets lies in the division of value added between themselves and the core. In all food commodity chains (or *filières*: Bernstein 1996) the setting of prices at the various points in the production, processing and marketing chain is not a matter of "real" value added or of supply-demand interactions, but reflects more the relative social/political bargaining strength of the parties involved. Contract farming, through institutionalizing monopoly/monopsony relations between farm and agribusiness, can reflect this property of "real" markets (Mackintosh 1990) in exaggerated ways.

The only thing that binds all contract-farming and outgrower schemes together as an analytical category is the existence of a contract. Here, it is important to bear in mind that "the contract is a representation of a relationship rather than the relationship itself, and the divergence between the two may prove crucial". This perspective helps us to move beyond narrow, legalistic and mystifying notions of the contract as a bargain freely made by two equal parties as found in neoclassical and agribusiness literature, towards a more useful view of the contract relation as potentially a "social relation of domination" and "an attempt to naturalize an unequal social relation and to represent that inequality as just" (Clapp 1994:79, 92f.).

This brings us to a second general point: whatever organizational form of contract farming is chosen, its implementation takes place in specific and concrete social and political environments, in which various actors and groups may exercise sufficient local powers to subvert and manipulate the scheme in their own interests. In short, contract farmers like any other small farmers are potential prey for whatever social-political predators may be present in a particular national or local context. Such predators may be located either within or outside the agribusiness core. At the same time, contract farmers are not merely potential prey, but also social agents who may engage in various forms of resistance, negotiation and protest in relation to both agribusiness projects of technology transfer and control and predators' attempts to find a niche in these projects.

So far we have discussed only problems which may arise in the relations between "core", contracting smallholders and various potential predators. The actual structure of production relations in many or most contract-farming formations is more complex than this, and research must also be alert to other sets of relationships below the level of the nucleus-smallholder relation, for example those between smallholders of different scale and power in a differentiated rural society, and also between the so-called "smallholders" and those who work for them (see for example Porter & Phillips-Howard 1995). In many cases, "smallholdings" in contract farming schemes are not actually "family farms" (family-labour based production units) but small or medium-scale enterprises based mainly on wage labour. In cases where the smallholding is indeed a family-labour based production unit the relations within household units also require study; particularly the internal division of labour, decision-making and control of earnings between household members based on hierarchies of age and gender. The position of women within contract farming systems, for example, seems to have received little attention to date (exceptions are Carney 1988 and 1994; Mackintosh 1989), and also the position of children and youth.

Contract farming, then, is one way in which West Java's (and Indonesia's) upland cultivators[2] are being "captured" or incorporated into wider economic circuits, often involving a shift from mixed-farming to monocrop cultivation. But just as there is a great variety of paths of transformation in both uplands and lowlands, within the broad category of "contract farming" there is also a great variety of situations and processes at work. As Little notes, the diversity of contract farming is so great that it is more useful to focus on the actual content (rather than the formal structure) of contracting relations in specific cases and the motives and power relations of those involved, rather than looking for blanket conclusions about contract farming as a generic institution (Little 1994:218f.).

INDONESIA AND WEST JAVA: THE AGRARIAN AND POLICY CONTEXT

West Java, Jakarta's agrarian hinterland, has a long and complex history of agro-production for export. It has a relatively large percentage of its agricultural land area in large (government and private) estates compared to the rest of Java and also, as recent discussions in the national and local press have documented, a high proportion of so-called "neglected estates" (*perkebunan terlantar*) which are considered not to be fulfilling their economic and social function in terms of productivity, export revenues and employment. Thus the search for an appropriate future for West Java's estate lands is a topic of high current policy relevance. At the same time, West Java's location close to the large urban centres of Jakarta and Bandung and to export outlets through Jakarta, makes it a preferred location for investment (both foreign and national) in various forms of commercial agro-production of high-value crops (vegetables, fruits, flowers), livestock and dairy production, poultry/eggs and various kinds of fisheries and shrimps production for export and/or urban markets. Many if not most of these dynamic forms of agro-production are based on contract-farming lines. Since the mid-1970s a series of Presidential Decisions and other national regulations have established institutionalized contract farming or outgrower ("PIR") systems as the preferred — and in some cases the only permitted — form of production in many branches of commercial agro-production, including sugar cane, tree crops (such as tea, rubber, oil palm and coconuts), dairy, poultry and egg production and coastal brackish-water shrimp ponds (*tambak*).

The contracted smallholder form (besides reflecting the new orthodoxy of the World Bank, which since 1975 has preferred outgrower schemes over plantations) responds not only to neo-populist ideology and nationalist sentiments (the distaste for large-scale capital in general and foreign capital in particular), but also to democratic rhetoric. It holds out the vision of an agrarian society based on small-farm units, using household labour, modernizing, prosperous, homogeneous and democratic, particularly where cooperatives are involved (as they nearly always are, if only by fiat, in Indonesian contracting schemes); in short, a Chayanovian dream. There is also, however, a darker underside to this vision of a sturdy modernizing peasantry, which reflects instead a project of control, in both its technical and political aspects. On the technical side, contract farmers (at least in government-promoted schemes) are totally dependent on the nucleus for all inputs, and all production decisions except those relating to labour. On the political side, at least in politically sensitive regions with a history of separatist movements (as in West Irian and Aceh) or leftist movements (as in many plantation regions), where tree crop nuclear estate schemes are now commonly promoted, there is undoubtedly an agenda for political control. The passive, disembodied and uncreative character of the ideal Indonesian contract farmer is reflected (perhaps accidentally) in the term commonly used in official documents to refer to outgrowers:

petani plasma (plasma farmers), which to English-speaking ears at least connotes a formless, homogeneous and pliable mass:[3] something even more remote from notions of citizenship and social agency than the Department of Industry's official symbol for the model Indonesian industrial-sector worker, the black ant (*semut hitam*).

The hilly southern region of West Java, stretching from Pandeglang and Lebak to the West to Tasikmalaya and Ciamis in the East is the least irrigated, the least densely-populated and the relatively least accessible part of West Java. Much of it was devoted to shifting cultivation until the opening-up of the region to plantation production, mainly rubber and tea, in the early twentieth century (van Doorn & Hendrix 1983). The national revolution (1945–49), subsequent nationalization of Dutch enterprises (1957–1958) and the fundamentalist *Darul Islam* rebellion (defeated in 1961) led to widespread collapse of plantation production and the virtual depopulation of much of the region. During the 1960s and 1970s, the nationalized plantation corporations began the rehabilitation of the larger plantations, but many of the smaller ones remained abandoned and were gradually re-occupied by peasants (those who had formerly fled the region, or new in-migrants from the West Java's Central plateau and from the Western parts of Central Java) in search of land for mixed upland garden cultivation. The region thus contrasts with much of the rest of Java, in that it has had until the 1980s much of the character of a pioneer settlement region, and access to land has been relatively easy. An important related feature is the relatively insecure status of land tenure. Many peasant holdings lie on what is officially classified as state land (*tanah negara*) and while those who occupy it may have "purchased" it from local officials and paid taxes on it (with certificates to show it) they do not have formal ownership title.

Although rehabilitation of the larger plantations was a focus of government intervention and World Bank lending from the late 1960s onwards, it is only since the early 1980s that major attention has been paid to the (re-)incorporation of the peasant sector of this southern zone into more intensive commercial production. This has been stimulated not only by the relative economic backwardness of the region, but also by the desire to establish greater political control in a region which was both the basis of the Darul Islam rebellion and (on and around the plantations) a site of considerable support for the Communist Party (PKI) and its labour-union and peasant affiliates (SARBUPRI and BTI). One aspect of this re-incorporation has been the stimulation of Nucleus-Estate Smallholder (NES) schemes for tree crop production (*Perkebunan Inti Rakyat Tanaman*

Perkebunan or *"PIR-BUN"*) under the aegis of the State Plantation Corporations PTP XI and PTP XIII with a major World Bank loan (1980) in the more remote regions bordering on the southern coast.

HYBRID COCONUTS ON CONTRACT: THE CISOKAN NUCLEUS ESTATE[4]

> Frequently the trading machine, concerned about a standard quality in the commodity collected, begins to interfere in the organization of production, too ... and turns its clients into technical executors of its designs and economic plan (Chayanov 1966: 262)

From Upland Mixed Gardens to Contracted Monocrop Farming

In 1980 the World Bank (which had adopted the smallholder-nucleus system as a general policy priority in the mid-1970s) signed an agreement with the Indonesian government for a series of seven 20-year loans to support the development of Nucleus-Estate Smallholder (NES) schemes under the aegis of various existing state plantation corporations. Two of these loans included five hybrid-coconut NES projects in the southern region of West Java, covering a total of almost 50,000 ha (about 20% of it reserved for the "nucleus" or *inti* and 80% for the *"plasma"*) and aiming to involve somewhat over 10,000 households as contract farmers. These contract farmers (the *petani plasma*) were to be assigned a total of 2.0 ha of land: 1.5 ha already cleared and planted with ca. 214 hybrid coconuts for monocrop cultivations, 0.3 of cleared land for establishment of a food crop plot, and 0.2 ha for a house and garden plot, with a house provided — or, in the case of existing housing, suitably improved — by the project. On receiving their plots the farmers would begin a 15-year period of repayment of a bank loan booked against their names for approximately Rp 3,600,000 and an additional loan of up to Rp. 1,300,000 for house construction or improvement. The basic format of all these NES projects is the same, and can be summarized in simplified form in four phases, as follows:

1. Staff of the National Land Board (Office of Agrarian Affairs) first map and mark with posts the area of land to be converted to the NES project.
2. (Years 1–3) The PTP clears the land; plants the coconut seedlings in both nucleus and smallholder land; prepares the home-garden (0.2 ha) and food-crop (0.3 ha) plots; builds the necessary storage facilities and processing plants and installs machinery (for copra and coconut oil

production); builds the necessary local infrastructure i.e. roads, contract-farmer's housing, school, places of worship, marketplaces and health clinics (for this, working together with district government and the relevant district offices of government departments); assists the various relevant government agencies in preparing procedures for the issuing of land titles, and in collecting data on candidates for participation in the scheme; provides technical guidance for the candidate-farmers (*calon peserta* or "*capes*"); stimulates the formation of farmers' groups (*kelompok tani*, "*poktan*") and (where not yet existent) co-operatives (*KUD*). All work on the nucleus and smallholder plots during these years is carried out by wage labour, with those selected as "*capes*" being given priority for employment.

3. (Years 4–5) At about the time when the young coconut trees should have flowered and begun to bear their first fruit, the PTP allocates the home, food crop and coconut plots to the participant farmers (by lottery organized together with village and local government); assists in completion of the issuing of land titles (which, although formally issued to individual farmers, will be physically kept in the BRI bank until the credit has been repaid in full); supports the KUD's development as the institution which eventually will be responsible for input supplies; calculates the volume of credit required by each farmer; and transfers responsibility for cultivation and credit repayment to the farmers while retaining responsibility for the processing plant, cultivation on the nucleus estate itself and infrastructure. This is known as the "conversion phase".

4. (Years 6–20) After "conversion" the PTP continues to be responsible for processing and marketing, and for the nucleus and other infrastructure; it assists the KUD with transport of harvested coconuts, and (for the 15 years when farmers are repaying their bank loans) assists the bank in monitoring repayments. After conversion the farmers are formally owners of the land, although as noted above they do not receive their land titles until their credit repayment is completed. The time when farmers should have repaid all their debts (about 20 years after planting) in fact will coincide with a sharp drop in the trees' productivity, necessitating re-planting and a period of 4–5 years' waiting until the new trees bear fruit, which presumably would be financed by a second round of long-term bank loan and repayment.

The choice of the crop, and of the NES form of production organization, were based on a number of considerations. The choice of crop relates to Indonesia's coconut-oil crisis of the late 1970s, in which coconut production failed to meet the domestic demand for cooking-oil and (rather than importing from the neighboring Philippines) the government diverted a part of Indonesia's growing palm-oil production from export to domestic markets. In the early 1980s the government began promoting hybrid coconut production as a means of satisfying domestic demand and thus returning oil-palm to its original role as foreign exchange earner. In fact, the market projections for hybrid coconut oil have proved incorrect, and prices in both export and domestic markets have been weak and fluctuating. Local consumers prefer to use the traditional varieties of coconut (*kelapa dalam*) which are purer-tasting and less oily, and crude coconut oil (CCO) from hybrid coconuts is regarded as inferior to palm-oil by the various factories producing cooking-oil for the domestic market.

The idea to introduce nucleus-estates into this region was seen as a means of opening up this undoubtedly backward and deprived region (through commercialization and the infrastructure of roads etc. linked to it), but the element of political control is also occasionally mentioned in documents,[5] besides being frequently raised in conversations by officials at various levels. This helps to explain the choice of Cisokan, which our researchers were told was regarded as a basis of Darul Islam separatism and very difficult to control due to its geographical isolation.

The most controversial of these projects was Cimerak (to the west of our case-study region), in which 2,000 ha of peasant mixed-garden land, including valuable clove and other trees, were appropriated without compensation from farmers who had been cultivating them for up to 30 years, and bulldozed for the planting of monocrop hybrid coconuts. This occurred in the face of intense peasant protest supported by various NGOs including the Bandung Legal Aid Institute (LBH Bandung 1985) which reached both the provincial and national parliament and continued throughout the 1980s, with considerable attention from the national and regional press. In 1990 the regional newspaper *Pikiran Rakyat* published a series of features on Cimerak, with photos of the ramshackle wooden houses provided by the project and a copy of a farmer's monthly settling of accounts with the nucleus PTP, showing that after various deductions for credit repayment and other purposes this farmer received a net monthly income of only Rp 2,831 (in a project whose feasibility study had projected farmer incomes of more than Rp 150,000 per month) (*Pikiran Rakyat* 17/10/90). A decade of protest, however, does not seem to have resulted in any redress or improvement for either those excluded from or those included in the project.[6] The Cisokan project, to which we now turn, did not receive any major publicity, although its problems seem to have been no less than those of Cimerak.

The area now occupied by the PTP XII's Cisokan rubber plantation and adjoining hybrid coconut Nucleus Estate lies somewhat more than 100 km south of Jakarta, in a 30 km stretch of low but often steep coastal ridges running East-West close to the southern coast with altitudes between 0–150 meters. Many parts of this area were deserted of people during the last years of the Darul Islam rebellion, and much land was abandoned. From the defeat of this rebellion (1961) to the late 1970s people were encouraged to return or settle there, with many coming from the more differentiated and land-hungry parts of West Java and western Central Java, attracted by the easy availability of land for cultivation. Relatively small payments to village officials would secure them plots of what was formally classified as state land (generally of around 2 ha) with official permission to cultivate *(Surat Izin Menggarap)*. After regularly paying land taxes and obtaining proof of it from the village *(kikitir)* or sub-district office *(Tanda Bukti Pembayaran Pajak)*, peasants felt relatively secure in their control of this land and the prospect of eventual formal "ownership" at such time in the future as the Department of Agrarian Affairs would issue ownership certificates.[7] Some had been paying land taxes for more than 20 years when the project began.

By the late 1970s, a pattern of mixed shifting agriculture and horticulture had developed, with smaller amounts of rainfed *sawah* and grazing land. Main dry-field crops were upland paddy, maize, soya and ground nuts. In home-gardens and mixed-gardens, many kinds of productive tree crops were grown in addition to bamboo stands, hardwoods and the ubiquitous coconut and banana (durian, pineapple, avocado, mango, mangosteen, *petai*, guava and cashew nuts). Informants characterize this as a time when "money was scarce": crops were more often exchanged than sold, and when traders came at the main harvest times they would often simply exchange rural crops for urban products without the medium of cash.

The main non-farm activities were linked to this pattern of agriculture and local raw materials: small-scale production of coconut oil, *tahu* and *tempe* (fermented soya curd and cake), construction materials (bricks, timber) and furniture, and various items woven from bamboo or pandanus leaves (*bilik*, mats and woven hats), but most importantly the tapping of coconut and aren palm for palm sugar *(gula kelapa / gula aren)*. Those with access to trees for tapping would generally produce 4–8 kg of palm sugar per day, sold (in 1989) at Rp 700 /kg on the free market or Rp 500 to a merchant who had previously given credit (*borsom*). Tapping of others' trees on a product-sharing basis would still provide a regular daily income somewhat higher than prevailing agricultural wages in either the peasant sector or on the nearby plantation. Circular outmigration to Jakarta, Bandung or other urban centres was also common, generally of younger household members, with men going mainly into construction labour and women into domestic service.

Although we know little about the forms and processes of differentiation before the project, this was certainly not a homogeneous society of small-scale mixed farmers. In the five villages included in the Cisokan project, a number of villagers (mainly from the first wave of settlers in the mid- and late 1960s) had acquired control of relatively large tracts of state land (though not enormous amounts: 7.0 ha was mentioned as a large holding) and rented out parts of it to others. Many of these and some others were by now relatively wealthy merchants (*tengkulak*) in the trade in agricultural products, often combining this with processing (copra, coconut oil etc.).

The project was initiated in 1982 and the first trees planted in 1983. As may be expected in such an extremely top-down project (which draws its staff mainly from a nearby plantation specializing not in coconuts but in rubber, and which requires the coordinated inputs of more than 18 government agencies at national level and 25 at provincial level),[8] the project has run into many serious technical problems. Although these are not the main subject of our study they deserve some discussion, particularly in view of the supposed superior role of outgrower schemes in effecting "transfer of technology" to smallholders.

The mandatory feasibility study for the project was undertaken only when the project had already been in operation for a year, and after pressure from the World Bank. Before this study was even completed, project officials were under pressure to identify land for planting, as hundreds of thousands of seed coconuts had been delivered before land preparation was completed, and some of them were already sprouting. Land was arbitrarily allocated for coconut planting and for house and food-crop plots by project staff, without proper land mapping and apparently by simply walking from one hill-top to another and planting markers in the ground. After planting, many plots of young hybrid coconut trees were ravaged by wild pigs, and sometimes completely destroyed despite more than one attempt at replanting. Although these various replantings cost Rp 1.6 billion no funds had been allocated for this purpose, since the report on the damage had been sent only to sub-district level, when decisions on such matters are made in the central office of the NES Coordination Team in Jakarta. For several years after the young trees had begun to produce (1986), the conditions of the project's access and feeder roads were so bad that they were only passable by tractor (particularly in the rainy season). Many thousands of coconuts lay for weeks and even months beside the feeder roads, waiting for collection vehicles which never came, until they sprouted or rotted.[9] Although the first ripe nuts were harvested in 1986, until 1992 there was no coconut oil processing factory in operation. The impassable roads and high transport costs made sale of fresh coconuts relatively unprofitable, the only remaining outlet being the Copra Drying Station at Karanganyar. Despite the relatively simple technology and design, the CDS too ran into many technical problems, including several fire outbreaks during night-time drying in 1989. At times mountains of up to a million coconuts waited for processing beside the CDS.

Due to these and other problems, coconut production is far behind the original projections, with the further consequence that farmers will be unable to repay their loans within 15 years according to the schedule based on these projections. More interesting for our purposes are the social and political dynamics underlying the emergence of the contract-farming community, its internal structure and its relations with external forces and the state. Some insight into these dynamics is provided by consideration of various areas of conflict and tension observable during the first years of the project. First we may consider the process of land appropriation for the project, and the subsequent process of selection and exclusion reflected in its reallocation to contract farmers. We will then turn to issues of agrarian differentiation, labour process, and the insulation of contracted producers from "free" markets.

Land Alienation

There is a major hiatus between local ideas of land possession rights (which does not necessarily mean "traditional", in a society composed mainly of recent in-migrants) and those embodied in the project's policy of appropriation without compensation. The project (perhaps because essentially borrowed from the "pioneer (re) settlement" NES model) assumed that the land was basically "empty" and available state land. Anyone occupying it had merely to be persuaded to surrender it to the project, induced if necessary by the prospect of selection as a NES participant and subsequent reallocation of land. For that reason, the project budget did not include any funds for compensation, either for the land or for the valuable trees or buildings on it. Local cultivators, in contrast, felt themselves to be in fairly firm right of occupancy, based not only on customary "pioneer's rights" of those who had cleared land and planted trees on it, but also on their formal rights, backed up by written proof from village government (Letter of Authorization to Cultivate) or sub-district government (certificates of assessment and payment of IPEDA tax). Furthermore they felt justified in the expectation of later acquiring ownership title.[10]

There was thus widespread unrest and significant opposition to the alienation of peasant holdings and their clearance for hybrid coconut planting. The local élite was among those who protested: in the five villages, some 20 of the most wealthy villagers opposed the NES from the beginning, not least because they had relatively the most to lose, having gained control of large tracts of state land and rented them out to small farmers. One woman had only recently bought 1.5 ha of land from the village head, paying Rp 800,000 and an additional Rp 60,000 for the proof of purchase letter (*Surat Jual Beli Tanah*) from the village office. Village-level certification was no proof against alienation and her land was taken without compensation. After the project was first announced in 1982, it was widely expected that it would be many months or even years before land alienation, and still longer before any land clearance would begin. Everyone seems to have been taken by surprise by the very rapid appearance of the land-clearance teams, who immediately met with physical confrontation. Farmers with knives and cutlasses drove project officials off their fields, holding up the land-clearance phase for two months, after which it was continued under military protection. Many of those who opposed land clearance were detained at the sub-district military command and predictably threatened with accusations of communist-party sympathies. Of the first 20 peasants to surrender their land, apparently only two were eventually accepted as NES participants.

Inclusion and Exclusion

These confrontations made many local farmers reluctant to enlist as candidate NES participants (*Calon Peserta*). Many explained that they were initially afraid of becoming involved in the scheme, recalling how others had previously become involved in the communist party with disastrous consequences. This, plus the fact that several hundred household heads in the five affected villages actually had been declared involved in the communist movement 20 years previously and were thus ineligible for participation,[11] meant that at first the project had considerable difficulties attracting participants.

Initially many local leaders and village officials were persuaded to sign up as participants (although not meeting the formal criteria), after which others followed. Because of the volatile situation and also for the longer-term agenda of security and control, priority in selection was often given to sub-district government officials, local schoolteachers and civil servants in various departments, while many of those whose land was appropriated were not selected. This led to a situation in which many "insiders are excluded, and outsiders included" in the project. Some of those who lost land and did not enlist as participants now regret it. Restricted to land outside the project, which is less fertile because it must now be cultivated continuously, they look back with regret to the time when land was easily available, one could still move from one plot to another, and yields of rainfed *sawah* and mixed-gardens were good.

Once the NES participants have been selected, the allocation of NES plots among them (when the trees are about four years old) is supposed to take place by lottery. In this case however the "lottery" systematically allocated the most fertile and conveniently located plots to village and sub-district officials and their wives, officials of the KUD cooperative, field staff of the Family Planning program and other relatively well-off or well-connected persons, while ordinary participants received the least fertile plots, far from the settlements and on the edge of the forest. In the joking and punning mode characteristic of situations where open protest is curbed, informants agreed that success in the NES project was a matter of "3–D": *duit* (money), *deukeut* (closeness, i.e. connections) and *deuheus* (lit. "to pay a visit to an official or social superior").

The food crop (0.3 ha) and home-garden (0.2 ha) plots and the houses built on them were allocated by project officials, without lottery. Each block of land was simply allocated to a group of farmers who were then supposed to arrange the individual allocations among themselves. Many problems arose from the careless way in which these plots had been surveyed and mapped by staff of the National Land Board. Boundaries were unclear (leading to disputes between NES farmers and "outsiders", as well as between NES farmers whose land titles showed them as owners of the same land). Often the land was so steep as to be effectively uncultivable, with very thin layers of humus washed away with the first rain, exposing the limestone underneath. About one-third of a sample of 99 smallholders surveyed in 1989 did not know where their food-crop plots were located or had seen them taken by other farmers. Others left their plots uncultivated or used them only for extensive cultivation of upland rice and groundnuts. Some had announced they were willing to abandon any claim to food crop land, so long as their debt could be reduced accordingly, but received no response from the project. NES participants' existing houses (if they came from the area) were assessed by village and project officials, and if designated below standard the farmer was provided with a new house (and an additional debt of Rp 1.3 million) or a packet of materials and cash for home-improvement (valued at around Rp 0.6 million).

As in so many land (re)allocation schemes of this type throughout the world, women were almost completely excluded from land allocation in their own right, regardless of whatever rights to land they may have exercised before the project. Our researchers did find a small number of women who were participants in their own name, but these were all cases where for some reason the relevant male was not available for participation: one whose father's land had been alienated for the project, one whose husband was too old to be registered himself, others whose husbands had died since the project began, etc. In a region where divorce and remarriage are as common as they are in West Java, claims of divorced women to land resources acquired in marriage are bound to arise. The project officials' creative solution, when confronted with this problem, was to ask the local Office of Religious Affairs to discourage candidate-NES farmers from divorcing!

Differentiation and Labor Process

Social differentiation in Cisokan can thus be seen in terms of distinctions and relations between four main groups: the project officials, many of whom also were allocated NES plots (and whose sturdily-built houses and lifestyles everyone envies, contrasting starkly with the ramshackle housing built by contractors at inflated costs for participant farmers);[12] other local officials and civil servants, and members of village élites who registered as NES participants but also retain land outside the project and often many mercantile or other off-farm interests as well, and who recruit farm servants (*pembantu*) to farm the NES plots for them, although this is formally not permitted; thirdly, the "genuine" NES participants who depend for their livelihoods on their NES plots, worked mainly or partly with their own family labor, and on whatever wage-opportunities are available at the nucleus or outside; and finally those who lost land to the project but are now excluded from it, and who are generally regarded as the poorest group.

As may be expected given this differentiation, the majority of the NES farms deviate considerably from the "family farming" model on which

NES project planning is based.[13] A sample survey of 99 households in 1989 found that about 60% of all labour inputs on the coconut plots, and 25% of those on food crop and homestead plots are hired (nearly all of this is male labour). It is not only the wealthy "armchair" NES farmers who hire labour. A surprising number of "genuine" NES participants have found themselves unskilled off-farm work in the region (59% of adult men and 42% of adult women reported having supplementary off-farm employment, see Grijns 1995). Many of these at times hire labor for work in the coconut groves so as not to lose their off-farm jobs, even though the daily wage given to the coconut labourer may be higher. In these cases, then, the farm is worked by women and children from the household and (male) wage labor. The practice of exchange labor (*liliuran*), common in the region's upland farming outside the project, is unknown in hybrid coconut farming; groups of 4–6 wage laborers however hire themselves out on contract to NES farmers at harvest time. Such a group can pick and husk 3,000–6,000 nuts a day, earning about Rp 2,500–5,000 per person. Wage labour opportunities on the coconut farms themselves are mainly restricted to men. The project's two Copra Drying Stations offer work to quite a large number of adult men and women, boys and girls, at relatively attractive piece-rates (around Rp 4,000/day for men, Rp 3,000 for women, Rp 2,500 for boys and Rp 1,250 for girls).[14]

Some NES farmers have also effectively opted out of direct involvement in hybrid-coconut farming by the (officially forbidden) practice of selling their unripe coconut crop in advance by pawning their Certificate of Payment. The wealthier farmer, trader or official who takes the certificate in pawn is then responsible for harvesting and husking the nuts, and for paying the 30% credit installments in the name of the farmer. Some others (especially those whose plots are frequently ravaged by wild pigs) have opted out completely, abandoning their plots and simply refusing to pay any credit installments.

Insulation from Markets

In the past few years the price paid to farmers for hybrid coconuts has fluctuated between Rp 67 and Rp 101 per nut. Price-setting is based on a formula whose main components are the market price of processed coconut oil (and its by-product *bungkil*, an ingredient in livestock feed) and the factory's production costs. The nucleus' profit margins are thus cushioned from price fluctuations, which are passed on to the smallholder. More serious from the farmer's point of view are the many additional deductions made at the time of payment. Besides the automatic 30%

deduction for credit and interest installments, these include deductions for fertilizer, mandatory deposit to the KUD, transport fees, and an additional 6.5% in fees to various officials. Taken together these deductions meant that when coconut prices were around Rp 75, farmers often received as little as Rp 45.

If price setting is one thing over which farmers have no control, getting or even knowing exactly what they have paid for is another. Although, as we have seen, deductions are made for fertilizer, the amounts distributed through the KUD are often way below the recommended dosage (about 200 kg for the whole farm of around 200 trees). Farmers begin their contract period with debts to the bank of between Rp 3.6–5.0 million, (depending on how much was allocated in the form of house construction or rehabilitation). This is a very large amount: an indication of what it means is that an adult male would have to work full-time for 8–10 years at the prevailing local wage rates to pay off the principal, let alone the annual interest payments of 10%.[15] Most farmers know the amount of the debt that was initially booked to their name, but not exactly what it was for. After some years of regular deductions they do not know how much still remains to be paid, or how to calculate it. Farmers do not receive written receipts for installment payments or statements of the outstanding debt.

The establishment of the local KUD cooperative (since 1990) has meant no more than another building-block in the institutional construction separating farmers from the market. All farm inputs and outputs pass through the KUD (with the accompanying increases in cost) on their way between farmer and core. No farmers are represented in the KUD administration, which is staffed mainly by local civil servants and plantation staff.

Another major area of tension and conflict surrounds the project's claim to monopoly of purchase and processing rights of NES produce and the effort to enforce this monopoly, which has at times included banning certain economic activities based not on NES produce at all but on the sale or processing of nuts from the still-numerous stands of "village" (non-hybrid) coconut trees in home-and mixed-gardens in and around the project. Among the various broken promises which NES farmers claim were made to them (besides the promise of asphalt access roads, rural electrification and other infrastructure) was the undertaking at the beginning of the project that NES participants would be allowed to sell their produce on the open market if prices were higher outside, so long as they maintained their credit repayments at the level of 30% of the value of cash sales. In practice, those who sell outside are accused of theft; a notion of "theft" quite alien to local understandings, as expressed in the typical irony of a participant farmer: "it's confusing, when (since conversion) we are told that we are now the owners of our NES farms, but if a farmer takes a few of the coconuts he grew on it to sell or even to make a bit of coconut oil, he'll be arrested and called a thief —but he has title to the farm, even if the land title is still retained by the bank".

In the first years of production (1986 onwards), when piles of harvested coconuts often lay rotting or sprouting by farmers' plots because of the project's inability to transport them to the Copra Drying Station, the project gave farmers permission to make their own copra at home. Later, however, this led to problems when farmers who had set up small copra-drying enterprises were accused of "stealing" project coconuts. In 1989, some local entrepreneurs making copra had their copra confiscated and were summoned to the sub-district military command, even though they claimed to have the proofs-of-purchase to show that they were using only "village" coconuts. Sales even of fresh village coconuts were supposed to be made only to the KUD although this had not yet begun its purchasing operations, the making of coconut oil was forbidden, and even the makers of the manual coconut graters (*parutan kelapa*) used in preparation of coconut oil were warned by the sub-district government to take them off the market. Meanwhile, as already noted, the project's own inability to absorb the NES harvests for sale or processing was resulting in large amounts of wasted produce.

By the 1994 rainy season this unsustainable situation had been partly resolved by the stipulation that NES farmers could make their own copra with conventional (village) technology before selling it through the KUD to the project for processing into oil, the Crude Coconut Oil (CCO) factory having come into operation in 1992. This, however, was to be allowed only in the rainy season while roads were impassable, while in the dry season farmers would still have to sell their fresh nuts to the project's Copra Drying Stations. We see here a good example of the insistence of agribusiness nuclei on restricting farmer involvement to primary production, retaining all upstream and downstream activities (and the value-added generated by them) for the nucleus.

CONCLUSIONS: SMALLHOLDERS, CONTRACTING AND LOCAL POWERS

In the case we have examined the contract-farming project has not worked out precisely as intended. The gap between the assumptions on which smallholder contracting schemes are based, and the actual conditions that have emerged, is quite large and the project of technology transfer and standardization has led to more technical problems than it has solved. While state regulations and decisions (as in other Indonesian development ventures) have initiated the project, whatever process of accumulation has been set in motion is not entirely within state control.

Given the enormous technical and organizational problems presently faced by this project, one may ask whether it is really working to anyone's benefit except a few project and KUD officials and "armchair" NES farmers. Even if some of these problems are overcome in future, this will not resolve the fundamental contradiction between the interests and aspirations of state and "nucleus" on the one hand, and peasant on the other, in terms of what each hopes to achieve. The project is driven by a desire to control, to incorporate, to modernize, to remove rural people from backwardness and "subsistence" through the imposition of contracted monocrop cultivation. Meanwhile, for Cisokan's peasants who previously had locally-sanctioned access to upland farms for mixed-cropping, the project has meant that they now find themselves owing large sums of money plus interest, for land which is often less productive than what they had before (and some which is literally worthless), in many cases also for housing which they could have built much better and cheaper themselves. They pay this imposed debt by cultivating a low-value tree monocrop which they would not have chosen to plant, with a weak position in both domestic and export markets, whose prices are fixed in ways they do not understand, but include many sizable deductions for inputs which they do not get and services and institutions which they do not need.

Profits and some degree of enrichment through hybrid coconut contracting are possible, but mainly for the "armchair" coconut farmers for whom the activity is an investment rather than a source of day-to-day earnings, whose farms are worked by others in wage or product-sharing relations, and for whom the schemes were not intended. For "real" smallholders it seems that contracting can provide only part of household incomes, and farming is supplemented with external employment of at least one family member, normally adult males. The day-to-day activities of smallholder coconut farming are mainly in the hands of adult women and children, who have no formal right of ownership to the smallholding, of membership in the cooperatives through which inputs and outputs are channeled, or of control over the incomes generated. One cause of smallholder impoverishment is the parasitic activities of cooperative officials.

For many peasants, the project seems more to represent an unwelcome but necessary and *temporary* interlude of forced monocropping, low productivity and *plasma* status. Out of this they hope to emerge after 15–20 years with at least more secure ownership rights, not as the project envisages to enter another round of debt for hybrid coconut replanting and contracting but to return to the kinds of mixed, part-commercial farming they had practiced previously. This is reflected in the question often put to our researchers:

> Is it true that after 15 years I will really be the owner of this land? And does that mean I'll be allowed to sell the land, or to sell my coconuts wherever I want, even to cut down the hybrid coconut trees and grow whatever I like again?

In such cases, one might say, the institutional framework surrounding contract farming is in serious need of democratization. The formal structure of rural cooperatives, farmers' groups etc. *is* of course democratic. Given prevailing structures of local power and privilege, however, institutions which on paper appear to foster "participatory", egalitarian forms of development tend in practice to be dominated by the wealthy and powerful and are subverted to their interests. Independent mass organizations as a possible countervailing force have been banned since the early 1970s and their state-sponsored monolithic substitutes do not offer an alternative.[16] "Everyday" forms of resistance and protest, some of which we have mentioned, have their limits. The problem is not one that can be solved by bureaucratic tinkering with the formal design of the institutions and processes involved, since it is a problem not of formal structures but of the actual function and substance of real relationships, which reflect the nature and exercise of power in rural society.

ACKNOWLEDGEMENTS

This chapter draws on research carried out under the collaborative project 'Rural Productivity and the Non-Farm Sector in West Java, Indonesia' (1987–1991) by the Institute of Social Studies, The Hague; the Centre for Development Studies, Bogor Agricultural University and the Centre for Environmental Studies, Bandung Institute of Technology and financed by the Netherlands Ministry of Development Cooperation (Section for Research and Technology). An earlier version, incorporating an additional case-study of cooperative dairy farming, was published as White (1996). Thanks for helpful comments to Henry Bernstein, Tania Li and an anonymous reader for the ISS Working Papers series.

NOTES

1. In the absence of a standard terminology for the different types of contract farming arrangements, "contract farming" is used here as a generic term covering all types of farm production on contract (other authors sometimes use "satellite farming" as the generic term, reserving "contract farming" for private-sector contracting schemes, cf. Glover 1992:3); "outgrower scheme" refers to government contacting schemes on which public enterprises purchase crops from farmers; "nucleus estate smallholder" (NES) schemes are a sub-type of outgrower schemes, in which the corporate nucleus administers a plantation as well as processing plants, and where contract purchases supplement plantation production.
2. Also some of those in lowland irrigated areas where since 1975 smallholder sugarcane "TRI" *(Tebu Rakyat Intensifikasi)* schemes have largely replaced the system of rental of peasant sawah by the sugar factories for sugarcane cultivation using wage labour, which had prevailed for the previous century (see for example Mackie and O'Malley 1988; Hartveld 1995).
3. See the dictionary definition at the beginning of the chapter.
4. Except where otherwise stated this section is based on the report of Gunawan *et al.* (1995) and about 250 pages of field notes of Nunung Sulastri and Titi Setiawati (who carried out the main field research in July and September 1989) and Rimbo Gunawan (who made a brief follow-up visit in August 1994). I follow Gunawan *et al.* (1995) in giving the project the pseudonym Cisokan.
5. For example Departemen Pertanian (1978).
6. In a curious twist, the involvement of LBH in the Cimerak case resulted in 1990 in a well-publicized delegation of NES farmers to the national parliament in Jakarta, in a truck provided by the PTP, whose leader — no doubt under various pressures — announced, to the surprise of some members of the delegation "we don't know anything about the LBH, we have had nothing to do with it, their activities make us nervous, and we just want to go on with our farming in line with the government recommendations" (*Editor* 29/9/90: 34)
7. State Law no 1 (1958), enacted to handle the land question after nationalization of foreign plantations, specifies that all *"tanah partikelir"* formerly leased on *erfpacht (Hak Guna Usaha)* basis to foreign enterprises would have the status of state land, and should be redistributed to peasants. Presidential Decision 32 (1979) on Basic Policies in Granting New Status to Land Originating from Conversion of Western Land Rights, following the provisions of the 1960 Agrarian Law which specified that all land under *erfpacht/HGU* land leases would revert to direct state control at latest by 24 September 1980, states that all such land that has been devoted to village settlement and smallholder cultivation will be assigned to those occupying it (Suhendar 1994).
8. At national level, besides the Department of Agriculture, the project requires coordination of various inputs and/or authorizations from the National Planning Agency, the State Secretariat, the Departments of Home Affairs, Forestry, Transmigration, Cooperatives (the latter two have since been merged), Manpower, Industry, Communications, Public Works, Justice, Foreign Affairs, Trade and Finance; the *Bank Rakyat Indonesia* and the Inspectorate of Development Finances (BPKP). At provincial level and below, almost all government agencies are involved.
9. At the time of our 1989 field study the main 46 km long access road, totally impassable except by tractor in certain places, could not be repaired according to project officials, because it still had the status of 'material evidence' in the project's court case claiming restitution for non-fulfillment of contract from the road-building contractor. By 1994 the problem still had not been resolved.
10. That the Agrarian Affairs Department was capable of rapidly issuing ownership titles for land of this type is evidenced by the experience of the occupiers of some 300 ha of land immediately south of the project and adjacent to the sea coast, which a consortium of five Jakarta-based companies led by the "Hari Kader Group" purchased in 1989 for Rp 3.75 million/ha for the development of an export-oriented brackish-water shrimp farm. For the sale to take place, land titles had to be (and were) rapidly issued to the occupants by the

Agrarian Affairs department for a fee of Rp 0.2 million/ha. Many of those who sold land in this way are now building the best houses in the local sub-district town, in stark contrast to those whose (much better-quality) land was alienated without compensation for the NES project.

11. Although accounts of the formal selection criteria differ and not all were followed in practice, the criteria most often mentioned in project documents and meetings were: age younger than 50 years (or, if older, with children old enough to help); local inhabitant; married; not wealthy (*kehidupannya minim*, literally "at minimum level of living"); politically clean (*tidak terlibat organisasi terlarang dan lingkungan bersih*, literally "not involved in any banned organization and having a clean environment", the latter being official newspeak for "politically clean, including one's close relatives"). Additional criteria mentioned and in many cases followed were: having surrendered land to the project, and having regularly worked as wage-labourer at the nucleus during the clearing and planting stage.

12. These are the people who can afford to spend a Sunday morning at the local fishponds, where for an entry fee of Rp 5,000 one can take away as much fish as one can catch. They are also criticized for the way they boss the farmers around and adopt the attitudes of rural "gentry" (in Sundanese, *juragan*).

13. This aspect is discussed in greater detail by Grijns (1995).

14. Boys and girls work less than a full day, and can combine this with school attendance.

15. Based on a daily wage of Rp 1,500 and a 6-day working week.

16. At the local level, the state-sponsored "Indonesian Farmers' Harmony Association" *Himpunan Kerukunan Tani Indonesia (HKTI)*, with the official mandate to promote the interests of both peasants and farm workers, is virtually invisible. Of all the state-sponsored monolithic substitutes for pluralistic interest-organizations (of women, workers, youth etc.) established in the early 1970s, the *HKTI* is one of the least active in Indonesia today. Interestingly, despite the enormous growth in NGO activities in rural areas of Indonesia in the past 10–15 years, they have not generally involved themselves in the support of smallholders in contracting schemes.

* Reprinted by permission from the Journal of Peasant Studies, Vol. 24, No. 3 published by Frank Cass & Co. Ltd, 900 Eastern Avenue, Ilford, Essex IG2 7HH, England. Copyright © Frank Cass & Co. Ltd.

References

Bachriadi, D., 1995, *Ketergantungan Petani dan Penetrasi Kapital: Lima Kasus Intensifikasi dengan Pola Contract Farming*, (Small-farmer dependency and capital penetration: five cases of intensification based on contract farming). Bandung: Akatiga Foundation.

Bernstein, H., 1996, "The Political Economy of the Maize *Filière*". *Journal of Peasant Studies*, 23, 120–145.

Carney, J., 1988, "Struggles over Crop Rights and Labour Within Contract Farming Households in a Gambian Irrigated Rice Project". *Journal of Peasant Studies*, 15(3), 334–349.

Carney, J., 1994, "Contracting a Food Staple in the Gambia" in *Living Under Contract: Contract Farming and Agrarian Transformation in Sub-Saharan Africa*, edited by P. Little and M. Watts, pp. 167–187. Madison: University of Wisconsin Press.

Chayanov, A., 1966, "Peasant Farm Organization" in *A.V. Chayanov on the Theory of Peasant Economy*, edited by D. Thomas, B. Kerblay and R. Smith, pp. 29–269. Homewood, Illinois: American Economic Association (orig. 1925).

Clapp, R., 1988, "Representing Reciprocity, Reproducing Domination: Ideology and the Labour Process in Latin American Contract Farming". *Journal of Peasant Studies*, 16(1), 5–39.

Clapp, R., 1994, "The Moral Economy of the Contract" in *Living Under Contract: Contract Farming and Agrarian Transformation in Sub-Saharan Africa*, edited by P. Little and M. Watts, pp. 78–94. Madison: University of Wisconsin Press.

Departemen Pertanian, 1978, *Pengembangan Daerah Pegunungan Selatan Jawa Barat dengan Pola Nucleus Estate: Suatu Gagasan Menghadapi Pelita III* (Development of West Java"s southern hill region on the Nucleus Estate model: a concept in preparation for the Third Five-Year Plan). Bandung: Department of Agriculture, West Java Regional Office.

Doorn, J. van and W. Hendrix, 1983, *The Emergence of a Dependent Economy: Consequences of the Opening Up of West Priangan, Java to the Process of Modernization*. Rotterdam: Erasmus University, CASP Series no 9.

Elson, R., 1994, *Village Java under the Cultivation System 1830–1870*. Sydney: Allen and Unwin.

Glover, D., 1992, "Introduction" in *Contract farming in Southeast Asia: Three Country Studies*, edited by D. Glover and Lim Teck Ghee, pp. 1–9. Kuala Lumpur: Institut Pengajian Tinggi, University Malaya.

Glover, D. and Lim Teck Ghee, 1992, *Contract farming in Southeast Asia: Three Country Studies*. Kuala Lumpur: Institut Pengajian Tinggi, University Malaya.

Glover, D. and K. Kusterer, 1990, *Small Farmers, Big Business: Contract Farming and Rural Development*. London: Macmillan.

Goldsmith, A. 1985, "The Private Sector and Rural Development: Can Agribusiness Help the Small Farmer?" *World Development*, 13(10/11), 1125–1138.

Grijns, M., 1995, "From Blueprint to Actual Households: Coconut Farmers in West Java" Paper prepared for the 3rd WIVS/KITLV Conference on Indonesian Women"s Studies, "Indonesian Women in the Household and Beyond: Reconstructing the Boundaries". Leiden, 25–29 September (draft).

Gunawan, R., J. Thamrin and M. Grijns, 1995, *Dilema Petani Plasma: Pengalaman PIR-Bun Jawa Barat*, (Dilemmas of contact farmers: the experience of tree-crop nucleus estates in West Java). Bandung: Akatiga Foundation.

Hartveld, A., 1995, "Transformations of the Sugar Sector in Malang District" in *Steps Towards Growth: Rural Industrialization and Socioeconomic Change in East Java*, edited by B. Holzner, pp. 219–253. Leiden University, DWSO Press.

Hefner, R., 1990, *The Political Economy of mountain Java: An Interpretive History*. Berkeley: University of California Press.

Hüsken, F. and B. White, 1989, "Java: Social Differentiation, Food Production and Agrarian Control" in *Agrarian Transformations: Local Process and the State in Southeast Asia*, edited by G. Hart, A. Turton and B. White, pp. 235–265. Berkeley: University of California Press.

LBH Bandung, 1985, "The People Plantation Scheme in Cimerak, West Java: Why the People Reject". *Human Rights Forum*, 1, Jakarta.

Little, P., 1994, "Contract Farming and the Development Question" in *Living Under Contract: Contract Farming and Agrarian Transformation in Sub-Saharan Africa*, edited by P. Little and M. Watts, pp. 216–247. Madison: University of Wisconsin Press.

Little, P. and M. Watts, (eds.) 1994, *Living Under Contract: Contract Farming and Agrarian Transformation in Sub-Saharan Africa*. Madison: University of Wisconsin Press.

Mackie, J. and W. O"Malley, 1988, "Productivity Decline in the Java Sugar Industry from an Olsonian Perspective". *Comparative Studies in Society and History*, 30(4), 725–749.

Mackintosh, M., 1989, *Gender, Class and Rural Transition: Agribusiness and the Food Crisis in Senegal*. London: Zed.

Mackintosh, M., 1990, "Abstract Markets and Real Needs" in *The Food Question*, edited by H. Bernstein, B. Crow, M. Mackintosh and C. Martin, pp. 43–54. London: Earthscan.

Payer, C., 1980, "The World Bank and the Small Farmer". *Monthly Review*, 36(6), 30–47.

Peluso, N., 1992, *Rich Forests, Poor People: Resource Control and Resistance in Java*. Berkeley: University of California Press.

Porter, G. and K. Phillips-Howard, 1995, "Farmers, Labourers and the Company: Exploring Relationships on a Transkei Contract Farming Scheme". *Journal of Development Studies*, 32(1), 55–73.

Suhendar, E., 1995, *Ketimpangan Penguasaan Tanah di Jawa Barat*, (Inequality in land control in West Java). Bandung: AKATIGA.

Watts, M., 1990, "Peasants under Contract: Agro-Food Complexes in the Third World" in *The Food Question: Profits Versus People?*, edited by H. Bernstein, pp. 149–162. London: Earthscan.

Watts, M., 1992, "Peasants and Flexible Accumulation in the Third World: Producing Under Contract". *Economic and Political Weekly*, 25 July 1992, 90–97.

Watts, M., 1994, "Life under Contract: Contract Farming, Agrarian Restructuring and Flexible Accumulation" in *Living Under Contract: Contract Farming and Agrarian Transformation in Sub-Saharan Africa*, edited by P. Little and M. Watts, pp. 21–77. Madison: University of Wisconsin Press.

White, B., 1996, "Agroindustry and Contract Farmers in Upland West Java". *Journal of Peasant Studies*, 24(3), 100–136.

WIM, 1994, *PIR, Anugerah atau Bencana? Sengketa Agraria antara Negara dan Rakyat pada Proyek PIR*, (Nucleus Estates, blessing or calamity? Agrarian conflicts between state and people in NES projects). Medan: Wahana Informasi Masyarakat.

Williams, S. and R. Karen, 1985, *Agribusiness and the Small-Scale Farmer: A Dynamic Partnership for Development*. Boulder: Westview Press.

Chapter 9

FROM HOMEGARDENS TO FRUIT GARDENS: RESOURCE STABILIZATION AND RURAL DIFFERENTIATION IN UPLAND JAVA

Krisnawati Suryanata

> Agricultural development is no longer bound by irrigation networks. By the year 2000, we will convert the degraded uplands of Purworejo to a thriving center of fruit production.
>
> The Regent of Purworejo
> (Suara Karya 9/14/89)

INTRODUCTION

Since the late 1970s, there has been an increasing concern over high soil erosion rates in Java's uplands. Development planners have constantly searched for ways to stabilize the upland's environment while still supporting the growing population. One of the initiatives, the national 'regreening' *(penghijauan)* program, promotes integrating trees into the upland farming system through various measures that include government subsidies and extension services (Mackie 1988). Unfortunately, adoption of tree cropping in response to government-sponsored programs was modest at best. Conversion of upland farming systems depended heavily on government subsidies (McCauley 1988; Huszar and Cochrane 1990), and peasants often reverted to old practices soon after a project ended.

Meanwhile, dramatic economic and land use changes have occurred since the 1980s in many upland villages in Java. Increased demand from the growing Indonesian urban middle class and international markets have improved the comparative advantage of upland commodities such as cloves, coffee, and most importantly, a variety of fruits traditionally grown in homegardens. Commercial fruit production is not a novel phenomenon in Java's uplands. In villages close to urban centers, homegardens have been the primary suppliers of the urban fruit market since at least early in this century (Ochse and van den Brink 1931; Tergast and de Vries 1951; Stoler 1978; Penny and Ginting 1984; de Jong and van Steenbergen 1987). An inquiry in the 1920s estimated the total annual commercial value of homegardens around Jakarta at a value of 6,000,000 guilders (Ochse and van den Brink 1931: v).

Although homegardens have long been integrated with markets, their distribution remains more egalitarian than wet rice lands (Stoler 1978; Penny and Ginting 1984). Rocheleau (1987) demonstrates how women mobilize agroforestry strategies to make the best use of the minimal landholdings allotted to them (cf. Leach 1994). Dove (1990: 159–160) notes that the ecological diversity and complexity of homegardens promotes equity by enabling smallholders to resist the extractive propensities of dominant classes. The tenacity of Javanese middle peasantry in non-rice upland production was also noted by Robert Hefner (1990: 233), who suggested that ecological obstacles to capitalist accumulation (cf. Mann and Dickinson 1978) have contributed to the ability of most upland peasants to be "at least marginally involved in farming their own land".

Homegardens are also known for their ecological properties which reduce soil erosion and shelter a diverse range of plants (Anderson 1980; Wiersum 1982; Nair 1989). These prodigious capacities of homegardens from the production, equity, and environmental stabilization standpoints make the promotion of agroforestry systems an attractive strategy, one that echoes throughout the developing world (Rocheleau and Ross 1995; cf. Schroeder 1995) in the context of the 1990s environmentalism, developmental populism and global market integration (cf. Peet and Watts 1996). The market-induced expansion of fruit-based agroforestry is therefore hailed as the answer to the problems of upland farming in Java (Roche 1987).

In this chapter I will argue that the adoption of fruit-based agroforestry in Java is, first and foremost, a strategy of private accumulation. Under these conditions, the new land use has social and economic implications that are entirely different from the 'traditional' homegardens. To appreciate the changed significance of tree planting in this case, it is necessary to look beyond the technical question of land management to examine differential impacts on the livelihoods of upland peasants.

Tree planting brings about new rights and pressures on existing property relations. Many societies including the Javanese recognize the separation of tree tenure from land tenure (Hill 1956; Dove 1985; Fortmann and Bruce 1988). The development of multiple tenure associated with tree planting often results in competing claims over resources in a tree garden (cf. Berry 1987; Mizuno 1985; de Jong and van Steenbergen 1987). Indeed, a growing literature has identified agroforestry and tree-based systems as sites of contentious political struggle with overlapping property and labor claims (Peluso 1992; Schroeder 1993; Bryant 1994; Rocheleau and Ross 1995). While the planting of trees produces a potential for claim multiplicity, their tenurial institutions are the result and embodiment of social relations of production and are always in a dynamic state (Dove 1985; Riddell 1987). This chapter examines local processes involved in the intensification of fruit production in two of East Java's upland villages. This particular agrarian transformation poses upland communities with new moral dilemmas, and their resolutions have been varied. The two case studies show how changing social values as well as ecological and crop dynamics mediate the processes of economic differentiation and resource stabilization that result from the fruit boom in upland Java.

THE SETTING

The Fruit Boom in Java

Beginning in the early 1980s, for the first time since Indonesia's independence, upland Java became the site of a dynamic agriculture based on commercial production[1]. Rapid growth in Indonesia's GDP[2] has boosted urban income and further improved the market for fresh fruit which is income elastic. Throughout the 1980s domestic demand for fruit increased at a rate of 6.5% per annum (*Pelita*, 1/9/91), while the export markets are still growing. Upland farmers were quick to respond to the opening of market opportunities, and the 1980s were marked by dramatic land use transformation in many upland regions. Dry fields, many of them already eroded, were planted with high-value fruit trees in combination with traditional annual crops. In many villages, this land use change has drastically increased incomes[3] (KEPAS 1985; Roche 1987; KEPAS 1988).

The province of East Java, where distinct dry seasons favor the development of several fruit crops, has been one of the primary sites of the fruit industry. Production of apples, mangoes, citrus, and pineapples in this province accounts for at least half of total national production. Table 1 shows that growth has been particularly high since the 1980s. Apple production increased tremendously until the mid-1980s but has since slowed down due to the limited possibility of further area expansion. By contrast, the production of citrus, mangoes, and rambutans which have a wider ecological spectrum, is still growing steadily. Between 1983 and

TABLE 1 ANNUAL GROWTH RATES OF FRUIT PRODUCTION IN EAST JAVA, 1974–1989.

YEAR	ORANGES	APPLES	MANGOES	RAMBUTANS	PAPAYAS	BANANAS	PINEAPPLES
1974–79	–7.8%	44.8%	–4.8%	1.8%	11.7%	19.5%	N.A
1979–84	27.1%	95.9%	–3.3%	30.3%	10.7%	–1.8%	N.A
1984–89	46.7%	17.8%	36.0%	17.2%	9.4%	14.5%	11.4%

SOURCE: KANTOR STATISTIK JAWA TIMUR, *JAWA TIMUR DALAM ANGKA* (VARIOUS YEARS).

1987 the production of *jeruk siam*, a type of mandarin orange, increased almost ten-fold (Mackie 1993). The two case studies examined in this chapter concern apples and oranges.

The Study Sites

The two study villages lie in the upper watershed of Brantas river, one of the 22 watersheds in Indonesia that have been assigned super-priority status by the government, mainly in view of the large investment in infrastructure development. The Brantas watershed is the second largest in Java, and the largest in the province of East Java. Siltation rates in several sub-watersheds of Brantas river rank among the highest ever recorded in the world, resulting from both accelerated topsoil erosion and mass wasting[4] (Carson 1989:51). Annual flooding in the lower watershed since prehistoric times has created one of the richest agricultural land resources in Java, and this area is a major producer of rice, sugarcane, corn and soybeans. The government has built several dams and reservoirs in the watershed to mitigate the flood problem and develop irrigation networks for the lowlands. For this reason, control over accelerated soil erosion in the upper watershed is one of the most pressing concerns of the Brantas Watershed Development Plan.

Land use classification put most lands in the two villages under a single category of upland dry field (*tegalan*), denoting the absence of irrigation for agriculture. Both villages suffered from accelerated topsoil erosion resulting from the cultivation of annual crops, but they both have received government recognition[5] for rehabilitating their degraded lands and becoming more productive in the past 15 years. However, the similarity ends there, as the two villages have distinctly different resources and social histories. Gubugklakah represents the typical volcanic highlands, with deep, fertile soils and the potential for growing high-valued fruit and vegetables of temperate origins. Tumpakpuri, on the other hand, represents areas with shallow soils and marginal fertility, traditionally growing low-value staple crops such as corn and cassava.

Gubugklakah has been integrated into the market economy since the late nineteenth century through its vegetable production. Its agricultural history is characterized by boom-bust cycles of commodity production coupled with ecological breakdowns. In 1991, average landholding per household was 0.53 hectare, with 31% of the village's population owning less than 0.1 hectare. Tumpakpuri, on the other hand, was largely based on subsistence production since its settlement around the turn of this century. Prior to 1970, difficult roads and marginal land productivity prevented it from being fully integrated into the market economy. Although farmers have grown coconuts, chili peppers and various beans as cash crops for a few decades, the importance of these cash crops was relatively small. In 1991, the average land ownership was 1.01 hectare, and its distribution was fairly egalitarian with almost all households having access to land in one form or another.

The spread of fruit-based agroforestry in the two villages has significantly transformed their farming systems as well as their market orientations. By 1991, fruit was the primary commodity, accounting for more than half of the farm income in both villages. Tree planting has increased the adoption of soil conservation practices which include mulching, terracing and low tillage. The impact on economic stratification, however, is less uniform. In Gubugklakah a class of capitalist fruit growers has quickly emerged, while there is no such class developing in Tumpakpuri. Examination of these two different cases will shed light on the different influences that affect processes of agrarian change in upland Java.

APPLE-BASED AGROFORESTRY

The Agricultural Ecology

The challenge in growing temperate fruit trees in the tropics lies in simulating mechanisms that can prevent or break bud dormancy in the absence of daylength and temperature variation. In Java, these mechanisms include shaping the trees' architecture and manually defoliating the trees. In addition, farmers must apply a large amount of pesticides and fungicides to overcome problems of pests and diseases common to crops of exotic origin. As a result, apple production in Java is extremely labor and capital intensive. The average density of apple trees ranges between 800 and 1100 trees per hectare. In the first few years of apple cultivation, all farmers grow vegetable crops between rows of apple trees. Scallions and leeks do not require additional chemical input when they are intercropped with apples. In contrast, the more valuable cabbages, potatoes and garlic require additional pesticides, manure and chemical fertilizers in addition to those used for apple cultivation. Apple trees in Gubugklakah begin producing three years after planting, and can be harvested twice a year.

Apple planting has increased the adoption of soil conservation practices in most of these farms. Unprotected sloping soils in this volcanic highlands could erode at the rate of 2 cm/year (Carson 1989), exposing and destroying roots within the lifetime of apple trees. Construction of bench terraces is thus a prerequisite for apple farming. By the time apple seedlings are planted, the completion of backsloping terraces and closed ditches have accounted for roughly 1000 person days per hectare of labor investment.[6] During heavy rainfall, virtually all mud carried by water runoff collects in the ditches of each terrace bench. After the rain, farmers return the mud to the terraces, thus minimizing the loss of topsoil and fertilizers.

Approximately three quarters of the land in Gubugklakah has been converted into terraced apple orchards or apple-based agroforestry. Of the remaining lands, about half have already been terraced. Overall, close to 90% of lands in the village have been stabilized in this manner within the last two decades. Government officers both at district and provincial levels, who had been struggling in their efforts to reduce soil erosion from Java's upper watersheds, have applauded this development, and Gubugklakah has often been cited as a model of successful upland development practices (KEPAS 1988; Carson 1989).

Social Classes and the Changing Economy

> There is no land-lord (*tuan tanah*) in Gubugklakah, but we have plenty of apple-lords (*tuan apel*). This is a good arrangement because nobody loses all means to make a living. A small farmer can still grow vegetables even when the trees on his land are leased-out.
>
> *Former Village Head, 1991*

More than a century of settlement and market production in Gubugklakah has resulted in a relatively more differentiated society than is commonly found in Java's rural uplands. Yet due to chronic labor shortages and environmental obstacles to farming in this sloping highlands, controlling land had never been the main strategy for wealth accumulation in this village. Up to the 1970s, livestock ownership was the primary form of wealth accumulation. Cattle were share-raised with poor households using a 'halving' system known as *gaduh*. Chemical fertilizers began to substitute livestock manure in the 1970s, and at about the same time other investment opportunities such as vegetable cropping and trade began to expand. As a result, the attractiveness of livestock keeping as a wealth accumulation strategy declined substantially. In 1991, none of the richest ten villagers owned more than a few heads of livestock.

The new chemical-intensive vegetable production led to a brief flourishing of a sharing institution (*maro*) in the 1970s, linking capital-poor peasants and the village's wealthy class. Some vegetables, such as potatoes, cabbages and garlic, require considerable outlays of capital and labor.[7] Capital-poor land owners promised half of their vegetable harvests to those who would finance the purchase of seeds, fertilizers and chemicals (Hefner 1990:102). This arrangement quickly lost its importance, mainly because capital-poor land owners could turn to growing leeks and scallions, which are less capital intensive.

The recent apple boom has accelerated the process of economic differentiation in Gubugklakah. Yet the pattern of land distribution has not significantly changed. About half of the village population belong to the middle peasantry with access to plots that range in size from 0.2 to 1 hectare. In 1991, 94% of the lands in Gubugklakah were owner-operated, with an average holding of 0.53 hectare and the richest 6% controlled about a quarter of lands in the village, which is fairly typical of the midslope Tengger region (cf. Hefner 1990: 98). A process of accumulation, however, occurred very rapidly over the valuable apple trees.

The apple boom has reinforced the distinction between land and tree tenure customarily recognized in homegardens. Claimants of different resources of a homegarden, however, are usually related by family ties.[8] By contrast, the high commercial value of apples has reproduced this tenure multiplicity among unrelated households through market mechanisms. Apple trees constitute a valuable asset with higher marketability than land itself, and are often exchanged independently of land. In times of emergency, rights over trees, especially mature trees at fruit-bearing stage, can quickly be liquidated to raise cash. The mechanisms for the transfer of these tree assets vary. Tree seedlings themselves are sometimes sold and transplanted, but the transfer of right to trees *and* the space they occupy is more common. By transferring only the tree tenure, a land owner retains the right to grow vegetables between the trees. In 1991, eight years after the first apple garden was established in this village, close to 20% were operated under some form of tree tenancy, and the number appeared to be growing.

Apple lords accumulate fruit-bearing apple trees through tree transactions and two forms of tree tenancy. Tree sharecropping (*maro apel*) resembles the vegetable sharecropping of the 1970s, in that capital-poor landowners seek credit to finance the labor and capital-intensive apple production. Apple sharecroppers, however, provide not only the capital, but in most cases the labor and skills necessary for the cultivation of apple trees. Landowners provide the land but retain the right to grow annual crops underneath the trees until it is prohibitively difficult to do so. Sharecropping agreements specify how profits from apple production are to be divided, as well as rules on access to the land surrounding the trees. In contrast to vegetable sharecropping, the longevity of apple trees and their permanent tenure precludes terminating the contract at a season's notice, unless land owners compensate their tenants for the trees, a practical impossibility in most cases given their high value.

Just like the vegetable sharecropping, the apple sharecropping institution quickly lost importance, in this case to a new tenurial institution: tree leasing (*sewa apel*). As apple sharecroppers acquired management skills and reduced production risks, they preferred fixed-rent leasing to sharecropping apple trees. Meanwhile, the persistent cash liquidity crisis of smaller scale owner-operators has created a rental market for apple trees. The typical arrangement involves those with capital, who are former merchants or apple sharecroppers, leasing apple trees from land-owning, capital-poor peasants.

Reasons for the liquidity crisis range from life crises such as illness and death of a family member, to basic demands of household reproduction such as children's education and house building expenses, to the increased consumerism that has accompanied new prosperity (cf. Lewis 1992). Most often, however, the need to lease out apple trees arises from the inability to maintain young trees that have absorbed investment capital, but have not yet produced any return. If a farmer owns several fields, tree leasing of one plot may be a way to raise capital to finance the operation costs for another. The rent is typically negotiated and payable in advance, albeit within the context of a renter's market.

The duration of tree lease contracts ranges from one harvest to as long as fifteen years (thirty harvests under a double crop regime). A lessor who needs extra cash before the contract expires can choose to extend the contract in return for an agreed sum of money. The lessor's bargaining position at that point, however, is far weaker than when the contract was first established. The lessee is in a position to negotiate a lower rent, impose more restrictions on growing field crops, or claim permanent tenure of the trees. Unfortunately, with the reduced amount of resources available to a lessor household after it enters into the contract, the likelihood of needing further credit extensions before the lease term expires is fairly high. Of the 29 cases of tree leasing in the study, more than half the lessors have renegotiated their contracts before original terms expired, resulting in reduced access to their own lands.

Tree transfers under such circumstances have contributed to rapid economic differentiation without apparent land accumulation. The richest 15% in the village controlled only 50% of the land in the village, but they controlled 80% of the apple harvest. Similarly, although only 21% of the village's households were landless, 68% did not have any access to apple harvest. Despite the fact that the largest land holding was only five hectares, the largest apple farmer operated close to 15,000 trees growing on 20 hectares of land.

Inter-crop Dynamics and Social Relations

As the rules and agreements of tree-leasing contracts become more complex, written documents increasingly replace oral agreements. A few apple lords have used the service of the village administration office to authenticate their leases. The formalization of the contracts emerged as a response to the conflicts that arose from tree leasing. These conflicts, however, are rarely caused by disagreements about the rent amount. Rather, they arise from the overlapping claims to land-based resources which result from the multiple tenures of apple gardens.

As apple trees mature, expanding canopies and intensive maintenance of apple trees begin to interfere with vegetable growing. In fields under tenancy contracts, this conflict becomes pronounced. Frequent trampling by apple workers damages vegetable crops. Conversely, apple lords blame lessors' activities while tending their vegetables for blemishes in apples. In such struggles, tree lessees invariably come out as winners. As a result, many fields have effectively turned into monoculture apple orchards which deprive lessors/land owners of access to their own lands (Suryanata 1994). The lessees' advantages are exercised either in the fields through imposing an environment hostile to the vegetable crop, or legally through formal terms in the contract extensions or conflict resolutions. The legal resolutions, however, are carried out in a political context increasingly dominated by a new village elite.

In apple farming, access to information outside the village regarding apple markets, cultivation methods and input supplies for this practically new crop proved to be critical. A majority of those who succeeded in this endeavor had worked outside the village, usually as vegetable merchants or middle-men during the 1970s. Even though they are originally from this mid-slope village, they have outward-oriented knowledge and value systems and have emerged as a new, dominant economic class. These people, all men, are actively changing the characteristics of this upland community, assuming new social roles in addition to their economic ones.

The upper slope of the Tengger mountains is one of the few enclaves in Java where a majority of the population are non Muslims, having escaped the wave of Islamization in the sixteenth century (Hefner 1985). Yet many Muslim lowland immigrants in the transitional, mid-slope regions such as Gubugklakah, have settled since the early twentieth century

(Hefner 1990). The most recent economic transformation has increased the dominance of Islam. Many apple lords have had close ties with the strongly Muslim lowlanders through their trade activities and have increasingly adhered to the lowlanders' values, as well as to lowland forms of symbolic capital, in building their local prestige. For most Gubugkalakah villagers, for example, it is the pilgrimage to Mecca which has come to epitomize upward mobility, connoting social and religious rank as well as wealth.[9] While there were at most two or three pilgrims per year, from other villages in the region in 1990 there were sixteen from Gubugklakah, continuing the trend that began with the apple boom. All but one of the apple lords had the "Haj" title and regularly serve as *Imam* at Friday prayers.

In addition to adopting religious roles, apple lords legitimize their social and economic rank by relating to their dependent wage-laborers in patron-client mode, a form of relationship which used to be absent in the uplands. Due to the steady labor supply necessary for apple cultivation, apple lords employ workers on a permanent basis, and additional daily workers as needed. As patrons, they offer benefits that include loan provisions with low or no interest, access to fodder from their fields, or year-round guarantees of employment.

About a quarter of lessors worked as paid laborers on their own lands. These arrangements provide lessors with the opportunity to personally ensure that apple maintenance does not cause trampling damage to their own vegetable crops. The lessors' residual rights are thus appropriated by the apple patron and *returned* to them as part of a labor contract. While this labor relation may partially mitigate the effect of lost control over resources, it does so only under terms which increase the dependency of lessors on their creditors, deepening the imbalance of power between them. Despite their formal land-owning status, they have formed a new class of "propertied laborers" (cf. Watts 1994).

From the resource stabilization point of view, the expansion of apple-based farming in Gubugklakah has significantly improved the conservation scene. The conflicts that have stemmed out of tenurial multiplicity, however, limit the choice of cropping strategies available to many landowners. Instead of benefiting from inter-crop dynamics that could create a stable and resilient environment, a majority of apple-based farming systems are increasingly simplified, relying heavily on chemical inputs. Although soil erosion is no longer a problem, there are new environmental hazards threatening the resource base.

From the equity point of view, land use change has restructured village society, transformed labor relations and contributed to a process of rapid economic differentiation. Although land distribution patterns have changed little over the past decade, a new class of capitalist growers has emerged as the village's dominant power. The general sentiment across the village population was that although the apple boom has increased economic inequality it has, nevertheless, improved many livelihoods. In 1991, more than 62% of the village households reported some income derived from agricultural wages. Although the pay scale discriminates against women and older workers, employment opportunities are usually available to most villagers. At the same time, outmigration by Gubugklakah villagers, a primary source of remittances only a decade earlier, has virtually disappeared.[10] Naively perhaps, many lessors view their loss of status as independent peasant producers as a temporary feature of this transition period and look forward to resuming control after the lease ends.

ORANGE-BASED AGROFORESTRY

The Agricultural Ecology

Since the mid-1980s, siamese oranges have been the most important commodity in Tumpakpuri. These oranges adapt well to soils of marginal fertility, and start bearing fruit by the third year after they are planted. When exposed to drought, however, branch tips and shoots of orange trees can dry and die prematurely. As a result, orange trees in drought-prone areas such as Tumpakpuri often suffer significant decrease in productivity and death by the eighth year. Despite the shorter tree life, oranges became extremely popular in this village. As in many regions where the land quality is marginal, the opportunity cost of growing oranges is fairly low. On average, fields growing solely traditional dry field crops such as cassava, corn, and beans only yield between 10% and 50% the harvest value of those with mature orange trees (Winarno 1987).

Most farmers utilize the space between trees for planting annual crops, facilitating the transition before the trees start to produce. Due to their narrow canopy structure, young orange trees cast minimal shade, and are an ideal tree crop for agroforestry. Moreover, the short life expectancy of orange trees in Tumpakpuri ensures that tree canopies remain narrow, and that orange-based agroforestry never shifts to a complete monoculture. Annual crops growing between the trees not only provide an interim benefit before the fruit trees start producing, they remain a major part of the system throughout the life cycle of the trees.

At the time of study in 1990–91, orange trees were a component of roughly half of the agricultural plots in Tumpakpuri. Farmers were also engaging in other improvements such as the addition of organic materials, construction of bench terraces, and extraction of surface rock, usually during the slack season (September-October). Overall, close to three-quarters of the sloping lands in Tumpakpuri had been improved, leaving only the rocky areas unsuitable for agricultural production.

Tree density in orange-based farms ranges widely between the recommended 400 trees per hectare to more than 1000 trees per hectare. Rice, corn, cassava, chili peppers, cotton and a variety of beans are the most common secondary crops grown between the trees. Oranges grow best when they are planted in combination with short, non-aggressive crops such as rice, chili peppers, cotton or mungbeans. By contrast, corn, polebeans and cassava easily outcompete oranges in securing soil nutrients and sunlight.

Owner-operators usually design their cropping patterns to secure the best possible growing environment for the trees, selecting the short, non-aggressive crops. Unfortunately, each of these crops has its shortcomings. A majority of lands in Tumpakpuri consist of sandy loam and rocky soils, in which wet rice cannot grow well. Farmers can grow upland rice on these soils, but the crop is vulnerable to drought and its productivity is much lower than wetland rice. Chili peppers and cotton have been plagued by problems of pests and diseases, and beans have been suffering from fairly low market prices. Corn is the preferred crop on upland fields when there are no orange trees but, once orange trees are planted, it is replaced by upland rice. In my sample (see figure 1) upland rice became the primary annual crop on 60.6% of these fields while an additional 11.5% of the fields were planted with mungbeans, also a 'good' intercrop. The strategy of combining rice or mungbeans with orange trees assures the least competition for orange trees. Farmers adopting this strategy, however, risk either a failure of rice production or a low market value of mungbeans and thus a poor return from their annual crop sector. They can reduce this risk by growing other annual crops in rows at a further distance from orange trees. The selection of these "row crops" includes the normally-favored corn, cassava, and polebeans.

Social Classes and the Changing Economy

In 1990, distribution of land ownership in Tumpakpuri was still fairly egalitarian. Farms larger than 2 hectares belonged to the richest 6%, and covered less than 20% of the village land. Up to the early 1980s, agri-

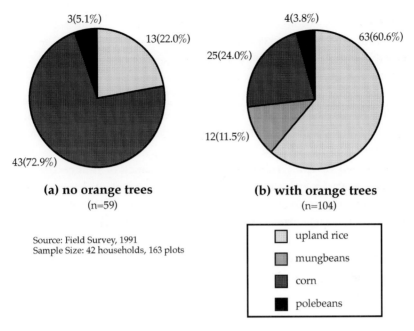

FIGURE 1 DISTRIBUTION OF UPLAND FIELDS IN TUMPAKPURI BY CROPPING PATTERNS.

cultural production was largely for subsistence purposes. The economic return to the few cash crops that could grow there was so low that it usually paid to migrate to other areas and the cities.

Most of the lands were owner-operated, but about 15% were operated under some forms of tenancy contracts. Sharecropping was the most important form of tenancy. In Tumpakpuri, sharecropping was primarily a means for securing labor to work the land. From 21 cases of sharecropping in the study sample, almost all of the land owners cited lack of household labor — due to old age, illness and out-migration of family members — as the reason for sharing-out their lands. About one-third of the share-tenancy contracts in Tumpakpuri were between parents and their children. The decision to share out lands is thus not based upon the owner having too much or even enough land, nor does it normally improve his or her social standing.

The development of the agricultural labor market in Tumpakpuri was fairly recent. Up to the 1960s the most common way of mobilizing non-household labor was through exchange. In the 1970s the increase of out-migration, especially of young males (to Blitar, Surabaya and Malaysia) caused shortages of labor for many households. Hiring labor for part of the farm work became increasingly common. By 1990, it was common to hire workers for activities such as soil preparation, sowing, weeding, and harvesting. The increased economic opportunity brought by orange farming had not attracted migrants back to Tumpakpuri, and the agricultural labor market remained tight.[11] As a result, share-tenancy remained prevalent, and tenants had relatively strong leverage in spite of the increasing commercialization due to orange planting.

Compared to apple cultivation, the maintenance of orange trees requires a relatively low capital input. Nonetheless, for subsistent peasant households, the start-up capital for orange farming can be a major hurdle. In addition, orange trees require more chemical fertilizers and pesticides than traditional crops. Unlike the case of apple-based farming in Gubugklakah, however, the introduction of orange trees did not result in transactions over tree tenure. From an investor's point of view, obtaining long-term tenure of orange trees poses a high risk. Year-to-year performance of orange crops is fairly unpredictable, depending not only on the level of management inputs, but also on the year's rainfall. On the supply side, the economy of orange cultivation[12] rarely traps peasants in a debt cycle that would force them to sell the tenure of their trees.

Also, in contrast to Gubugklakah, tree planting has not created new tenancy institutions in Tumpakpuri. It has, rather, reshaped existing ones. In reaction to the high profitability of orange farming, many landowners asked to renegotiate the terms of share tenancy contracts with their tenants. Because tree cropping requires lower labor inputs than food crop cultivation, it provides a new option to labor-poor landowners and thus increases their leverage against their tenants. A few landowners decided to retrieve their share-cropped lands and operate them on their own. A majority of landowners, however, prefer to reclaim partial access in order to plant trees while tenants continue to grow annual crops between the trees. This option is especially attractive for landowners who had shared-out their lands due to a shortage of labor. Although tree owners are primarily responsible for the maintenance of their trees (i.e. pruning, fertilizing and spraying), keeping the share-tenancy contract intact eliminates the need for weeding[13] and thus makes it possible for such owners to partake in orange production even with their limited labor resources.

In contrast to the case of apple-based agroforestry, however, controlling a booming tree commodity did not confer a means to accumulate power and has not caused economic differentiation. Inter-crop conflicts in orange-based agroforestry are distinctly different from those in apple-based agroforestry. Most importantly, these conflicts occur in the context of a tight labor market and relatively unchanging social values.

Inter-crop Dynamics and Social Relations

As in the case of apple-based agroforestry, inter-crop conflicts are intensified in fields subject to multiple claims. Many landowners who have planted orange trees attempt to redefine share tenancy contracts with respect to crop choices. In order to minimize competition for orange trees, the owners invariably forbid the planting of corn, cassava, or polebeans close to the trees. Tenants who have rights only over annual crops, on the other hand, prefer these crops to rice or mungbeans. The cropping strategies that actually prevail on lands under share tenancy contracts are the result of ongoing negotiations between landowners and their tenants.

The survey in 1991 showed that rice or mungbeans dominated as primary annual crops in all fields under share-tenancy in Tumpakpuri. The redefinition of contractual terms has apparently been able to force tenants to plant primarily these non-aggressive (and less productive) crops on these fields. Nevertheless, there is a wide variation in the density of other, more aggressive, crops which grow in rows crossing the fields. In cases where the tree/land owner has an ability to enforce the rules on cropping strategies, aggressive crops only grow on the fields' borders or every few rows between the trees, consequently reducing the tenants' share.

Continuous, actual supervision on cropping patterns, however, is not always possible for landowners/ orange growers. Some tenants use a few tactics to increase the proportion of land planted with corn or cassava. A tenant could phase in the planting of corn, planting a later batch after the landowner inspected the field for the season. He could also grow more of the aggressive crops in a concealed part of the farm. In one case, a field inspection by a tree owner resulted in the uprooting of the "forbidden" crops, causing a major conflict that was brought to the village authorities. Under the direction of the *lurah*, both parties agreed to renegotiate the contractual terms. The tenant received a larger share of the annual crops to compensate for the reduced space in which the favored crops were permitted. In return, he would stop growing these crops close to the trees.

More often, de facto field conditions rule. Landowners/ orange growers make an effort to discourage tenants from breaking the new rules on crop choices, but once the crops are planted, they generally are left standing for the season. Due to the short growing season of the annual crops, it is usually too late to replace uprooted crops with rice or mungbeans. In addition, the general feeling is that landowners do not want to make their

TABLE 2 SHARE-TENANCY TERMS AFTER TREE PLANTING IN TUMPAKPURI.

TENANCY TERM	LANDOWNERS' REASON FOR SHARING-OUT			
	PRE-INHERITANCE	LABOR SHORTAGE	TO HELP RELATIVES	TOTAL
	NUMBER OF SHARE-TENANCY CASES			
ORANGES PLANTED, MANAGED BY LANDOWNER	2	6	0	8
ORANGES PLANTED BY LANDOWNER BUT SHARED BY TENANT	2	1	1	4
ORANGES PLANTED AND SHARED BY TENANT	1	2	0	3
ORANGES PLANTED, MANAGED BY THIRD PARTY AND SHARED WITH LANDOWNER	2	1	1	4
NO ORANGE TREES PLANTED	0	1	1	2
TOTAL	7	11	3	21

SOURCE: FIELD SURVEY, 1991.

tenants, who are mostly close neighbors or kinfolk, lose what they have already invested in the annual crops. In contrast to the case in Gubugklakah, growers of these fruit trees do not adhere to alternative value systems, and they still seek their neighbors' approval and respect to maintain their social standing. More importantly, they are dependent on their tenants' labor in the context of a tight labor market.

In extreme cases, such as where absentee[14] or sick landowners are involved, sharecroppers have full control of the actual operation of the fields. They continue to make their cropping decisions as if there are no orange trees present. On some fields, aggressive field crops crowd orange trees to the point of no growth. Through cropping strategies, the sharecroppers have effectively prevented the tenancy rules from shifting to their disadvantage.

About one-third of the share-tenancy cases in Tumpakpuri have included both trees and annual crops in their contracts (Table 2). All landowners who are part of this arrangement have in common their lack of household labor due to old age. This contractual arrangement has been the least problematic. The absence of conflicting claims on this land allows for tenants to act like owner-operators and design an optimum cropping strategy, taking into account inter-crop dynamics as well as other resources available to them.

It appears that after about a decade, the introduction of oranges in Tumpakpuri has not involved a thorough reorganization of production. Multiple tenures and economic differentiation did not proliferate as they had in Gubugklakah. Labor organization and property relations changed little following the orange boom. The role of sharecropping remains important as the means for securing labor for landowners and of gaining access to land for landless peasants. This mutual dependence has to a certain extent kept the social value systems fairly stable in spite of the increased degree of commercialization.

The high commercial value of oranges in this agroforestry system has not led to simplified tree-based gardens as in Gubugklakah. Instead, the cropping strategies remain complex, minimizing the risks that derive from reliance on too few commodities. Ironically, the reasons why the system has remained relatively egalitarian are the same ones that make it less favorable from the resource stabilization point of view. Short life cycles and high vulnerability of orange trees mean that orange-based agroforestry requires repeated start-up capital. The fruit boom of the 1980s and the early 1990s has provided the fuel for the present transformation, but the future is less certain.

CONCLUSION

The 1980s were marked by dramatic land use changes in many upland regions in Java. After a long string of failures in reducing soil erosion rates from Java's sloping uplands, improved market incentives for tree products have presumably given peasant producers economic incentives to plant fruit trees, and encourage land rehabilitation and soil conservation practices. The shift to fruit farming was made possible by adopting fruit-based agroforestry practices, which allowed smallholders to maintain benefits from annual crops before the trees started to bear fruit.

The two case studies I presented in this paper show the contradictions that result from efforts to stabilize the environment through the market, as commoditization leads to shifting patterns of access and resource control. In each place this process takes on distinct local characteristics, and produces different forms of social friction; these frictions in turn shape the developing land use. In both cases, commercial fruit trees embody all investments in land rehabilitation that farmers make to accommodate the planting of trees. Accumulation of the most valuable asset (i.e. the fruit trees) becomes the primary mechanism in the economic differentiation process, invariably leading to multiple claims in fruit gardens.

The fruit boom accelerated social and economic differentiation in Gubugklakah, but it has not created a capitalist class in Tumpakpuri. In Gubugklakah apple lords can appropriate the ecological interactions between apples and vegetables, facilitating the accumulation of apple trees and the polarization of resource control. By contrast, the tight labor market and the high risk of orange cultivation in Tumpakpuri deters the formation of 'orange lords'. Tenants can appropriate the ecological interactions between oranges and annual crops, defend their use rights, and prevent the gradual dispossession of resources by tree growers that occurs in Gubugklakah.

Where it occurs, rural differentiation in upland Java has been reinforced by the legitimation given to it by the Indonesian state. Trees represent 'good' and stable environments, and commercial cultivation represents economic development. The discourse that trees are good, and commercialization equals development, dominates the academic community that champions agroforestry as the appropriate technology for marginal, sloping uplands, as well as policy makers interested in economic integration and protecting the lower watershed. At the village level, new institutions of tenancy, labor practice and actual land use have also taken shelter under the discursive shade of trees as symbols of 'green' goodness (cf. Rocheleau and Ross 1995: 408). They all favor large growers who are able to make use of the modern agricultural inputs and produce high-quality, market-bound fruit and thus can be integrated into the national economy.

The case studies show that these commercial fruit gardens share few ecological and social characteristics with popular homegardens. Incorporation of apple trees certainly does not protect the disenfranchised from extraction by the dominant classes. Neither do the cases show a high degree of control by women producers of the kind that arises in homegardens. The recognized apple lords are all men, and formal contractual arrangements in the commercial fruit sector (leases, share-crop tenancies) are in men's names, even when women are active in day to day management tasks or supply field labor. Unlike the nutritionally efficient homegardens, both case studies show heavy reliance on a high level of external inputs, a condition that was made possible in the 1980s only by the highly favorable market conditions of apples and oranges resulting from trade protection. Increasing trade liberalization and removal of market protections for domestic fruit will likely affect the demand for apples and oranges, threatening the economic viability of these fruit gardens. At this point one can only speculate on the impact of a possible economic downturn in fruit prices upon the resource stabilization efforts and the relations of production in upland Java.

ACKNOWLEDGEMENTS

One of the two case studies presented here was published in *World Development* 22(10), 1994. I would like to acknowledge the generous support from the Osborn Forestry Policy Research Grant, jointly funded by the World Wildlife Fund and the Conservation Foundation, and the Ford Foundation for research in Indonesia. Thanks to Michael Watts, Louise Fortmann, Jeff Romm, Nancy Peluso, Marty Olson, Vinay Gidwani, Mark Baker and Keith Mattson who helped in the dissertation writing.

NOTES

1. Upland Java was the fountainhead of colonial revenue in the 19th century, particularly when coffee was the primadonna export crop. Between the 1950s and 1970s, however, the focus was on increasing rice production in the irrigated lowlands.
2. Averaged more than 7% per annum since 1967 (*Nota Keuangan*, various years).
3. Since the late 1980s, newspapers regularly reported on land improvement and the emerging *nouveaux riches* of the fruit industry. For example: "thanks to Fruit, 30% of Segeran Population Made Pilgrimage to Mecca" (*Kompas* 1/2/88); "Orange Boom in Ponorogo: From Critical Land to Gold Mine" (*Pelita* 2/3/89).
4. Mass movement of earth materials, including *lahar* and mud flows, landslides, and gully formation, usually independent of land use.
5. Tumpakpuri won the provincial *Kalpataru* award in 1988, the national regreening award in 1988, as well as 3 other prizes in village development contests between 1987 and 1990; Gubugklakah received commendations from the Brantas Watershed Project Office in 1985. Both villages regularly became destinations for official field trips as show cases in uplands development. They were selected for my study because of their successful transition to fruit-based farming.
6. Terrace construction requires 700–1000 person days per hectare, depending on slope gradient. Planting apple rootstock and grafting requires an additional 100 person-days. Large growers usually hire workers while smallholders use unpaid household labor to build terraces gradually over a few years.
7. Hefner (1990: 89) reported that the total cost of non-labor inputs for growing potatoes and cabbages is 15 to 20 times that for corn.
8. Parents commonly retain the rights to harvest the trees while bequeathing their child(ren) cultivation rights prior to full inheritance. They may grant the trees to a daughter, while passing on the land to a son.
9. The trip cost Rp. 6,000,000 (US$3,000) in 1990.
10. Gubugklakah has begun to draw labor migrants from neighboring villages. State forest officials also report a massive decline of poaching in bordering forest lands, a phenomenon they attribute to the increase of wage labor opportunities within the village itself.
11. Wages in Tumpakpuri were about 50% higher than in Gubugklakah.
12. Cultivation of orange trees requires a modest amount of capital to purchase seedlings and a small amount of pesticides and fertilizers.
13. Owner-operators reported that between December (the beginning of the rainy season) and August (orange harvest) they need to weed three times, each requiring about 50 person-days.
14. Absenteeism is not common in Tumpakpuri; the two cases in my sample were men who resided with wives originating from outside the village.

References

Anderson, James N., 1980, "Traditional Home Gardens in Southeast Asia: A Prolegomenon for Second Generation Research" in *Tropical Ecology and Development*, edited by J.I. Furtado, pp. 441–446. Kuala Lumpur: The International Society of Tropical Ecology.

Berry, S., 1987, *Property Rights and Rural Resource Management: The Case of Tree Crops in West Africa*. African Studies Center, Boston University.

Bryant, Raymond L., 1994, "The Rise and Fall of Taungya Forestry". *The Ecologist*, 24(1), 21–26.

Carson, Brian, 1989, *Soil Conservation Strategies for Upland Areas in Indonesia*. Occasional Paper 9. Honolulu: East West Center Environment and Policy Institute.

de Jong, Wouter and F. van Steenbergen, 1987, *Town and Hinterland in Central Java*. Yogyakarta, Indonesia: Gadjah Mada University Press.

Dove, Michael R., 1990, "Review Article: Socio-Political Aspects of Home Gardens in Java." *Journal of Southeast Asian Studies*, XXI (1), 155–163.

1985, "The Kantu System of Land Tenure: The Evolution of Tribal Land Rights in Borneo" in *Modernization and the Emergence of A Landless Peasantry: Essays on the Integration of Peripheries to Socio-economic Centers*, edited by G.N. Appell, pp. 159–182. Williamsburg, Virginia: College of William and Mary.

Fortmann, Louise and John W. Bruce, 1988, *Whose Trees? Proprietary Dimension of Forestry*, Rural Studies Series of the Rural Sociological Society. Boulder and London: Westview Press.

Hill, Polly, 1956, *The Gold Coast Cocoa Farmer: A Preliminary Survey*. London: Oxford University Press.

Hefner, Robert W., 1985, *Hindu Javanese: Tengger Tradition and Islam*. Princeton: Princeton University Press.

Hefner, Robert, W., 1990, *The Political Economy of Mountain Java*. Berkeley: University of California Press.

Huszar, Paul C. and Harold C. Cochrane, 1990, "Subsidization of Upland Conservation in West Java: The Citanduy II Project". *Bulletin of Indonesian Economic Studies*, 26(2), 121–132.

KEPAS, 1985, *The Critical Uplands of Java: An Agroecosystems Analysis*. Jakarta, Indonesia: KEPAS, Badan Penelitian dan Pengembangan Pertanian.

KEPAS, 1988, *Penelitian Agroekosistem Lahan Kering Jawa Timur*. Bogor, Indonesia: KEPAS.

Leach, M., 1994, *Rainforest Relations: Gender and Resource Use among the Mende of Gola, Sierra Leone*. Washington DC: Smithsonian Institute Press.

Lewis, Martin, 1992, *Wagering the Land : Ritual, Capital and Environmental Degradation in the Cordillera of Northern Luzon, 1900–1986*. Berkeley: University of California Press.

Mackie, Cynthia, 1988, *Tree Cropping in Upland Farming Systems: An Agroecological Approach*. USAID, Upland Agriculture and Conservation Project.

Mackie, Jamie, 1993, "Plantations and Cash Crops in East Java: Changing Patterns" in *Balanced Development: East Java in The New Order*, edited by Howard Dick, James J. Fox and Jamie Mackie, pp. 187–213. Singapore, Oxford, New York: Oxford University Press.

Mann, S. and J. Dickenson, 1978, "Obstacles to the Development of Capitalist Agriculture." *Journal of Peasant Studies*, 5, 466–481.

McCauley, David S., 1988, *Citanduy Project Completion Report, Annex V: Policy Analysis*. US Agency for International Development.

Mizuno, M., 1985, *Population Pressure and Peasant Occupations in Rural Central Java*. Canterbury: Centre of South-East Asian Studies, University of Kent.

Nair, P.K.R., 1989, *Food Producing Trees in Agroforestry Systems*. Symposium on Agroforestry Systems and Technologies. Bogor: BIOTROP.

Ochse, J.J. and R.C. Bakhuizen van den Brink, 1931, *Fruits and Fruitculture in the Dutch Indies*, English edition ed., translated by C.A. Backer. Batavia and The Hague: G. Kolff & Co. and Martinus Nijhoff.

Peet, Richard and Michael Watts, 1996, "Liberation Ecology: Development, Sustainability and Environment in an Age of Market Triumphalism" in *Liberation Ecologies: Environment, Development, Social Movements*, edited by Richard Peet and Michael Watts, pp. 1–45. London and New York: Routledge.

Peluso, Nancy Lee, 1992, *Rich Forest, Poor People and Development: Forest Access Control and Resistance in Java*. Berkeley: University of California Press.

Penny, David H. and Meneth Ginting, 1984, *Pekarangan, Petani dan Kemiskinan: Suatu Studi tentang Sifat dan Hakekat Masyarakat Tani di Srihardjo Pedesaan Jawa*. Yogyakarta: Gadjah Mada University Press.

Riddell, James C., 1987, "Land Tenure and Agroforestry: A Regional Overview" in *Land, Trees and Tenure*, edited by John B. Raintree, pp. 1–16. Nairobi, Kenya and Madison, Wisconsin: ICRAF and University of Wisconsin Land Tenure Center.

Roche, Frederick C., 1987, *Sustainable Farm Development in Java's Critical Lands: Is A Green Revolution Really Necessary?*. Cornell University Division of Nutritional Sciences. Unpublished.

Rocheleau, Dianne, 1987, "Women, Trees and Tenure: Implications for Agroforestry Research and Development" in *Land, Trees and Tenure*, edited by J. Raintree, pp. 79–121. Nairobi, Kenya and Madison, WI: ICRAF and University of Wisconsin Land Tenure Center.

Rocheleau, Dianne and Laurie Ross, 1995, "Trees as Tools, Trees as Text: Struggles over Resources in Zambrana Chacuey, Dominican Republic". *Antipode*, 27(4), 407–428.

Schroeder, Richard A., 1993, "Shady Practice: Gender and the Political Ecology of Resource Stabilization in Gambian Garden/Orchards". *Economic Geography*, 69(4), 349–365.

Schroeder, Richard A., 1995, "Contradictions along the Commodity Road to Environmental Stabilization". *Antipode*, 27(4), 325–342.

Stoler, Ann, 1978, "Garden Use and Household Economy in Rural Java". *Bulletin of Indonesian Economic Studies*, 14(2), 85–101.

Suryanata, Krisnawati, 1994, "Fruit Trees under Contract: Tenure and Land Use Change in Upland Java". *World Development*, 22(10), 1567–1578.

Tergast, G.C.W.Chr. and Egbert de Vries, 1951, "Indonesia and Western New Guinea" in *The Development of Upland Areas in the Far East*, pp. 45–95. New York: Institute of Pacific Relations.

Watts, Michael, 1994, "Life Under Contract: Contract Farming, Agrarian Restructuring and Flexible Accumulation" in *Living Under Contract: Contract Farming and Agrarian Transformation in Sub-Saharan Africa*, edited by P. Little and M. Watts, pp. 21–77. Madison: University of Wisconsin Press.

Wiersum, K.F., 1982, "Tree Gardening and Taungya in Java: Examples of Agroforestry Techniques in The Humid Tropics". *Agroforestry Systems*, 1, 53–70.

Winarno, M., 1987, "Program Penelitian Jeruk" in *Lokakarya Implementasi Rehabilitasi Jeruk conference in Batu*, edited by A. Supriyanto, A.M. Whittle, Nurhadi and Roesmiyanto, pp. 30–37. Malang: Sub Balai Hortikultura Malang.

Chapter 10

AGRARIAN TRANSFORMATIONS IN THE UPLANDS OF LANGKAT: SURVIVAL OF INDEPENDENT KARO BATAK RUBBER SMALLHOLDERS

Tine G. Ruiter

INTRODUCTION

This chapter concerns the formation of a peasantry in the uplands of Langkat regency (*kabupaten*) on the east coast of the province of North Sumatra. The area is dominated by state and privately owned rubber, oil-palm and tobacco plantations. Karo Batak inhabited these uplands before Western entrepreneurs began plantations there around 1870. The Karo, living literally on the "margins" of the plantations, survived the booming plantation economy as peasants. A majority of them now earn a living as independent rubber smallholders.

The modernization paradigm of the 1960s envisaged a dual economic system in which a "modern" plantation sector co-existed alongside but quite separate from a "traditional" peasant sector. This paradigm has been strongly critiqued in the academic literature, but it is still used by some development economists, and it continues to inform the agrarian policy of the Indonesian government. For example, recent state sponsored projects in North Sumatra which aim to develop the rubber smallholder sector are based on such a vision. Rubber smallholders are seen as small scale "traditional" producers, whose productivity is low and who are unable to replant with high yielding varieties without the help of the state.

Contrary to the idea of separate economic spheres, this chapter will examine interconnections in the spheres of land, labor and production which were central to the formation of Karo peasant society. It will also examine the impact of both the state and plantation sector upon the formation of communities as political arenas, and their role in shaping the forms of differentiation which occur in "traditional" settings. A case study of Bukit Bangun, a village involved in the production of rubber since the 1920s, will serve to illustrate processes of change at the local level.[1] I pay special attention to agrarian differentiation, defined by White (1989:20) as: "a cumulative and permanent ... process of change in the ways in which different groups in rural society — and some outside it — gain access to the products of their own or others' labor, based on their differential control over production resources and often, but not always, on increasing inequalities in access to land." Changes in surplus extraction by groups outside and inside the village are thus central to my analysis.

The state plays a crucial role as surplus extractor; equally important is its role in setting the conditions for rural differentiation. Bringing the state into the analysis, according to Hart "... entails understanding how power struggles at different levels of society are connected with one another and related to access to and control over resources and people" (Hart 1989:48). My analysis will demonstrate how administrative policies of the Dutch colonial state interacted with kinship based village polities to shape local processes. Village "tradition" was used and molded by powerful groups for their own purposes. Patterns of differentiation were especially affected by state regulations concerning peasant access to land and smallholder rubber production in both the colonial and the postcolonial periods. My study thus explores how the Karo have refashioned their economic and social relations in the context of new constraints and considers the consequences of their encounters with particular state regimes.

After a short introduction of the village of Bukit Bangun, I give an overview of state policy and agrarian changes in the uplands of Langkat in colonial times followed by a description of village level changes in access to land and in labor relations. I then deal with changes in the period after independence in the uplands generally and in the study village. Because of the supposed relationship between commercialization, i.e. a growing market involvement, and agrarian differentiation (White 1989:26), I pay special attention to trends and changes in the degree of commercialization in the village over time. I consider whether Langkat experienced a cyclical pattern of commercialization and "decommercialization" during the first decades of this century similar to that observed in Java's lowlands (Hüsken 1989:303) and mountain Java (Hefner 1990:79,80).

In order to examine patterns of differentiation, I identify factors which might foster polarization (the concentration of land and a growing landlessness), or which might, alternatively, cause a leveling of differences in landed property. A growing inequality would contradict Karo kinship ideology, which stresses "help" and "unity", making kinship a possible countervailing factor to polarizing tendencies. I consider the significance of a special characteristic of rubber cultivation, namely the need to plant new groves or to replant old rubber to compensate for the decreasing productivity of the trees after twenty-five years. Some suggest that because of the high costs of replanting, rubber production itself tends to result in a bifurcation of strong peasants and landless sharecroppers or wage-laborers (Lee 1973). Other factors which are relevant to understanding the specific forms which agrarian differentiation has taken in Langkat include market trends, demographic pressure, the investment priorities of richer households, and income earning possibilities for poor households. I make brief mention of the Javanese households that have recently settled in the village as a new and especially disadvantaged landless class, but in this chapter I focus primarily upon the Karo.

The Karo Batak Rubber-Producing Village of Bukit Bangun[2]

The village of Bukit Bangun is in the uplands of Langkat regency some sixty kilometers by road to the west of the regency capital of Binjai. The village is in the district of Salapian and from the district capital of Tanjung Langkat, a small road, partly unpaved, leads to the village across extensive rubber and oil-palm plantations. The huge palm trees and durian trees of the village are a sign that the village has been there for a long time. It was originally a Karo Batak settlement. At the time of my research in 1986/1987, there were seventy households, of which 40% were Javanese former plantation laborers or their descendants. The houses of the Karo families were built around two yards, reflecting the existence of two villages in colonial times. Most of the Javanese built their houses in a cluster around a mosque, on the land of the village headman, several hundred meters outside the Karo settlement. Other Javanese lived dispersed in the rubber fields. Whereas the Javanese were all Muslim, most Karo were Christian and some followed the old Karo animistic religion now called *Agama Pemena*. Intermarriage between the two groups was minimal. The Karo were especially disapproving of marriages between Karo women and non-Karo men because, as they said, "the clan will disappear."

The principal product of Bukit Bangun is rubber, sold once a week at a market not far from the village. Other crops of importance both for home consumption and for sale are kapok, cassava, peanuts, fruit and pepper. New cash crops include cloves and coffee. Rice is no longer grown on dry land and the former paddy fields along the river have been planted with cash crops. Residents buy their rice at the weekly market in Tanjung Langkat. Commercial production is controlled by the landowners, who employ sharecroppers for their rubber gardens and day-laborers for their coffee and clove fields. Some people are also employed in petty trade, for example in durian. Women sell some of their vegetable and fruit produce at the market in Tanjung Langkat. There are four shops in the village, where kerosene, salt, dried fish, pepper, cigarettes and so forth are sold. One of the shops is also a coffee shop (*kedai*), where the men gather in the afternoon.

The distribution of land is highly uneven. Half the population has no land, and this half is mainly Javanese. Stratification is clearly visible from the houses they live in. Poor peasants have houses of bamboo, with roofs of palm leaves (*atap*), and a floor of earth. Middle and rich peasants have houses made of wood with a roof of zinc and a floor of cement, and the richest ones have houses made of stone. The two richest Karo peasants, who each own around twenty hectares, belong to the clans of Sitepu and Sembiring.

The Karo, who belong to the linguistically and culturally related ethnic groups in North Sumatra called "Batak," have five clans (*merga*)[3] and each of these clans contains from thirteen to eighteen subclans (also known as *merga*). Their kinship ideology, as a moral order, is characterized by several interrelated principles: patrilineal descent, patrilineal inheritance, clan exogamy and an asymmetric marriage system with a preferred matrilateral crosscousin marriage (Kipp 1976:2). Clans are seen as united wholes, whose members (*senina*), should treat each other as brotherly equals and provide mutual support. Affinal relations are assumed to be unequal. The superior wife-givers (*kalimbubu*), also called "visible God" (*dibata ni idah*) give blessings, fortune, health and the underling wife-receivers (*anakberu*), referred to as "the tired ones" (*si latih*), are the servants. The inferior position of the wife-receivers stems from a perpetual debt, because they received women, the source of new life (Singarimbun 1975:139). Reality, however, contradicts this model. The Karo clans are not "united wholes" but noncorporate groups (Singarimbun 1975:72). Although the model states that the Karo should always honor their kalimbubu, in practice only the close kalimbubu are important in daily life. Similarly, only with some senina and anakberu do effective relations exist (Kipp 1976:272). Nowadays, the duties of the latter include organizing and serving at the feast of their kalimbubu, helping in times of crisis in their kalimbubu's family and mediating (Singarimbun 1975:115–137).

The village of Bukit Bangun is, like other Karo villages in the uplands, associated with the clan of the village founder although all five Karo clans are present. The older descendants of the village founder can trace their genealogical connection with him. Others, however, seldom know their ancestry beyond three generations and effective kinship is of a situational nature. Villagers can usually link themselves to others in more than one way, but choose the closest relationship as the basis for interaction. Since a person could be in one context an anakberu and in the other a kalimbubu, it is the case, as Kipp (1976:87,88) remarks, "that a system of hierarchical status serves an egalitarian end".

THE KARO BATAK UPLANDS OF LANGKAT IN COLONIAL TIMES

In pre-colonial times, the uplands of the present-day Langkat regency became a frontier area when Karo Batak from the highlands moved to the fertile hilly area downstream. They made a living as shifting cultivators, developed trade relations with the lowland Malays, and their pepper and tobacco were exported overseas. In the course of time, Karo village unions developed simultaneously with an administrative structure that was more or less independent of the highlands.

The tribal political structure of the Karo Batak lacked central authority, and could be called republican-democratic in contrast with the more monarchic-autocratic state structure of the lowland Malay. There was a hierarchy of village headmen: a village (*kuta*), the smallest administrative and juridical unit, formed part of a larger village union consisting of a mother village (*perbapaan*) and colonies (*dusun*), and several village unions together formed a federation (*urung*). Dispute settlement could take place at any of these levels. Actions of village headmen were controlled by their kinsmen, thus contributing to the democratic element of the system while the fact that the position of village headmen was hereditary supplied its somewhat oligarchical element. The unions were seen as genealogical units, and they did not have fixed boundaries prior to the annexation. Villages regularly split and new ones formed. Warfare contributed to the lack of stability of these unions. It was a means for the leaders to show who was the strongest and in general it did not cost many lives (Westenberg 1914:457–465, 496).[4] The flourishing market in pepper and tobacco in the beginning of the nineteenth century probably had only a minor effect on the then existing social and economic system, because money was not yet capital and was used, in the form of silver dollars, primarily for marriage payments (Steedly 1993:90–94).

"Langkat" was an Islamic Malay harbor state. Control over people, rather than land, was the key to the power of the Malay chiefs, whose income came from taxing trade (Milner 1982). For the Malay Sultan who settled in the center of this state near the mouth of the Langkat river in the lowlands (*hilir*), the uplands (*hulu*) formed the periphery of his realm.[5] Before the interference of the Dutch, the Malay chiefs exercised some influence in the uplands, but their relations with the main Karo Batak chiefs were those of alliance rather than subordination. Colonial rule changed this situation, as the Karo Batak uplands were incorporated into the Sultanate of Langkat after a political contract was signed between the Malay Sultan and the Dutch in 1865.[6] The Sultan informed the Dutch that he owned all the land up to the highlands, and described the Karo as "inferior jungle people" (Westenberg 1914:416). The political implication was that the Karo in the uplands, or *dusun*, had become subject to Malay chiefs.[7]

State Policy and Agrarian Change in the Uplands

The Dutch authorities were forced to strengthen their administrative control shortly after the annexation of Langkat and the other Malay Sultanates along the east coast such as Deli, Serdang and Asahan. At the time Western entrepreneurs, most of them Dutch, opened tobacco plantations and later, in the early twentieth century, rubber plantations. In a few decades, the predominantly forested area had been transformed into a huge plantation region with tens of thousands of contract laborers from China and Java working on the estates.[8] While law and order was their primary concern at the start of their rule in the area, the Dutch later switched to the "ethical" policy directed toward the "development" of the people. In the uplands of Langkat, an especially troublesome area at the beginning of Dutch rule, the policy adopted was a protectionist one which aimed to make the Karo into allies. The Dutch authorities took measures to regulate the incomes of the Karo chiefs, to preserve Karo customary law (*adat*) and to defend their land rights.

The Karo opposed moves by the Sultan to lease land to Westerners for the expansion of tobacco plantations. In 1872, the "Batak War" broke out in the Karo uplands of Langkat and Deli, and lasted for several months. Military help was requested by the Dutch Resident from Batavia to suppress the revolt (Schadee 1920:17). The Karo then continued to demonstrate their disapproval of the use of their ancestors' lands by burning the tobacco sheds. Dutch officials subsequently instituted measures to promote the acceptance of plantations in their area, ordering the Malay Sultan to give the Karo chiefs part of the rents paid by the Western planters to him. They also took measures to preserve the Karo *adat*, endangered by the wish of the Muslim Malay chiefs to extend their power in the Karo area. Thus they installed a special official, the *Controleur* for Batak Affairs in 1888, and undertook the codification of customary law (adat). At the same time, Dutch policy towards the Karo aimed to defend Western planters' interests. The colonial government feared a possible attack in the plantation area led by the Acehnese (Westenberg 1914:461). Because of the same fear Western planters started in 1890 to give financial support to the Dutch mission in the Karo area of the neighboring Deli. The idea was to create a Christian "buffer" against so called fanatic Muslims (Kipp 1990:41,63).

The Dutch also undertook measures to defend the land rights of the Karo against the expansionism of the Western planters. They designed "Model Contracts" stipulating the rights and duties of the Sultan of Langkat and the planters in relation to the Karo dwelling on the estates in 1876 and revised them in 1884 and 1892. According to the contract of 1876 the planters were required to reserve four *bouws* (2.8 hectares) for each Karo household on their estates as swidden land. The contract of 1884 further stipulated that land should be reserved for the Karo around the villages in the tobacco areas (the *tanah seratus)* and that Karo could use harvested tobacco fields, *jalurans,* for rice cultivation. To win the sympathy of the Karo, the planters had allowed them to use the harvested tobacco fields for rice and gave the village headmen extra shares. According to the contract, however, the practice of distributing jalurans became a requirement. Tobacco, an annual crop, was cultivated in a rotation of eight years and the planting and harvest times of tobacco and rice did not compete. Karo peasants were eager to use the jalurans, benefiting from fertilizers the planters had applied and from the low labor required to grow rice on these ready-made fields when compared to shifting cultivation. They were, however, restricted from growing crops other than rice on this land. When rubber replaced tobacco on the estates after 1910, there was again a requirement to reserve swidden land for the Karo. It was on these reserve lands that the Karo themselves started to grow *Hevea* rubber in the 1920s.

Dutch protection of the Karo was half-hearted, and many changes occurred in Karo territory as a direct or indirect result of Dutch interventions. The Karo became more aware of the power of the state or the *Kompeni,* as they called it, when it showed its military strength in restless times. Changes were made in the administrative system. Indirect rule in Langkat implied administrative autonomy (*zelfbestuur*) for the Malay Sultan, but the Dutch took measures to simplify, rationalize and centralize the existing administrative structure. As the process of modern state formation proceeded, a Malay-Karo bureaucracy was formed with a hierarchy of salaried officials ruling bounded territories. The main Karo chiefs became subordinate to the Malay ones, and the administrative structure of Karo Batak society became more autocratic. Karo village headmen were made responsible for the levying of taxes and acted as representatives of the colonial government. Around 1908 the Sultan of Langkat levied both a labor tax (*rodi*) and a monetary one and it was the latter which, despite the small sums involved (Fievez de Malinez van Ginkel 1928), stimulated the Karo to produce for the market. The position of "traditional" leaders was strengthened by Dutch recognition. Their receipt of salaries accentuated their privilege and the salaries themselves grew in importance when, in the course of the twentieth century, money became treated by the Karo as investment capital. Dutch respect for Karo adat, and their policy of preserving it, contributed to a sharper consciousness of adat in village society (De Ridder 1935:55). As the example of Bukit Bangun will demonstrate, traditional leaders whose position was bolstered by the Dutch were able to use the adat for personal gain.

Dutch protection of Karo in agrarian matters was also half-hearted, as officials had to take account of the interests of Western entrepreneurs in the area. Although they defended the land rights of the Karo, the four bouws of land reserved for them were not enough to sustain shifting cultivation. Moreover, regional Dutch officials knew that Western tobacco planters did not in fact reserve the four bouws specified, eager as they were to keep the land at their own disposal (rubber planters were more compliant in this respect). Officials did not even enforce the swidden land requirement in the 1930s when there was a serious shortage of jaluran land (caused by the reduction of the total area planted with tobacco) as well as population pressure. Instead, Western tobacco planters offered the Karo compensation for the loss of jaluran at the rate of one *bouw* (0.7 hectare) uncleared estate land per household per year. For these *rabians* or *bosjalurans* as they were called, the same restrictions applied as for the jaluran, i.e. only the cultivation of dry rice was permitted. After one or two harvests, the land had to be given back to the estate owner and new rabians were distributed. The rabians were located on the steeper, less desirable slopes not well suited for plantation tobacco. These were the parts of their concession which the tobacco planters planned to return to the Sultan when plantation and peasant land were divided under the terms of new long term leases. The unsatisfactory allocation was opposed by the Karo, who demonstrated and asked for more and better land.

Did colonial officials want the Karo to become rice producers rather than rubber cultivators? The Dutch wanted the Karo to increase the rice production but they never took the structural measures necessary to achieve this. In times of crisis, as in 1918 and in the 1930s, the government ordered the planters to let peasants and laborers plant rice on the estates, but only on a temporary basis. Reports indicate that regional Dutch officials had a low opinion of the Karo as peasants. Resident Grijzen, for example, in a report in 1921, criticized the Karo for not being sawah cultivators, and for producing dry rice only for home consumption. He *also* criticized their strong interest in planting cash crops such as rubber and coconut since the reserved swidden land on the rubber plantations was, in his view, intended for rice cultivation.[9]

Western planters often used the image of the Karo as bad farmers prone to destroy fertile land in order to justify keeping most of the estate land at their own disposal. Irrigation was possible on parts of the estate land, but the planters refused to give up estate lands for this purpose, even though a large part was not cultivated but kept "in reserve." The colonial government did not begin to criticize this situation, nor to demand a permanent renouncement of the "reserve" land for irrigation purposes, until the end of the 1930s. However, plans to divide plantation land from peasant land and to reorganize the whole area were not implemented because of the Japanese occupation. In any case, these plans demonstrated a wish to intensify peasant food production and not to support extensive rubber production by smallholders (Pelzer 1978:104–112).

In fact, Dutch colonial policy together with the practices of the planters strongly reduced and restricted the commercial activities of the Karo. The shrinking pool of land available to the Karo was one factor. Rubber plantations expanded rapidly after 1910, and the land reserved for Karo dwelling on estates was inadequate. Official concern about the scarcity of land for the peasant population led to a decision in 1917 to reject proposals for new plantations in Langkat,[10] just a year after the creation of forest reservations had restricted Karo (and planter) access still further.[11] Ironically the planters, themselves major forest destroyers, had urged the government to create forest reservation to prevent the erosion they felt was caused by shifting cultivators (van Zon 1915:438).

Due to their limited access to land and the regulations concerning land use, such as the ban on planting cash crops on the jalurans, the Karo could not fully profit from the growing demand for rubber and the high rubber prices in the 1920s. But they did plant rubber on their own initiative wherever it was permitted, namely around the village on the tanah seratus lands on the tobacco estates, on the free land between the concessions, and on the reserved land on the rubber plantations. They also continued to cultivate dry rice. In the Depression of the 1930s the owners of rubber plantations urged the government in the Hague to restrict production of both plantation and smallholder rubber. New planting was restricted (by area), but replanting was permitted. The plantation owners replanted as much as possible with high yielding varieties. Smallholders, however, who could not afford the loss of income from a replanted garden during the gestation period (7–10 years) and who were hindered from planting new trees to compensate for the decreasing production of older ones, lost their competitive position in the long run.[12]

Colonial rule and the expanding plantation economy also caused Karo to cease cultivation of tobacco and pepper, the cash crops they had grown in pre-colonial times. Western tobacco planters, afraid that smallholders would steal plantation tobacco that they could sell as smallholder produce, successfully took measures to stop Karo production.[13] Pepper cultivation was no longer possible on a large scale because of the restrictions on land use in the area of tobacco plantations.

Thus the colonial period saw the transition of most of the Karo in the uplands from mobile shifting cultivators (and cash crop farmers) to smallholders living in permanent communities and cultivating tree crops such as rubber in addition to dry rice. Dutch policy on the reservation of land around the Karo villages located on the tobacco estates strongly contributed to this change. It was a policy intended to settle the rural people in a well-defined space (cf. Scott 1995:21). It was pursued even though the Karo refused to work as permanent plantation laborers. Living in close proximity to the plantations, their confrontation with the working conditions of the indentured coolies seems instead to have strengthened their identity as autonomous Karo smallholders with an aversion to subservience.[14]

BUKIT BANGUN IN COLONIAL TIMES: COMMERCIALIZATION AND THE RISE OF A VILLAGE ELITE

Colonial rule implied an interaction between Karo kinship-based village polities and the Dutch colonial state. The somewhat hierarchical structure at the village level was formed by the descendants of the alleged village founder who formed a ruling lineage, called the *bangsa taneh* or the "owners" of the land, and the governed people (*ginemgen*). The bangsa taneh claimed the right to the position of village headman, which was hereditary in the male line. Village headmen were administrators and judges at the same time. The position of their main assistant was claimed by those Karo recognized as the "traditional" wife-receivers (*anakberu taneh*) of the ruling clan, and was also hereditary. Mediation between the governed people and the village headmen was part of their role (Westenberg 1914:473). Colonial rule re-enforced this hierarchical structure by recognizing "traditional" leaders and paying them salaries. It thereby strengthened their position in the village.

The village (kuta) of Bukit Bangun belonged to the federation of villages (urung) of Salapian. It was a "mother village" (perbapaan) with four

FIGURE 1 KARO BATAK GROUP IN KAMPONG LAU TEPU, C. 1885
(G.R. LAMBERT, COLLECTION ROYAL TROPICAL INSTITUTE AMSTERDAM).

subordinate villages (*dusun*). All the village headmen belonged to the same clan. This federation was part of a larger union, with a court located in Lau Tepu, where the village headman bore the title of *pengulu balai*. In total, there were four main Karo chiefs in Salapian, a genealogical unit which Dutch regulations turned into an administrative unit, a district with fixed boundaries.[15]

The ruling lineage in Bukit Bangun was formed by Karo belonging to the Sitepu clan and the anakberu taneh by those of the Sembiring clan. According to the story of village founding a Karo of the Sitepu clan, whose native village was Penampen in the Karo highlands, founded Bukit Bangun. He had mystical powers, his body was hairy and he was a good fighter. His successor, his oldest son, chose as his main assistant one of his close wife-receivers, namely his sisters' husband (SiHu), a Sembiring. Since then the Sembirings occupied the position of main assistant of the village headman. Another assistant of minor importance was one of the village headman's clanmates (senina), but he was considered a distant relative. Bukit Bangun was, according to a former village headman who kept a written genealogy, founded some six generations ago. A striking element in this story, told by the Sitepus, is that clan names of the wife-receivers and wife-givers of the village founder have been forgotten. Other stories note that a village is always founded by a man together with his anakberu and kalimbubu (Singarimbun 1975:89). In this way the position of the ruling lineage and of the anakberu taneh could be legitimized. In my view, the story told by the Sitepus in the village illustrates the importance of the Sembiring clan.

Karo leaders were able to take advantage of Dutch administrative regulations to strengthen their position. Around 1900, the village headman of Bukit Bangun ordered one of his sons to found a new village so that he could receive a government salary. The new village, Kuta Mbaru, was very close to Bukit Bangun and the existing kinship based structure was repeated there with a Sembiring as assistant to the village headman. The villages were small: in the 1930s there were ten households in Kuta Mbaru, twenty in Bukit Bangun, and all five Karo clans were represented.

The expansion of tobacco plantations in the uplands also stimulated the political elite of these two villages to strengthen their economic position. The village headmen claimed to control the reserved land around the village, the tanah seratus, on the basis of Karo adat and their position as bangsa taneh. They also became brokers between the villagers and the plantations.

It was not until 1922 that the Karo in Bukit Bangun and Kuta Mbaru got access to jalurans. The tobacco plantation of Tambunan opened nearby around 1880, but the villagers were never given permission to use the plantation jalurans because the village was situated outside its boundaries. This situation changed when a German planter opened the tobacco plantation Glugur Langkat in 1922. The Dutch Resident had just forbidden the opening of new plantations in 1917, but for this one he made an exception "because the people liked to receive jaluran," he wrote in a report.[16] Tobacco was then planted close to both of the villages on former village land. As required in the contracts, land around the two village settlements was reserved for the peasants. The boundaries were stipulated under the supervision of a Dutch official, the *Controleur*. The reserved land had to have a radius of 200 meters equal to 100 *vadem*, measured from the house of the village headman. Because of this measurement, the bounded village land was called tanah seratus, or hundred land. The tobacco of Glugur Langkat was planted right up to the borders of the tanah seratus and no swidden land was reserved on the estate. The Karo did not know about their right to swidden land according to the contracts.

From then on, the Karo of Bukit Bangun and Kuta Mbaru started to grow rice on the harvested tobacco fields of the plantation. In the past, they had practiced shifting cultivation on fields allocated annually by the village headman. The communal nature of land tenure was evident from the rule that village land could not be sold or used by residents of other villages, and reverted to village control if a resident departed. Newcomers who settled in the village and took part in specified communal labor tasks were given permission to open fields for rice cultivation. However, land planted with permanent crops such as fruit trees and wet rice paddies (*sawah*) were considered individual property; inheritable, but not for sale.[17] As we will see, land practices changed with the expansion of the tobacco plantation.

The village headmen, making use of the stronger position they had acquired as a result of Dutch backing, distributed the tanah seratus unequally and also made stipulations regarding its use. Referring to Karo adat, they claimed to be the "owners of the land", entitled to use it to their personal advantage. They kept the largest part of this land with a total of around fifty hectares for themselves and their brothers, gave their two assistants a reasonable part, and gave smaller parts to their relatives and others who settled in the village, thereby creating allies. The "newcomers", or those Karo not closely related to the Sitepus, were excluded. The tanah seratus became valuable land, consisting of sawah along the Salapian River and gardens with durian, coconut and palm sugar trees (*aren*). Under such uses it also became individual property according to Karo adat. Emerging stratification in the village closely reflected the unequal division of this land.

The economic elite consisted of the village headmen, their assistants and the brothers of the headmen, all of whom belonged to the Sitepu and the Sembiring clan. In the 1920s, members of this elite profited from the growing market for rubber in the area. They were the first to plant rubber on the tanah seratus, while the village headmen prohibited others from following suit. The two village headmen became the most powerful members of this village elite. They were salaried and put in charge of the payment of money taxes to the government. With the opening of the tobacco plantation, they were responsible for the annual distribution of the jalurans and were given an extra share by the plantation owner. They organized the seasonal work done by the Karo for the tobacco plantation, and were paid for this with a percentage of the wages. The headmen also retained several of their traditional privileges. They could ask all the villagers to donate a number of days of agricultural labor every year, especially valuable in view of their extensive tanah seratus fields. They ordered the villagers to perform other chores for the village itself, such as guarding the village and maintaining the village road. They could claim a tenth of the monies exchanged in civil law suits, such as divorces, and a tenth of the slaughtered animals. They asked for small amounts of money at weddings or funerals and for granting permission to move out of the village. On certain occasions, gifts in the form of rice and chickens also had to be presented to them.[18]

The proximity of the plantations together with need to pay taxes (from around 1910) directly and indirectly furthered the monetization of the rural economy. Roads were constructed in the area for the use of the plantations but they also facilitated local commerce. In the 1920s, a period of general economic growth, the Karo sold their agricultural products such as rice and fruit to the plantation laborers, they delivered materials such as *atap* to the tobacco plantations and they started to sell rubber to Chinese traders. Karo men and women also earned money at the nearby tobacco plantation as temporary workers. But within the village, land and labor were not commoditized. Rice cultivation as well as house construction were undertaken with cooperative labor, and fruit and vegetables were commonly shared.

Differences grew between the village elite and the non-elite with respect to access to land and wealth, but new forms of surplus extraction such as sharecropping or wage labor did not emerge. Middle peasants using their own household labor remained predominant. Thus increasing inequality in access to land and to the possibility of rubber production were not caused by economic laws or the "cumulation of advantages" in the market (Harrison 1977:133), but by the actions of Karo with power to deploy kinship and adat to their own advantage. Over time, however, the elite did begin to treat money as capital, investing in trade and rubber seedlings. They also spent money on better housing and clothing, and on rituals. Access to education was still very restricted at the time, although it would later become a main object of Karo investment, as we will see.

Conditions for the Karo in the villages worsened in the Depression of the 1930s. Because Glugur Langkat switched from tobacco to rubber in 1930 they lost access to the fertile jaluran, and had reduced opportunities for seasonal wage work. The new contract for Glugur Langkat as rubber plantation pertained to half of the former area, leaving the other half — the hilly, poor quality land — for the peasants.[19] This form of compensation represented a decline for the peasants in comparison with their former situation. When the plantation manager (administrator), told the

Karo of Bukit Bangun and Kuta Mbaru that they would lose their jaluran they protested. They perceived the loss of this fertile land as a loss of their right to use their former ancestral lands. One of the Sitepus was so furious he wounded the administrator of Glugur Langkat. The conflict between peasants and planters in Bukit Bangun was settled by the Sultan of Langkat. At a meeting with the villagers, he promised to give them land, and sought thereby to uphold his prestige as a good ruler. But his intervention brought little improvement. The peasants were relegated to the poor quality land, forced into a "marginal" position from which they could produce only meager crops of dry rice.

Due to the Depression and the prohibition on new smallholder plantings which was in force between 1934 and 1944, rubber was not planted on the new reserved land in the thirties. Karo who already had rubber fields could still gain a modest income from them, and were therefore somewhat sheltered from the effects of the Depression. During the Japanese occupation the Karo returned to a subsistence economy. A renewed commercialization occurred after Independence. Members of the village elite with a firm base in the tanah seratus could then accumulate by means of their rubber lands, which formed a kind of reserve. New conflicts arose because of the competition among the peasants for access to land outside the tanah seratus on which they could expand their rubber gardens.[20]

POLITICAL AND ECONOMIC CHANGES IN THE UPLANDS AFTER THE JAPANESE OCCUPATION

After the Japanese occupation (1942–1945), a strong republican movement became active in North Sumatra. It was directed against the "feudal" Malay aristocracy and resulted in the Social Revolution in 1946. The movement had mass support and eventually led to the dissolution of the Malay dominated (and Dutch-backed) Negara Sumatera Timur (NST) (Reid 1979).

The uplands of Langkat became part of the non-republican area (formed after the Dutch military campaigns) during the period of the NST (1947–1950). There was much unrest, caused by repeated mass occupations of plantation land and military action against "illegal" squatters, including different groups of Karo, Javanese and Chinese. Actions by radical organizations in the rural area were primarily directed to the occupation of plantation lands. They were not aimed at an equal division of the village lands, the tanah seratus, monopolized by the Karo village elite. The Karo themselves were strong supporters of the Republican movement during the Independence War (1945–1950). Their involvement can be viewed as an expression of enduring anti-Malay feelings, fed by an abhorrence for subordination.

Under the land policy of the new republic, plantation land and peasant land was legally separated in 1954 (Van de Waal 1959). This implied that planters were no longer obliged to reserve land for peasants within the plantation borders. In the uplands of Langkat some plantation land remained — and still remains — occupied by peasants. The mass occupation of plantation land continued in the 1950s, after the integration of Langkat in the Republic of Indonesia and the nationalization of the Dutch plantations in 1957. Plantation land was seen by peasants as "the gift of the revolution". The occupations ended with stronger military repression in the 1960s.

The Indonesian government formally recognized the status quo in 1962 by ordering land with settlements, land permanently cultivated, and land in use by peasants within the plantation borders to be reserved for the people. However, peasants could be displaced for plantation expansion, and given compensation (Parlindungan 1981). For the peasants, this did not mean security, especially in times of economic growth.

The upland economy recovered slowly after the hardships of the Japanese occupation. During the NST (1947–1950), the Karo started to tap their rubber fields again and Chinese traders built eleven new remilling factories in Medan for smallholder rubber. Like the colonial government before it, the NST followed a policy of "betting on the strong", and offered material help such as tapping knives and sheet machines to the village headmen.[21] The Karo mainly produced sheet rubber at the time, a capital and labor intensive product. In the mid-1950s, the Korean War boom induced Karo peasants to produce more rubber and to engage in new planting. They planted the traditional varieties, notwithstanding the new republic's plan to promote high yielding varieties. The policy known as the "cess-scheme", which aimed to support the richest peasants, failed completely (Barlow and Tomich 1991).[22]

The trade in smallholder rubber was still in the hands of the Chinese but there was a strong anti-Chinese sentiment. Karo rubber smallholders in the uplands of Langkat organized rubber co-operatives in the beginning of the 1950s in an attempt to bypass them. In 1959, the government ordered the Chinese to move to the district capitals, with the intention of breaking their dominant trade position. When market conditions again became unfavorable due to high inflation at the end of the 1950s and in the early 1960s the Karo did not totally retreat from the market, as they had during the war. Those with rubber fields continued to tap them, while still cultivating rice. The 1970s saw a period of general economic growth under the "New Order" regime.

The state sponsored projects (*Bimas*) designed to make Indonesia self-sufficient in rice started in Java at the end of the 1960s, and were introduced in North Sumatra in the 1970s. The result of this policy was that in the uplands of Langkat, the agricultural extension officials concentrated upon wet-rice areas, and more or less ignored districts such as Salapian where the production of rubber was dominant. The mass of rubber smallholders in the uplands received no government assistance. Credit was only available for peasants who were involved in state projects aimed at a "green revolution" in the cultivation of rubber. These projects started in the 1980s after the official abolition of the "cess-scheme" in 1973. Their rationale was that smallholders needed state assistance to increase production. The projects included bureaucratic control, block planting with superior varieties of rubber and the processing of smallholder rubber at the national estates.[23] Their stated aim was to reach the landless poor by allocating them two hectares in the project area together with credit (Tomich 1991:253,254).[24] At the time of my research, these projects had just started on a small scale, and they were not part of my systematic observations. It appears, however, that headmen whose villages were included in these projects, as well as higher government officials, were employing landless Javanese to cultivate plots on a sharecrop basis. The contradiction between this outcome and the project's stated purpose of making the landless into smallholders was rationalized by officials in terms of the inability of the impoverished Javanese to take on the "risks" of project membership.

A recent phenomenon in the uplands of Langkat is the large scale settlement of Javanese former plantation laborers and their descendants in Karo villages. In 1986/87 Javanese were already present in the more remote Karo villages in the district of Salapian. They are invisible in government statistics, which do not record ethnic origins. A spontaneous land colonization by Javanese plantation laborers had occurred in colonial times in the Malay lowlands but not in the Karo uplands. This could be explained by the aversion of Muslim Javanese to pagan "pig-eaters", and also by the unwillingness of the Karo to admit the settlement of non Karo in their villages. The recent demand for labor by the Karo and the notion of belonging to one nation (*satu bangsa*) could have contributed to the willingness of both sides to accept an arrangement under which they live in one village, albeit in separate sections. In this area, as in other plantation zones (Stoler 1985), the Javanese form the lowest class. They are the poorest and the vast majority of them do not own land. Because of their marginal position, they are also excluded from full participation in state projects, as we saw above. Given their need for labor, the Karo value the Javanese as sharecroppers and as laborers for the same reasons as Western planters: for their diligence and reliability. But in cultural terms, the Karo view the Javanese as deficient. As the village headman of Bukit Bangun once said to me: "What a pity for the Javanese. They have no kalimbubu, no anakberu and no senina!"

The village of Bukit Bangun was not directly involved in any of the state projects described in this section. After the Japanese occupation, the position of the old village elite in Bukit Bangun remained strong. In the next section I identify the initiatives taken by this elite and other Karo villagers as they sought to solve the problem of the need to replant their rubber at a time when they were also eager to give their children a good education.

BUKIT BANGUN AFTER INDEPENDENCE: RUBBER PRODUCTION, AGRARIAN DIFFERENTIATION AND KINSHIP

At the time of the revolution, some members of the village elite were active in the guerrilla groups, the *laskyars*. The revolutionary spirit in the Karo area influenced the people's views on the administrative structure of the villages. The autocratic position of the village headman was challenged and viewed as "feudal". For the first time, in the 1950–1957 period the village headmen did not belong to the ruling Sitepu clan. The two chiefs in this period, who also united the two villages, both belonged to the wife-receivers (anakberu) of the Sitepus. In fact, the Sitepus asked them to help by representing them, because the former position of the village head had been discredited. After 1957, however, a member of the Sitepu clan again became village headman and the position of headman has since remained with this clan. In the general elections, the people in the village chose a man of the Sitepu clan.

The position of village headman changed after independence. Bukit Bangun became part of a new administrative union of six villages (*lorong*), together forming one administrative village (desa) of Naman Bukit. Of the village headmen (*kepala lorong*), one was recognized by the Indonesian government as the main village headman bearing the title of *kepala desa*. Up until now, the kepala lorong of Bukit Bangun has not been salaried, but has kept some of the adat-income for his services at marriages, divorces, and funerals. He no longer works with one of his anakberu and senina as assistants. Because he is responsible for the levying of a land tax, he represents the Indonesian state. Although his position as village headman is less autocratic than in the past, members of the old village elite and their descendants have managed to keep an important position. In the next section, I deal with the period of renewed commercialization and its decline during the "Old Order" (1950–1965) and the competition for access to land in the village.

Bukit Bangun in the Old Order: Rubber Production and Competition for Access to Land.

During the Korean War rubber boom of the 1950s, only the village elite in Bukit Bangun owned full grown rubber fields. As people rushed to plant new rubber, members of this elite attempted to maintain their advantageous access to land, shifting their attention to village land outside the tanah seratus. Some Karo also used uncultivated plantation land close to the village. On the basis of adat rules, land planted with rubber became individual property that could be inherited, contributing to its attractiveness.

There was fierce competition among members of the village elite, the Sitepus and the Sembirings, to plant as much rubber as possible outside the tanah seratus. They excluded the "newcomers" from this competition in a brutal manner, by burning down their young rubber trees while expanding their own fields. However, the "newcomers" also had opportunities to improve their position in the course of time. In the 1950s, they gained some access to the tanah seratus. In the 1960s, they were able to expand their rubber fields outside the tanah seratus because of the way the Basic Agrarian Law (1960) was implemented in the village. Although there was no distribution of the land of the richest peasants, which was one of the purposes of this law, land became individual property and the headman of Bukit Bangun started to distribute the remaining wasteland. Villagers could decide how much land they wanted, so long as they could afford the official land tax. They had to give a small sum to the village headman "for making boundaries". At the time, for example, two "newcomers", brothers of the merga Perangin-angin, together requested seven hectares, which they planted with rubber.

Since the 1960s Javanese former plantation laborers also tried to open land in the village and to plant rubber, obtaining permission from the Karo village headman on the same conditions as the Karo inhabitants. A few of them succeeded in becoming independent rubber smallholders by working temporarily as tenants for the Karo before they were able to live from their own rubber fields. Many other Javanese left the village after losing their land or because they thought prospects were better somewhere else. Non-economic factors also caused some flux in the Javanese population. Many Javanese feared the Karo because of their supposed magical power. The Karo were thought to be using black magic to cause sickness among Javanese would-be rubber smallholders, especially in the period before the mid-1970s when the Karo became Christian. The Javanese claimed that one would need to be really "brave" (*brani*) to stay in the village. It seems that the Karo did everything possible to reinforce their superiority over the Javanese immigrants, especially in relation to their control over ancestral village land.

In the period of inflation after the end of the 1950s, rice prices were high and rubber prices low. By stealing and selling rubber from the plantation, villagers could earn some extra money. This occurred on a wide scale in the uplands, and rubber traders from Medan came to the countryside to buy the stolen plantation rubber as if it were smallholder rubber. At the beginning of the 1960s, stealing diminished because of stronger military repression and because there was less need to steal once the young rubber fields became tappable.

How commercialized was the village economy in this period? The new agrarian regulations made land an individualized commodity but the Karo disapproved of land selling. If forced to sell, they looked for buyers among their close relatives. Sharecropping relations spread, but there was no shift to wage-labor relations in the village. Cooperative forms of labor continued in a very restricted form. The sharing of products disappeared or became the subject of contention. For example, the Sitepus had customarily shared the fruit of their durian gardens with families to which they were closely related. This sharing was called durian *kerin*, and the families who helped guard the ripening fruit could keep the durian that fell to the ground during their turn. When land became individualized, the Sitepus wanted to guard their own gardens and keep all the fruit for themselves. They argued that the custom of sharing was "feudal" and that they were now "modern" (*moderen*). The other families protested and the question was brought before the court in Binjai. Here it was decided that the gardens of the Sitepus were indeed individual property, and they had the right to do as they pleased with them.[25]

Some changes could be observed in the method of rubber cultivation. The Karo who expanded the rubber fields in the course of the 1950s belonged to the second generation of rubber smallholders in the village. Unlike the first generation, they planted their one-hectare plots in regular rows, plantation style, and were conscious of the higher productivity of good rubber seeds. They searched for strong seedlings growing spontaneously on the rubber plantations. The demonstration effect of rubber plantations, just like the stealing of rubber from the plantation by the Karo smallholders, again contradicts the view that plantations and smallholders existed in separate spheres.

During the *Orde Lama* (1950–1965), sharecropping relations spread in the village. This occurred not because of land concentration but because some villagers, including Karo "newcomers" and Javanese, became tenants or sharecroppers for the village elite while waiting for their own young rubber trees to yield. Their payment was half the market value of the product, and the sharecropping agreement was therefore called "to divide in half" (*bagi-dua*).

For the Karo, the education of their children became one of the new and prime objects of investment. Some members of the village elite even sold part of their land to pay for their children's education, a few sending them away to Jakarta because the quality of education there was thought to be higher than in Medan.

Bukit Bangun During the New Order (Orde Baru): Commercialized Village Economy

At the time of my fieldwork in 1986/87, the village of Bukit Bangun was mainly a rubber-producing village. After the period of general economic growth started in the 1970s, a renewed commercialization occurred with the following characteristics. Firstly rubber grew in importance in relation to the total production of the village. Higher rubber prices stimulated the Karo to expand their rubber cultivation more rapidly. By the mid-1970s, the village land outside the former tanah seratus was completely planted with rubber, reflecting the Karo decision to abandon the cultivation of dry rice. The fact that rice could be bought regularly at the weekly market in the nearby district capital helped make the change to rubber less risky. The rubber smallholders, the second and third generation of rubber owners in the village, now regularly tapped their fields regardless of the rise or fall of rubber prices. This was new and reflected the importance of rubber income in household budgets. The ideal way of tapping, in their opinion, was to tap the same trees for two weeks and then leave them alone for two weeks, a system they knew was used in the past on the rubber plantations. Besides rubber, new cash crops including cloves and coffee were planted on the tanah seratus lands situated close to the Karo houses.

Secondly, the production system became more labor and capital intensive. Fertilizers were used for the new high yielding varieties of cloves and coffee and also, starting in the 1970s, for rubber. The availability of fertilizer in the uplands was a by-product of the state-sponsored wet-rice projects (Bimas). After the mid-1970s, the Karo changed the form of their rubber production from sheets to slabs because the price difference between the two products became too small for them to continue the more labor intensive process.

Thirdly, land and labor became commodities, but not in the full sense of the word. Land selling was still disapproved of and occurred only on a small scale, while the lending of land without rent was still common among Karo. More wage labor was used for the cultivation of the new cash crops, but sharecropping remained the main method of extra household surplus extraction. Besides the cooperative labor (mainly for the upkeep of the village road) imposed by the headman upon all male household heads (*gotong royong*), cooperative labor in the Karo community only occurred in a very restricted form. It was used between close family members or neighbors to perform some small agricultural tasks. While experts from outside built the houses of stone using paid labor, the Karo still offered their help and were given a meal to build a simple house for family members. Fruit and vegetables could no longer be "borrowed", as was noted above. Vegetables were now sold in the village itself, although the price was lower than at the market. Food crops were valued in different ways than cash crops. When I passed older Karo women in the village selling and buying vegetables, they apologized and told me this was not the custom in the past. Although durians are now sold at a good price to traders, the Karo still think of this fruit as something you should share with others. In the durian season, the Karo still invited family members and friends living outside the village to come and eat durian with them.

The village frontier was closed in the mid-1970s. Since then it has only been possible to gain access to land through inheritance or purchase. Land tenure remains highly uneven and, at the time of fieldwork, 40% of the village population was landless. Landlessness was not the result of the proletarianization of the existing community, but of in-migration in a situation where waste land was not available. The majority of the landless were Javanese who recently settled in the village to work as sharecroppers on the extensive rubber fields of the Karo elite. The rubber planted after the beginning of the 1960s became tappable at the beginning of the 1970s and caused a high demand for labor.

The sharecropping relations between Karo landowners and Javanese landless were asymmetric labor relations between two agrarian classes, to which was added the element of credit, as Karo regularly lent their Javanese sharecroppers money, interest-free. However, beyond making loans, Karo did not feel a moral obligation to guarantee their subsistence. The Javanese, for their part, did not provide any free services to the Karo outside agriculture. This pattern contrasts with the diffuse and enduring patron-client bonds observed in wet rice areas of lowland Java (Hüsken 1989:311). The sharecropper was responsible for the daily tapping of the trees, the daily production of rubber, the weekly transport of rubber to the market and the upkeep of the fields twice in a year. Costs were shared with the landowner, and the sharecropper received 50% of the income from the rubber. The sharecropper worked quite independently and this was why the Javanese preferred to work in the village rather than on the plantations: "We like to work here because there is no superintendent (*tidak ada mandur*)".

Within the Karo community itself, land distribution was also highly uneven. Some of the richest Karo owned twenty hectares, including land outside the village, while the middle and rich peasants owned two to five hectares. Karo households without land were in this position only temporarily. Most were young married couples sharecropping the rubber fields of their parents as part of the process of land transmission between generations. To prevent land conflicts, the parents usually divided part of their land when they were still alive and rented it out to their sons. After all the children were married, they gave land to their sons with a right to its full income, and still reserved some land for their own use. Young Karo whose parents' land was insufficient had to tap the rubber of their close relatives, a sign of the growing pressure on land (also a reflection of the uncertain outcomes of investing in education). Sharecropping relations between Karo were in the first place relations between kin and contained the element of assistance, although the half-shares arrangement remained strict regardless of kin ties or the rubber price.

The Karo belonging to the "newcomer" group also improved their position in this period. Many became middle and rich peasants who rented land to others, mostly to the Javanese. In the course of time all of these newcomers developed family relations with the Sitepus so that, by the time of my research the Karo often told me, "we are all family, there are no others."

There were several factors countering the loss of landed property and the impoverishment of the poorer Karo households after independence. Firstly there was, until recently, a relative abundance of wasteland in the village. Secondly, the fragmentation of land owned by Karo families in the village was limited by the patrilineal inheritance system and by the predominance of out-migration over in-migration over a period of time. The village population grew only slowly. Some young men moved from their native village to their wife's village after marriage if there was a better opportunity to open new land there. Out-migration also occurred when young Karo found employment elsewhere, a result of the heavy investment their parents had made in education, especially for sons. Thirdly, thanks to the stability of rubber production, the sale of rubber did not lead to severe debt relations with Chinese traders,[26] and poor households were not forced to sell their land

As noted above, the closing of the land frontier made some Karo temporarily landless, but the kinship ideology stressing unity and help countered their potential impoverishment. Landless Karo were able to borrow land from relatives without payment and use it for growing vegetables, although not for permanent crops like rubber. They were also assured of paid work on the fields of their Karo relatives, undertaking tasks such as weeding or harvesting coffee and cloves. The position of the Javanese landless in the village was far less secure. Besides rubber tapping on a share-crop basis, their laboring opportunities were limited by the Karo preference to give work to their own people. None of the Javanese were solely day-laborers. All of them held rubber share-cropping contracts, but these could be terminated by the landowner without any special reason, forcing the Javanese to leave the village in search of work elsewhere. There was in fact a high rate of horizontal mobility in this class. They often described themselves to me as migrants, *orang merantau*. The political conditions in the "New Order" and the lack of free trade unions indirectly contributed to the continuation of unstable tenant contracts and to the vulnerability of this component of the upland population.

Kinship relations thus made it easier for the landless Karo to gain access to the products of the land and to earn extra income. More generally, kinship relations and underlying kinship ideology stressing "unity" and "help" lessened potential class tensions within the Karo community. Although I will not elaborate on kinship at the ritual level, it is important to note that the numerous rituals the Karo had to attend, especially weddings and funerals, which were time and money consuming, continuously re-emphasized the veracity of the kinship model of society (cf. Kipp 1976). However, it remains to be seen whether new land conflicts, including potential conflicts among kin, will arise in the context of growing land pressure.

Several factors limited the accumulation of wealth by the rich Karo. These included their investment in education rather than in agriculture and a consumption pattern in which prestigious objects (houses, furniture) and contributions to Karo rituals played an important role. But rich Karo in Bukit Bangun remained full members of the village community, and did not settle in town as absentee landlords. Even large landowners continued to tap part of their fields themselves until their retirement. Poor Karo expected a kind of solidarity from them. Being arrogant (*sombong*) was negatively valued, while diligence (*rajin*) was respected.

Karo continued to place great importance on their kinship system. "Our adat is strong," they said to me. Nowadays for the Karo their adat is, primarily, their kinship system with its specific norms and rules. This understanding of Karo adat is, as Steedly has observed, not a reflection of its timeless quality but rather an "outcome of quite complex historical processes of social negotiation" (Steedly 1993:50).

The priorities of rich and middle peasants did not differ, although some had to make choices. All of them invested in the education of their children, boys and girls alike. They bought land and made investments in agricultural production and trade. They also renovated their houses and bought motorcycles and some luxury products. However, none of the rubber smallholders invested in the rubber trade or in rubber processing (smokehouses or remilling). It seemed that they had no access to this part of the rubber industry. Also, none of them invested in high-yielding rubber clones as a means to increase production, for reasons which I address below.

Replanting of Rubber by the Karo Rubber Smallholders

When the village frontier was closed in the mid-1970s, smallholders were faced with the need to replant the old rubber fields which were decreasing in productive capacity. Because of its high costs, replanting rubber could potentially lead to a bifurcation between strong peasants and landless sharecroppers or wage-laborers (Lee 1973). This was not the case in Bukit Bangun as villagers chose to replant their fields not with rubber but with other cash crops. The oldest rubber trees planted in colonial times on the tanah seratus were cut down in the mid 1970s, and replaced with cloves. Coffee was the tree favored for replantings in the 1980s, since cloves had been affected by disease and were no longer profitable. Characteristic of the cultivation of both crops were: 1) the planting material was of a high yielding variety (*bibit-unggul*), 2) shorter gestation period than rubber, 3) market prices of both crops were high when the Karo started to plant them and 4) both crops were cultivated without state assistance.

Clove seedlings were obtained from Chinese and Karo traders who visited the uplands of Langkat and gave farmers information on how to grow the crop, the planting distance and the amount of fertilizer to be used. There were stories about clove-growers elsewhere who could afford to build houses of stone or "buildings" (*rumah gedung*), as they were called in the rural area. Dreams of rapid riches stimulated clove planting. At the time of my research, coffee "fever" had just hit the Karo village community. A Karo woman told me she used to dream about the young coffee plants she had just planted with her husband. The village headman and some of his close relatives went by bus to the lowlands of Langkat to buy plant material from small-scale Javanese producers, and receive instructions on how to grow it. Again, there were rumors of rapid riches. The first coffee growers in Bukit Bangun sold their young coffee plants to other Karo and a Javanese landowner in the village also started to trade in the young plants, spreading cultivation to the surrounding villages.

The Karo who started to replant their oldest rubber fields in the 1970s and 1980s belonged to the middle and rich smallholders in the village. They were able to afford the temporary loss of income from the replanted area because they still had a regular income from their stock of younger, productive trees. Rubber and its replacements took on different roles in the household economy, as a Karo woman explained: "rubber is to get food and coffee is to save money." Those whose rubber was planted since the 1960s had no need to replant but they too tried to cultivate coffee wherever they could to raise their income.

It is striking that the Karo chose to cultivate cloves and coffee of high-yielding varieties and never tried to cultivate the high-yielding variety of rubber, although they were already long familiar with the budgrafts used by the plantations. The reasons for this given by the Karo were the expense of budgrafts, susceptibility to disease, and their lack of capital. Whereas the plantations used expensive poisons to kill the roots of the cut rubber trees, smallholders could not afford to follow suit, and would therefore have to wait four years before rubber could be replanted. Other crops, however, could be planted immediately. Growing the high yielding variety of Arabica coffee was thus less costly and it more rapidly produced an income. Their calculation, however, was a short-term one, since it did not include the costs of replanting the coffee a few years later. The education of their children was an important factor in their investment decisions.

All the Karo in the village, irrespective of land ownership, expressed a negative opinion about the state projects for replanting with high yielding varieties of rubber. In their view, farmers who entered such projects became "coolies" (*kuli*) on their own land. They preferred to stand on their own (*berdiri sendiri*) as independent smallholders.

SUMMARY AND CONCLUSIONS

This chapter has described the formation of a peasantry on the margins of the colonial plantation economy in the uplands of Langkat. The Karo changed from being mobile shifting cultivators growing hill rice, tobacco and pepper to smallholders living in permanent settlements, producing rice and rubber.

Interconnections between the plantation and the peasant sector included Karo access to "prepared" land (jalurans) which they could use to grow rice before the next tobacco rotation and to swidden land reserved for them within plantation boundaries. Stimulated by a need for additional cash to pay taxes, they performed seasonal work for tobacco plantations and sold their products to the plantation laborers. But they refused to become full time wage workers. Instead, they sought to retain their independence while planting the profitable Hevea rubber on their smallholdings.

Dutch officials working in the plantation area and Western planters alike typified the Karo as bad farmers, and as small scale "traditional" or marginal producers. They neglected to note the extent to which the marginality of Karo farms was a product of their own policies. The Karo had been forced to abandon their commercial crops, tobacco and pepper, in the nineteenth century. They later planted rubber on their own initiative, but severely reduced access to land and the planters' insistence that only rice could be grown on the jalurans prevented them from profiting fully from the high rubber prices in the 1920s. During the Depression of the 1930s, they were further restricted by a ban on new rubber planting designed specifically to save the rubber plantations and reduce the competitive position of smallholders. Land suitable for irrigation was not made available to the Karo, and the reserved swidden land was insufficient. Under these conditions, the best the Karo could do was to maintain the cultivation of rice and rubber at low levels, and spread the risks posed by market involvement.

The village as a political arena was also formed and reformed in the shadow of the plantation economy and the colonial state. In Bukit Bangun, the state relied on headmen to collect taxes and carry out orders, and planters offered them various fees and favors. The elite which emerged used a language of "kinship" and adat to take advantage of the new economic possibilities while denying full access to others. They divided land unequally, and monopolized access to rubber production.

The involvement of Karo peasants with the market did not follow an unilinear trajectory. Partial retreat from the market in the Depression years was followed by a complete return to a subsistence economy during

World War Two. Up to the end of the colonial period, land and labor had not become commodities. Inequalities of access were a result of the exercise of power by the Karo elite. In colonial times the Karo re-invested money in rubber and trade and also spent money on houses, clothes and rituals. The elite maintained their advantage during the Depression, obtaining a small but significant income from their rubber.

The Indonesian government continued the Dutch practice of governing the Karo uplands by way of "traditional" leaders. "Adat" leaders elected by Karo villagers were duly confirmed as village headmen. In Bukit Bangun the village elite and its descendants remain important at the local level, and the position of village headman has remained in the Sitepu clan. However, since the village is not involved in state projects, he is not a powerful client of the state.

Showing more interest in the peasantry than its colonial predecessor, the Indonesian government has attempted to increase rubber production among Langkat smallholders. But both the replanting schemes of the 1950s and the renewed efforts of the 1980s have failed. The majority of rubber smallholders continue to produce outside these projects and without any state assistance. The programs have been based upon the assumption that the Karo are "traditional" and need to be guided and assisted if they are to change. Contrary to this assumption, the Karo have repeatedly shown themselves to be innovative farmers eager to adopt new practices and increase profits. In Bukit Bangun, they resolved the replanting problem by switching out of rubber and into new crops, cloves and coffee. They obtained both information and planting material through their own, multi-ethnic, regional network operating quite outside government channels. They selected these crops according to their own priorities, which included the need for a high income within a short period. With this they were able to pay for the education of their children, an investment of central importance to the Karo after independence.

Since the 1950s, new forms of agrarian differentiation have emerged. Production has become more capital intensive, as fertilizers are used on high yielding varieties of cloves and coffee. Land and labor have become commodities, although still in a somewhat restricted form. Land selling continues to be negatively valued and occurs only on a small scale. Sharecropping remains the predominant form of extra household surplus extraction although there is some wage labor. In Bukit Bangun, Karo "newcomers" (those without kinship relations with the ruling lineage) were able to improve their position and expand their landed property in the 1960s. They worked temporarily as tenants for their Karo fellow villagers while acquiring land and waiting for trees to yield. Young Karo couples waiting to inherit continue to do likewise, and are offered work by their kin who adhere to an ideal of kin-based solidarity and assistance. In contrast, most Javanese former plantation laborers who entered the village to work as share tappers remain landless.

A number of factors limited class differentiation among the Karo. The kinship ideology required rich Karo to contribute generously to rituals. The late closing of the frontier, patrilineal inheritance, stronger out-migration of Karo than in-migration, and the rubber marketing system (which does not cause debt relations), served to counter the possible loss of land and impoverishment of the poorer Karo households. The replanting of rubber fields did not lead to a bifurcation of stronger peasants and sharecroppers or wage-laborers, because the villagers who replanted, members of the village elite and "newcomers" alike, could afford the replanting. While some have profited more than others from the opportunities associated with rubber and other cash crops, the values of a middle peasantry and a preference for autonomous smallholding are as deeply entrenched now as they were when this peasantry emerged as a refractory component of the colonial plantation economy.

ACKNOWLEDGEMENTS

Data collection in various archives and anthropological field work in the Karo Batak area of *kabupaten* Langkat occured in 1985–1987, in the context of PhD research supervised by Professor O.D. van den Muijzenberg of the University of Amsterdam. The research was financed by PRIS (Programme of Indonesian Studies). See Ruiter (forthcoming b).

NOTES

1. Lack of previous research ruled out a re-study approach (Hüsken 1988; Van den Muijzenberg 1991) but changes in landed property, land use and population growth were traced through other methods.
2. I use pseudonyms for the village and its clans.
3. The merga are named Karo-Karo, Sembiring, Perangin-angin, Ginting, Tarigan (Kipp 1976).
4. Westenberg, the first *Controleur* of Batak Affairs, codified the Karo adat in the dusun of Langkat, Deli and Serdang, planned the centralization of the political structure of the Karo in the highlands, and played an important role in the negotiations with Karo chiefs. He spoke Karo and was married, unusually for that time, to a Karo woman.
5. Scott characterizes the *hulu-hilir* contrast in pre-colonial Malay harbour states as "non-state space" versus "state space" (Scott 1995:25).
6. *Acte van erkenning*, Raja van Langkat Tengku Pangeran Indradi Raja Amir, 21 Oktober 1865, in: "Overeenkomsten met de Zelfbesturen in de Buitengewesten." *Mededeelingen* (1929:87–89).
   For a description of this and other political encounters of the colonial era see Ruiter (forthcoming a).

7   The Karo used the term dusun for new villages founded by Karo from the "mother village" (perbapaan). The term dusun was also used for the whole piedmont area colonized by Karo from the highlands (Neumann 1951).
8   See Stoler (1985) and Breman (1987).
9   "Memorie van Overgave" of the Governor of Sumatra's East Coast, H.J. Grijzen. Medan, 1921. State Archive The Hague (ARA), Kol. AA 210 (afdeling 2), no. 186.
10 "Memorie van Overgave" of the Governor of Sumatra's East Coast, Van der Plas. Medan, July 1917. ARA, Kol. AA 210 (afdeling 2), no. 184.
11 "Memorie van Overgave" of the Governor of Sumatra's East Coast, Van der Plas. Medan, July 1917. ARA, Kol. AA 210 (afdeling 2), no. 184.
12 Production quotas and tax measures were also set to the disadvantage of smallholders. See Bauer (1948).
13 Broersma (1919:125–128) describes measures taken by the union of tobacco planters (1879) to make stolen tobacco unmarketable.
14 Dutch missionaries from Deli reported that "fear strikes the heart of every rightminded Batak who hears the word coolie. Being a coolie is the last resort for a Batak .... With unlimited contempt they look down at coolies; they are beast of burden" (op cit. in Kipp 1990:120).
15 Letter of the Controleur of Deli, Kroesen, 17 juni 1873. ARA, Mailreport (MR) 1874, no. 31.
16 "Memorie van Overgave" of the Governor of Sumatra's East Coast, L.H.W. van Sandick, Medan, August 1930, Appendix "Agrarische Aangelegenheden", p. 18. ARA, Kol. AA 210 (afdeling 2), no. 187.
17 Similar land rules applied to other Batak areas of the period (Enda Boemi 1925).
18 Adat-incomes of the Karo village headmen in Langkat, Deli and Serdang were mentioned in Fievez de Malinez van Ginkel (1928:138–9).
19 "Memorie van Overgave" of the Governor of Sumatra's East Coast, L.H.W. van Sandick. Medan, August 1930, Appendix "Agrarische Aangelegenheden", p. 18. ARA, Kol. AA 210 (afdeling 2), no. 187.
20 The Karo situation fits Wolf's model of a kin-ordered production mode in which chiefs were able to lay hold of power sources independent of kinship control, and become new surplus takers. Thus kin ordering, under conditions of closed resources, produces inequalities (Wolf 1982:94,99–100).
21 This assistance was paid by the NIRUB (*Nederlands Indisch Rubber Uitvoer Bureau*).
22 Reasons for the failure included understaffing, mismanagement, non-availability of high-yielding varieties of rubber material and corruption (Barlow and Tomich 1991).
23 In the 1950s, the Dutch economist Boeke had argued similarly that only the state could bridge the gap between plantation and smallholder by reorganising the "best" peasants in a plantation-like way (Boeke 1948:103).
24 For a critical discussion of these state projects, heavily sponsored by the World Bank and based upon Bank "models", see Tomich (1991).
25 A court case is not the usual way for the Karo to solve conflicts. Slaats and Portier (1981) describe *runggun*, the traditional institution of deliberation by Karo kin groups.
26 Lee (1973:445) makes this point with regard to Malaysian rubber smallholders.

References

Barlow, C. and T. Tomich, 1991, "Indonesian Agricultural Development: The Awkward Case of Smallholder Tree Crop". *Bulletin of Indonesian Economic Studies*, 27(3), 29–54.

Bauer, P.T., 1948, *The Rubber Industry: a Study in Competition and Monopoly*. Cambridge: Harvard University Press.

Boeke, J.H., 1948, *Ontwikkelingsgang en toekomst van bevolkings- en ondernemingslandbouw. Deel I*. Leiden: E.J. Brill.

Breman, J., 1987, "Koelies, planters en koloniale politiek; het arbeidsregime op de grootlandbouwondernemingen aan Sumatra's Oostkust in het begin van de twintigste eeuw". *Verhandelingen van het Koninklijk Instituut voor Taal-. Land- en Volkenkunde, 123*. Dordrecht: Foris Publications Holland.

Broersma, R., 1919, *Oostkust van Sumatra; deel I De ontluiking van Deli*. Batavia: Javasche boekandel & drukkerij.

Enda Boemi, 1925, *Het grondenrecht in de Bataklanden*. Leiden: Eduard Ydo.

Fievez de Malinez van Ginkel, H., 1928, *Verslag van den economischen toestand en den belastingsdruk met betrekking tot de Inlandsche bevolking van de gewesten Oostkust van Sumatra en Lampongsche districten*. Weltevreden: Kolff.

Harrison, M., 1977, "Resource Allocation and Agrarian Class Formation: the Problem of Social Mobility among Russian Peasant Households, 1830–1930". *Journal of Peasant Studies*, 4(2), 127–161.

Hart, G., 1989, "Agrarian Change in the Context of State Patronage" in *Agrarian Transformations: Local Processes and the State in Southeast Asia*, edited by G. Hart, A. Turton and B. White, pp. 31–50. Berkeley: University of California Press.

Hefner, R.W., 1990, *The Political Economy of Mountain Java; An Interpretive History*. Berkeley: University of California Press.

Hüsken, F., 1989, "Cycles of Commercialisation and Accumulation in a Central Javanese Village" in *Agrarian Transformations: Local Processes and the State in Southeast Asia*, edited by G. Hart, A. Turton and B. White, pp. 303–332. Berkeley: University of California Press.

Kipp, R.S., 1976, *The Ideology of Kinship in Karo Batak Ritual*. Ph.D. diss., University of Pittsburgh.

1990, *The Early Years of a Dutch Colonial Mission: The Karo Field*. Ann Arbor: University of Michigan Press.

Lee, G., 1973, "Commodity Production and Reproduction Amongst the Malayan Peasantry". *Journal of Contemporary Asia*, 3(4), 441–457.

Mededeelingen van de Afdeeling bestuurszaken der Buitengewesten van het Departement van Binnenlandsch Bestuur. Serie A no. 3. Weltevreden, 1929.

Milner, A.C., 1982, *Kerajaan: Malay Political Culture on the Eve of Colonial Rule*. Tucson: University of Arizona Press.

Muijzenberg, O.D. van den, 1991, "Tenant Emancipation, Diversification and Social Differentiation in Central Luzon" in *Rural Transformations in Asia*, edited by J. Breman, S. Mundle, pp. 314–338. Delhi: Oxford University Press.

Neumann, J.H., 1951, *Karo-Bataks – Nederlands Woordenboek*. Medan: Varekamp & Co.

Parlindungan, A.P., 1981, *Kapita Selekta Hukum Agraria*. Bandung: Penerbit Alumni.

Pelzer, K., 1978, *Planter and Peasant. Verhandelingen van het Koninklijk Instituut voor Taal-, Land-, en Volkenkunde 84*. 's Gravenhage: Martinus Nijhoff.

Reid, A., 1979, *The Blood of the People*. Kuala Lumpur: Oxford University Press.

Ridder, J. de, 1935, *De invloed van de Westersche cultures op de autochthone bevolking ter Oostkust van Sumatra*. Proefschrift (Ph.D. diss.), Leiden.

Ruiter, T.G., 1982, "The Tegal Revolt" in *Conversion, Competition and Conflict: Essays in the Role of Religion in Asia*, edited by D. Kooiman, O. D. van den Muijzenberg, P. van der Veer, Anthropological Studies Free University Amsterda, pp. 81–99. Amsterdam: VU Uitgeverij/Free University Press.

forthcoming a, "Dutch and Indigenous Images in Colonial North Sumatra" in *Images of Malay-Indonesian Identity*, edited by M. Hitchcock and V.T. King. Kuala Lumpur: Oxford University Press.

forthcoming b, *Rubberboeren, planters en de staat; sociaal-economische veranderingen in het Karo Batak gebied van Langkat, Noord Sumatra (1870–1987)*. (forthcoming: Ph.D. diss., University of Amsterdam).

Ruiter, T.G. and H. Schulte Nordholt, 1989, "The Village Revisited: Community and Locality in Southeast Asia". *Sojourn*, 4(1), 127–135.

Schadee, W.H.M., 1920, *De uitbreiding van ons gezag in de Bataklanden. Uitgaven van het Bataksch Instituut 19*. Leiden: S.C. Van Doesburg.

Scott, J.C., 1995, *State Simplifications: Some Applications to Southeast Asia*. Centre for Asian Studies Amsterdam, CASA Wertheim Lecture 6.

Singarimbun, M., 1975, *Kinship, Descent and Alliance among the Karo Bataks*. Berkeley: University of California Press.

Slaats, H. and M.K. Portier, 1981, *Grondenrecht en zijn verwerkelijking in de Karo Batakse dorpssamenleving*. Proefschrift (Ph.D. diss.). Nijmegen: Publicaties over Volksrecht IX. Universiteit van Nijmegen.

Steedly, M.M., 1993, *Hanging Without a Rope: Narrative Experience in Colonial and Postcolonial Karoland*. Princeton: Princeton University Press.

Stoler, A., 1985, *Capitalism and Confrontation in Sumatra's Plantation Belt*. New Haven: Yale University Press.

Tomich, T.P., 1991, "Smallholder Rubber Development in Indonesia" in *Reforming Economic System in Developing Countries*, edited by D.H. Perkins and M. Roemer, pp. 249–271. Cambridge: Harvard University Press.

Waal, R. van de, 1959, *Richtlijnen voor een ontwikkellingsplan voor de Oostkust van Sumatra*. Proefschrift (Ph.D. diss.), Wageningen.

Westenberg, C.J., 1914, "Adatrechtspraak en adatrechtspleging der Karo-Bataks". *Bijdragen tot de Taal-, Land-, en Volkenkunde van Ned. Indië*, deel 69, 453–601.

White, B., 1989, "Problems in the Empirical Analysis of Agrarian Differentiation" in *Agrarian Transformations; Local Processes and the State in Southeast Asia*, edited by G. Hart, A. Turton and B. White, pp. 15–31. Berkeley: University of California Press.

Wolf, E., 1982, *Europe and the People without History*. Berkeley and Los Angeles: University of California Press.

Zon, P. van, 1915, "Bijdrage tot de kennis der boschgesteldheid van de residentie van de Oostkust van Sumatra". *Tectona*, 9e jrg, deel 9, 251–269, 349–375, 429–449.

INDEX

Abu-Lughod, L., 205
adat (customary practice): and cultural difference, 83–87; and Dutch rule, 10, 113, 284, 286; and elite gain, 290–92, 296–97; inheritance, 120; as kinship system, 303; and Leiden School, 87–88, 100; marriage exchanges, 112; *rumah adat* (ceremonial house) 136. *See* Land
Agrarian Law (1870; Java), 13
Agrawal, A. and K. Sivaramakrishnan, 3
Agroforestry, 30–31, Chapter 9
Alatas, S., 14, 215
Alifuru (wilderness or forest people; Ternate), 4, 62, 66, 136
Alisjahbana, Takdir, 100
Anderson, B., 80, 94, 196
Anthropology: concepts of culture, 82–84, 90, 93, 124; and study of power, 205–6, 210, 220–23
Attenborough, D., 150

Backwardness, of upland people, 2, 14–15, 18, 24, 35, 88, 106, 108–9, 114–15, 123, 159–61, 167, 169, 175, 179–80, 196, 203, 210–18, 237, 240, 250. *See also* Exotic culture, State views
balai (multi-family dwelling; Meratus), 173–75, 191, 194–95. *See also* Longhouse
Bali, 13, 47–48, 50–52, 60–62, 81–82
Ballot, J., 89–91
bangsa (race, nation), 18, 83–85, 295
bangsa taneh (owners of the land; Karo), 288, 290
Barber, C., 10, 15–16, 27–28, 32
Barlow, C., 27, 207, 294

Barth, F., 82
Basic Agrarian Law (1960), 134, 190, 213, 297
Basic Forestry Law (1967), 15, 134, 190
Bastaman, H., 33–34
Berry, S., 24, 204, 258
Boeke J. H., 88–89, 125
Borneo. *See* Kalimantan
Bourdieu, P., 108, 110–11, 114
Breman, J., 13, 105, 125, 210
Brody, H., 189
Brookfield et al, 16, 24, 28–29, 32

Cacao. *See* Cocoa
Capitalism: and flexible accumulation, 232–33; international, 95–96, 207, 233, 236; late, 223; and peasant smallholders, 92–93, 105–8, 118–20, 123–25, 232, 236, 258, 261–64, 267, 274; as threat to upland people, 23–24, 165. *See also* Class differentiation, Markets
Cash crops: and Karo, 281, 287–88, 299–300, 303; and Nuaulu, 131, 134–35, 143–44, 149; as part of portfolio, 92, 212, 261, 269; in swidden systems, 25–26, 169. *See also* Tree crops, Plantations, Tobacco, Vegetables
Cassava, 25, 45, 50, 260, 267–68, 271, 281
Census, 12–13
Chayanov, A., 233, 236, 238
Chinese: merchants, 53, 57–58, 66, 108, 117–19, 122, 126, 132, 175, 292, 294, 302–3; tobacco, 52–53, 58; occupation of plantation land, 293

INDEX

Christianity, 108–9, 281, 284, 298
Class differentiation, 30–34, 83; and agroforestry, 31, 258–59, 261–74; and contract farming, 233, 235, 242, 246–51; and ethnic identity, 18–19; and Karo, chapter 10; and smallholders, 105, 107–8, 119–123, 127; and "traditional" peoples, 23–24. *See also* Patronage, Capitalism, Markets
Closed corporate community, 125–26
Cloves, 31–32, 132, 134, 158, 240, 257, 281, 299, 302
Coastal polities, 3, 4, 7, 8, 65, 86, 131–32, 135, 145, 284, 293. *See also* Malay
Cocoa, 27, 33, 134, 175
Coconut, 132, 232, 236, 238–51, 261, 287, 291
Coffee production, 7, 9, 25–27, 48, 55, 90, 134, 169, 175, 232, 257, 299–304, 306
Colonial rule, *See* Dutch
Communalism, 23, 100, 106–9, 114, 120–26, 161, 172, 215, 291
Communists, 18, 83–84, 237, 244–45
Community, concepts of, 3, 5, 10, 13, 22–24, 33–34, 93, 97, 125, 159–62, 165, 167, 176, 180, 194, 198, 265, 303
Conservation, 1, 2, 22, 30–31, 159–67, 175, 184–85, 196, 261, 266, 273
Consumption, 22; and class status, 264, 303; at Dayak feasts, 211; household as unit of, 114–16, 120; and identity, 24, 31, 33; of opium, 54; reducing cost of 118; of tobacco, 53, 56, 58, 63. *See also* Rice
Contract farms and nucleus estates, 17, 21, 23, 27–28, 207, 213–14, 217, 219, Chapter 8

Corn, *See* Maize
Corvee, 7, 13. *See also* Slavery
Credit: and agricultural inputs, 122; and agroforestry, 263–64 *et passim*; coerced, 7; and contract farmers, 208, 239–40, 242, 247–48; government, 31; and petty trade, 118–19; and rubber replanting, 295; and sharecroppers, 301; and tobacco production, 58–59, 66–67. *See also* Patronage
Cultivation System, 26, 49–50, 55–56, 232
Cultural difference, 3–5, 8–11, 24–25, 30, 34, 79–89, 99, 106–8, 142, 144, 146, 161, 165, 176, 184, 196, 203, 206, 211, 296; and economic conflict, 14, 89, 215, 216, 218. *See also* Backwardness, Exotic culture
Culturalisation, 3, 25, 80–89, 99
Culture, concepts of, 81–82, 84; of para-statal corporations, 210, 215, 221; in relation to nature, 139–141. *See also* Backwardness, Traditionalism, Anthropology, State

Darul Islam, 237, 240–41
Dayak (indigenous people of Kalimantan), 32, 48, 52, 81, Chapter 6, 210–13, 216–19
Departments, Government of Indonesia: Agriculture, 16, 20; Education and Culture, 179; Forestry, 15–16, 195; Social Welfare, 18, 169; Transmigration, 16–17. *See also* State
desa (village, administrative unit), 13, 17, 173, 296

development: assumptions or narratives, 1, 5, 9, 11–12, 14–20, 30, 35, 105, 108, 111, 114, 120, 122–27, 136, 204, 210, 220, 279; demand for, 21–22, 33, 176–80; green alternatives, 22–24, 30, 258, 262, 274, Chapter 6; Dutch views on, 88–89, 106, 284. *See also* Sustainable, State, Backwardness

Dobbin, C., 48, 54

Domain Declaration (1874), 90, 91

Dove, M., 1, 25, 27, 30, 137, 258, 259

Dryland cultivation, 26, 28, 63–64, 66, 92, 165, 218, 241, 259–60, 267, 281, 286–88, 293, 299. *See also tegal, gaga*, Shifting cultivation

Dual economy, 88, 125, 279

Dutch colonial period (Netherlands East Indies): agricultural policies, 9, 46–49, 58–60; 284–88, 290, 305; archival sources, 46, 132; in Bukit Bangun, 288–293; bureaucratic expansion and native officials, 86, 92, 97–98, 285; contact with Nuaulu, 131–32, 142–43; critiques of native communalism, 106, 127; culturalism and critiques of colonial liberalism, 85, 87–89, 99–100; ethical policy, 90, 93, 98, 284–85; and identity formation, 10; indirect rule, 10, 14, 108, 144, 280, 283–86, 288–91; and land alienation in West Sumatra, 89–91, 93; missionaries, 284; and nationalism, 80, 85; and peasantisation, 89–93, 105–8, 113, 120, 124–25, 127; plantations, 204, 206, 210, 215, 218, 237, 284, 286, 294; territorialisation strategies, 12–14; in upland Langkat, 283–288. *See also* State formation

Eastern Indonesia, 6, 7; and maize production, 45–48, 51–52; and tobacco production, 53, 62, 66. *See also* Sulawesi, Maluku, Irian

Education, 21, 33–34, 92, 98, 146, 171, 177, 179, 214, 264, 292, 296, 299, 301–4, 306

Edwards, I., 134, 138

Elson, R. 97–98, 210, 215, 232

Environment, Environmental: activism, 11, 21–24, 30, 258, Chapter 6; conditions for fruit production 261, 266–68, 274; conditions for maize and tobacco, 45; degradation, 24–30, 34, 61, 65, 67, 133–34, 260; determinism, in models of natural economy, 123–25; knowledge, Meratus 163–65, 184–88; knowledge and consciousness, Nuaulu 131, 134–35, 148–50; problems on nucleus estates, 207; schema, 2–3; stabilization program, Java, 257–58, 266, 273; struggles, 3, 24, 203

Epidemics, coastal, 65

Erosion, 25, 29, 61, 134, 148, 257–58, 260, 262, 266, 273, 287

Estates. *See* Plantations, Contract farming, Tree crops

Evolutionary schema, 4–7, 79, 87–89, 106, 113–14, 137, 165, 196

Exotic culture, 3, 159–60, 167, 180–83, 196, 222. *See also* Primitives, Tribes

Export: and contract farming, 234, 236, 240, 250; and Cultivation System 125, 231; of fruit, 259; of maize 45, 47, 51; of oil and gas, 207; and resource expropriation 22, 26, 166, 207; of tobacco 53–8, 61–3, 283. *See also* Plantations, Tree crops, Forest products, Rice

INDEX

Farmer, farming: concepts of, 2, 27–28, 34, 214, 287; innovation, 25, 27, 31, 259, 261, 306; relations with Penan, 6; within "forest" boundaries, 2, 13, 15–16, 23, 31–32, 164, 173, 194. *See also* Contract farming, Smallholders, Land, Backwardness, Shifting cultivation, Forest communities

Feasting, 9, 106, 109–13, 211, 282

Ferguson, J., 17, 20, 210

Fire, 203, 243

Forest: communities, 15–16, 21–23, 30, Chapters 5, 6; concessions, 2, 16, 23–4, 27, 29, 32, 92, 131, 134, 148–49, 151, 164, 174, 177, 185, 194, 195, 204; conversion, 9, 16–7, 19–20, 26–32, 47, 65, 132, 134, 212, 284, 287; non-timber products, 6–8, 18, 30, 132, 145, 184, 212; Nuaulu rights recognized, 135–36; reserves, 20, 91–92, 137, 190–91, 195, 287; as sanctuary, 6, 13, 15; state control over 12–16, 21–22, 27, 32, 145, 164, 166, 190–95; timber loss through swiddening 184; as wilderness, 2–4, 22–23, 137, 186–88. *See also* Land, State

Fruit, 26, 28, 138, 173, 177, 188, 236, Chapter 9, 281, 291–2, 298, 300

gaga (dryfields, Java), 45, 55, 57, 66. *See also tegal*, dryland

Geertz, C., 12–13, 81, 119, 125

Gender, 91, 93, 140–41, 235. *See also* Women

Gift exchange, 106–14, 119, 123, 292

Global: environmentalism, 24, 159, 164, 184, 198–99, 203; cultural understandings, 11, 145, 150; proletariat, 233; market, 258

Green agendas, 20, 22–24, 34, Chapter 6, 274

Green revolution, 3, 29, 107, 117, 119, 122, 231, 295

hak ulayat (communal property), 91. *See also* Communalism, Land

Hardjono, J., 26–31, 135

Hart, G., 19, 21, 31, 149, 280

Hefner, R., 24, 26, 29, 33, 50, 58, 65, 231, 258, 280. *See also* Tengger

Henley, D., 4, 25, 29, 64

Hoben, A., 11, 12

Homegarden (Java), 257–58, 263, 274

Households: colonial Java, 13; Karo, 292; Minangkabau, 91–3; To Pamona, 106–7, 114–16, 118, 120–22, 126; and contract farm labour 232–36, 242, 246–47, 250; and land leasing, 269–70, 272; Meratus, see *umbun*. *See also* Kinship

Identity, 9, 10, 19, 33, 67, 144–46, 149, 159, 179–80, 196, 288. *See also* Community

Illegality, 13–15, 27, 32, 92, 134, 164, 293. *See also* Land

Indigenous peoples, 3, 17–19, 21–23, 26–28, 88, 90, 98, 117, 134, 136, 138, 159, 164, 166, 180, 189–90, 196, 212. *See also* Tribes, Exotic, Backwardness

International Situationists, 163

Investment, 28, 31; in education, 299, 301–4, 306; foreign, 97; in fruit production, 236, 262, 264, 270–73; in green revolution technology, 122; in kin ties, 110, 122, 126; money treated as capital, 286, 292; non-agricultural, 33–34, 306. *See also* Class differentiation, Capitalism

INDEX

Irian Jaya, 54, 62, 79, 236
Irrationality, 106, 114, 117–19, 122, 215–17. *See also* Backwardness
Islam: Darul Islam rebellion, 237, 240–41; Dayaks converted to, 8, 175, 178–79; early conversion, 3; flight from, 6, 9, 65, 265; Haj pilgrimmage 33, 266; modernist, 84–86; in Tengger, 266

jagung (maize). *See* maize
jaluran (plantation land used for rice between tobacco harvests; Karo), 285–87, 290–93, 305
Jameson, F., 223
Java, Javanese: contract farms, 219, chapter 8; culture, 4; forests, 13, 15, 32; fruit agroforestry, chapter 9; idea of power, 94; inequalities, 94; labourers in Karo villages, 281–82, 295–302; maize 47–50; in New Order, 82; peasant agriculture, 125; plantation labour, 14, 32, 215, 281, 284; plantation managers, 209–10; pre-colonial rule, 13–14; sugar production, 208–9, 213–15, 220; tobacco, 52–53, 55–60 *et passim*; transmigrants, 211; upland farms, 26, 28–29; viewed as superior, 4, 19, 25, 204. *See also* Tengger

kacang (pulses), 61, 63. *See also* Pulses
Kahn, J., 9, 10, 13–14, 105–7, 119, 125
Kalimantan, 6–9; agricultural conversion, 16, 19, 20, 26–27, 29, 32; maize, 48, 52; mines, 94–97; plantation land and labour conflicts, 203, 207, 209–18; tobacco, 54, 63; Chapter 6
Keane, W., 11

KEPAS (agro-ecosystem study group), 23
Kinship: Karo, 280–92; 301–7; To Pamona, 97, 107–14, 119–23, 126. *See also* Household
Kipp, R., 282, 284, 302
KUD (official co-operatives), 239, 245, 248–50

Labour: and contract farms, 231–37, 246–47; exchange, 9, 106, 107, 112, 114, 116–22, 126, 291–92, 295, 297, 300; migration, 8, 28, 242, 247, 267, 269, 302; plantation, 14, 19, 28, 211–14, 218, 281, 284, 288; productivity of swiddens 6, 32, 63; requirements of fruit agroforestry 261–63, 270; requirements of maize and tobacco 45, 57, 63, 66; state control over, 13–14, 19, 25, 56, 231; waged, 15, 28, 31, 34, 92, 116–18, 212–13, 266, 270, 272, 300. *See also* Household, Class differentiation, Transmigration, Sharecropping
ladang (dry field; term used off-Java). *See* Dryland cultivation, Shifting cultivation
Land: *adat* (customary), 22, 26–27, 30, 90–93, 131, 135–36, 144, 146, 162, 164, 171, 189–95; inheritance, 120, 282, 291, 297, 301–3; marginal or degraded, 9, 14, 32, 260, 293; sales, 32, 131, 134, 136, 150, 298, 300; and social stratification, 94–95, 117–22, 126, 258, 260–75, 280 *et passim*; spontaneous settlement of, 9, 13–14, 19, 21, 26–27, 29, 34, 92, 134–35, 237, 241, 295; state control and expropriation, 13–16, 19–21, 29, 30, 33, 89–93, 134,

148, 150–52, 171, 174, 180, 191–95, 203, 207–9, 211, 213–14, 217, 220, 237, 240–45, 284–90, 294; tenure, 21, 23, 31, 33, 237, 258, 263–65, 273, 291, 300; titles, 15, 21, 239, 249, 251. *See also* Tree tenure, Forest
Landless, 9, 14–15, 32–33, 118–20, 171, 264, 273, 280–81, 295, 300, 302–3, 307
Leach, E., 215
Legal Aid Institute, 240
Livestock, 59–63, 66, 232, 236, 247, 262
Longhouse, 106, 114, 116, 120. *See also balai*

Maize, 6, 25, Chapter 2, 143, 241, 260, 267–68, 271
Malays, 54, 65, 283–85, 293–95. *See also* Coastal polities
Maluku, Moluccas, 6, 45–46, 51, 62, 131–32, 136, 138, 143, 145
Maps, 12–13, 20–21, 132, 135, 162, 175, 188–98, 238, 242, 245
Marginality, Chapter 1, 93, 99, 105–7, 119, 123, 126, 196, 223, 293, 296, 305,
Markets, and contract farming, 231–36, 239–43, 247–50; and moral economy, Chapter 4; production for, 5, 7, 23–24, 26, 28–29, 31, 33, 34, 46, 56–58, 62, 65–66, 94, 169, 257–63, 268, 273–75, 280–83, 285, 292, 294, 305. *See also* Capitalism, Export
masyarakat terasing (officially designated isolated or backward tribe), 14, 18, 21, 135–36, 145, 171
Mataram, 3, 6, 52
merga (clan; Karo), 282, 297
Migration. *See* Labour migration, Land settlement, Transmigration

Military, 15, 32, 34, 95–96, 132, 164–66, 220, 244, 249, 284–85, 293, 294, 298
Minangkabau, 3, 10, 51, 63, Chapter 3
Mines, 7, 35, 90–93, 96
Modernity, modernism, 5, 8, 10–11, 18, 21, 23, 28, 80, 83–89, 93, 97–99, 106, 108, 111–12, 114, 125, 170–71, 176, 196, 236, 250, 279, 298

Nader, L., 221
nagari (Minangkabau village communities), 83, 90–92, 98
Nationalism, 80, 84–85, 160, 196
New Order, 10, 80, 82–83, 85, 95, 99, 108, 114, 295, 299
NGO (non government organization), 23, 95, 160, 165–66, 198, 240. *See also* Yayasan Kompas Borneo
Nucleus estate schemes (NES). *See* Plantations, Contract farming

Obeyesekere, G., 183
palawija (non rice annuals; Java), 26, 49–50

Palte, J., 9, 31, 50
Pancasila, 150
Parry, J., 109–10, 113, 123
Patronage, 6, 12, 19, 20–21, 31–32, 58, 66, 85, 94–99, 110, 117, 123–26, 148, 162, 170–71, 245–50, 266, 280, 290, 295, 301
Peasantisation, 89–97, Chapters 4, 10. *See also* Smallholders, Farmers
Peluso, N., 7, 12–15, 20, 22, 32, 137, 189, 231, 258
Pemberton, J., 4, 10, 80, 82, 90
Penan, 6

Plantations, 3, 14–17, 19, 20, 23–27, 29, 46, 63, 90–93, 134–35, 164, 231–32, 236–68, 241–42, Chapters 7, 10. *See also* Land, Labour, Tree crops, Contract farms
Police, 15, 95–96, 132
Popkin, S., 124
Population (size and distribution), 5–7, 9, 12, 14–15, 25–26, 29, 34, 49, 64–65, 67, 108, 134–35, 171, 207, 212, 237, 257, 286, 298, 302
posintuwu (gift exchange; To Pamona), Chapter 4
Poverty, 2, 9, 15, 18, 25, 29, 30, 32, 34, 52, 66–67, 80, 82, 110–12, 118, 120–21, 127, 179, 233, 246, 250, 262–64, 281–82, 295–96, 301–3, 307
Power, Chapter 1, 80–81, 85, 89–100, 123, 142, 149, 159, 162–63, 168, 172, 182–83, 198, 204–6, 221–23, 234–35, 249–51, 280, 292, 298, 306. *See also* State, Patronage
Primitive, 8, 18–19, 22, 79, 80, 136, 169, 170, 175, 177, 180, 182, 284. See Backwardness, Tribes
Pulses, 49–51, 59–62, 66–67

Raffles, Sir Thomas Stanford, 49
Reid, A., 5, 6, 54, 293
Resettlement, state-sponsored settlement, 9, 12, 14, 16–18, 21, 22, 106, 108, 131–32, 144, 164, 171, 241. *See also* Transmigration, Land settlement (spontaneous)
Resistance and critique, 7, 20–24, 27, 89–90, 93, 96, 115, 123–24, 138, 145–53, 189, 196, 198, 208, 213–20, 235, 244–45, 251, 258

Rice: dry (hill or swidden), 5, 6, 48–51, 63–64, 92, 185, 218, 245, 268, 271, 285–88, 291, 293, 299, 305; wet (irrigated, *sawah*), 3, 7, 16, 25–26, 28–29, 31, 45–46, 50, 65, 91, 94, 106, 108, 116, 119–22, 208, 231, 241, 245, 286, 291, 295, 301; consumption, 33, 48, 52, 65, 111, 117, 143, 281; trade and tribute, 6, 62, 65, 67, 122, 292
Roads, 17, 19, 21, 29, 32, 135, 177–78
Roseberry, W., 2, 107, 124
Rumphius, 54–55, 60, 132
rupiah, Rp. (Indonesian currency)

Sago, 6, 48, 135, 137–38, 141, 143–44, 146, 149
sawah (irrigated or wet rice field). *See* Rice
Scott, J. C., 123–25, 217, 288
Sharecropping, 31–32, 92, 262–64, 269–73, 280–81, 292, 295–303, 306–7
Shifting cultivation, swidden, 5, 6, 13, 16, 18, 23–29, 50, 32–33, 105, 117, 121, 134, 137–40, 144, 163, 165, 171, 173, 184–88, 194–95, 212, 237, 283, 285–88, 290–1, 305. *See also* Dryland cultivation
Singarimbun, M., 282, 290
SKEPHI (Indonesian environmental NGO), 17, 22, 135
Slavery, 6, 7, 9, 65
Smallholders, peasants, 2–5, 8, 9, 16, 19, 21, 25–28, 34, 46, 49–50, 55–58, 66–67, 85, 90–96, 99, 105–10, 119–27, 149, 160, 207–8, 213, 218, 231–39, 242, 245, 247, 249–51, 257–58, 267, 273, 279–80, 287–88, 293–99, 303–7; in conflict with plantations, chapter 7. *See also* Peasantisation, Farmer, Shifting cultivation

INDEX

Social forestry, 16, 21–23, 32
State formation, 11, 13, 97–98, 285
State: ethnographic studies of, 205–6, 221–23; New Order government policies and programs, 5, 8, 11–18, 29, 67, 105–7, 111, 121–27, 150, 164, 166, 170–71, 178, 190–98, 203–4, 235–43, 250, 274, 295, 304; views of upland people, 11, 14–15; 18, 27, 98, 105–9, 111–14, 120–27, 136, 164, 169–71, 176–79, 203–6, 214–18, 279. *See also* Patronage, Military, Police, Dutch colonial period, Coastal polities, Territorialisation, Departments, Land
Stereotypes, 2, 19, 22, 28, 125, 136, 159, 169–70, 180–81
Stoler, A., 19, 196, 212, 215–16, 218, 257, 258, 295
Stratification, 7, 65, 94, 261, 282, 291. *See also* Class differentiation
Subsistence, 5, 9, 15, 25, 27, 34, 45, 105–7, 115, 117, 120, 123, 127, 131, 137, 139–144, 149–50, 169, 211, 231, 250, 260, 269, 293, 301, 305; insurance, 105, 107, 115–16, 118–19, 121, 126
Sugar, 9, 26, 48, 55, 204, 206–15, 220, 231–32, 236, 241, 260, 291
suku (matrilineal kin group; Minangkabau), 84, 91
Sulawesi, 4, 9, 29, 33; plantations, 209, 212; maize production, 45–48, 51; tobacco production, 55, 63; Chapter 4
Sumatra, 3, 9, 14, 25–27, 32, 45; maize production, 48, 51, 65; smallholder tobacco, 54–55, 63; plantations, 209, 214–16, 218; Chapters 3, 10
Sustainable development, 1, 22–23,
32–34, 66–67, 134, 137, 148–49, 159–61, 163, 165–67, 184, 198. *See also* Development
Swidden (dry fields used in long fallow rotation). *See* Shifting cultivation

Taman Mini, 80, 82
tanah seratus (land allocated by plantations to Karo peasants), 285, 287, 290–93, 297, 299, 303
taxation, 7, 9, 12, 49, 90, 92, 212, 283, 285, 291–92, 296–97, 305; of tobacco, 52–53, 55, 66; and land tenure, 237, 241, 244
tegal (dry fields; Java), 28–29, 45, 49, 50–51, 53, 55, 57–59, 62, 64, 66, 260. See also *gaga*, Dryland
Tenancy and tree lease, 94, 263–65, 269–72, 274, 297, 299, 302, 306. *See also* Sharecropping
Tengger Highlands, 3, 9, 34, 65, 263, 265
Territorialisation, 11, 12–17, 20–21, 164, 173, 188–95, 285
Timber industry. *See* Forest concessions
Timor, 7, 46–47, 54, 62, 132
Tobacco: smallholder, Chapter 2; under contract to Europeans, 55–56, 63; on estates, Chapter 10
Tourism, 10, 80, 108, 163, 174–75, 180, 183–84, 198, 211, 213
Trade. *See* Market
Traditionalism, 5, 8–11, 21–24, 27–28, 32–35, 45, 48, 79–82, 85–90, 93–94, 97–100, 105, 137, 142–46, 161, 164–70, 176–85, 190, 194, 196, 199, 212, 216, 257–58, 279–80, 286–88, 291, 305–6, Chapter 4
Transmigration, 14, 16–21, 23, 26, 32,